The Impact of Discovering Life Beyond Earth

The search for life in the universe, once the domain of science fiction, is now a robust research program with a well-defined roadmap, from studying the extremes of life on Earth to exploring the possible niches for life in the Solar System and discovering thousands of planets far beyond it. In addition to constituting a major scientific endeavor, astrobiology is one of the most popular topics in astronomy, and is of growing interest to a broad community of thinkers from across the academic spectrum.

In this volume, distinguished philosophers, theologians, anthropologists, historians, and scientists discuss the big questions about how the discovery of extraterrestrial life, whether intelligent or microbial, would impact society. Their remarkable and often surprising findings challenge our foundational concepts of what the discovery of alien life may hold for humankind. Written in easily accessible language, this thought-provoking collection engages a wide audience of readers from all backgrounds.

Steven J. Dick held the 2014 Baruch S. Blumberg NASA/Library of Congress Chair in Astrobiology at the John W. Kluge Center of the Library of Congress. In 2013 he testified before Congress on the subject of astrobiology. He served as the Charles A. Lindbergh Chair in Aerospace History at the National Air and Space Museum from 2011–2012, and as the NASA Chief Historian and Director of the NASA History Office from 2003–2009. He is the recipient of numerous awards, including the NASA Exceptional Service Medal and the Navy Meritorious Civilian Service Medal, and is author or editor of 20 books, including *The Biological Universe*. He was awarded the 2006 LeRoy E. Doggett Prize for Historical Astronomy of the American Astronomical Society. In 2009, the International Astronomical Union designated minor planet 6544 stevendick in his honor.

"Living in our Milky Way galaxy with its billions of habitable planets, we humans are aching to know something, anything, about our intelligent neighbors among the stars. *The Impact of Discovering Life Beyond Earth* offers thoughtful and beautiful notions for the coming breakthrough contact."

Geoff Marcy, *University of California, Berkley*

"Are we alone in the cosmos? If yes, we can never be sure. If no, we might well have an answer within the decade. The 22 authors of these fascinating and informative essays say no, we are not alone, despite the fact that so far not a shred of evidence has been found for the existence of life elsewhere in the universe. But just in case they are right, we need to start thinking about the possibility that we are not alone, and here is a good place to start."

Owen Gingerich, *Harvard-Smithsonian Center for Astrophysics*
Author of *God's Planet*

The Impact of Discovering Life Beyond Earth

Steven J. Dick

Former NASA Chief Historian

CAMBRIDGE
UNIVERSITY PRESS

University Printing House, Cambridge CB2 8BS, United Kingdom

Cambridge University Press is part of the University of Cambridge.

It furthers the University's mission by disseminating knowledge in the pursuit of education, learning and research at the highest international levels of excellence.

www.cambridge.org
Information on this title: www.cambridge.org/9781107109988

First published 2015

Printed in the United Kingdom by TJ International Ltd. Padstow Cornwall

A catalogue record for this publication is available from the British Library

Library of Congress Cataloguing in Publication data
The impact of discovering life beyond Earth / [edited by] Steven J. Dick, former NASA Chief Historian.
pages cm
Includes bibliographical references.
ISBN 978-1-107-10998-8
1. Life on other planets. 2. Exobiology. I. Dick, Steven J.
QB54.I47 2015
576.8'39 – dc23 2015014815

ISBN 978-1-107-10998-8 Hardback

To Baruch S. Blumberg
and
John Billingham
In The Spirit of Exploration

Contents

Contributors

Linda Billings, *NASA HQ*
Eric J. Chaisson, *Harvard University*
Carol E. Cleland, *University of Colorado Boulder*
Guy Consolmagno, SJ, *Vatican Observatory*
Steven J. Dick, *Former NASA Chief Historian*
Iris Fry, *Technion–Israel Institute of Technology*
Robin W. Lovin, *Center for Theological Inquiry, Princeton, N.J.*
Mark Lupisella, *NASA Goddard Space Flight Center*
Jane Maienschein, *Arizona State University*
Lori Marino, *The Kimmela Center for Animal Advocacy*
Carlos Mariscal, *Duke University and Dalhousie University*
Michael A. G. Michaud
Margaret S. Race, *SETI Institute*
Michael Ruse, *Florida State University*
Susan Schneider, *University of Connecticut*
Dirk Schulze-Makuch, *Washington State University* and *Technical University Berlin*
Seth Shostak, *SETI Institute*
John W. Traphagan, *University of Texas at Austin*
Julian W. Traphagan, *Lehigh University*
Douglas A. Vakoch, *SETI Institute*
Clément Vidal, *Free University of Brussels (VUB)*
Elspeth M. Wilson, *University of Pennsylvania*

Introduction: Astrobiology and society

STEVEN J. DICK

The search for life in the universe, once the stuff of science fiction, is now a robust research program with a well-defined roadmap and mind-bending critical issues (Des Marais *et al.*, 2008; Dick, 2012; Dick and Strick, 2004; Sullivan and Baross, 2007). The science of astrobiology – and there is no longer any doubt it is a science, simplistic slogans about "a science without a subject" notwithstanding – is funded by NASA and other institutions to the tune of tens of millions of dollars of ground-based research, not to mention the hundreds of millions spent on space-related missions. Biogeochemists study extremophile life on Earth, biologists study the origins of life, a bevy of spacecraft have orbited or landed on Mars, others have found potentially life-bearing oceans on Jovian and Saturnian moons as well as organic molecules on Titan, and the Kepler spacecraft has discovered thousands of planets beyond the solar system – all just a prelude to future studies. Recent US Congressional hearings on astrobiology indicate it is a hot topic in the policy arena (United States Congress, 2013 and 2014). And international interest is also strong, particularly within the European Space Agency. Although no life has yet been found beyond the Earth, the search for such life has arguably been a driver of the space program since its inception, has inspired multidisciplinary research on Earth, and is the subject of great popular interest that shows no signs of abating. As this volume illustrates, it is also a perennial theme in science fiction literature, igniting dreams of other worlds.

Given both scientific and popular interest in astrobiology it is important for scholars, practitioners, and policymakers to examine the societal implications of discovery in the event of success. Substantial studies have been undertaken on the societal impact of other scientific endeavors such as the Human Genome Project, biotechnology, nanotechnology, and spaceflight. Even closer to astrobiology's core interests are planetary protection protocols, which are certainly studies of potential impact since one of their goals is to prevent a catastrophic "Andromeda Strain" scenario, in the terminology of Michael Crichton's 1969 novel. We should be under no illusion that millions of dollars are going to be spent to study the implications of finding extraterrestrial life – not, that is, until it is discovered, in which case the floodgates may open as they did with the Human Genome Project, now in the form of a practical problem rather than a theoretical one.

But how to approach rationally such a "far out" problem as the societal impact of discovering life beyond Earth? That was the question posed at a symposium held at the Library of Congress in September, 2014, for which this volume is the elaborated and fully referenced record (Library of Congress, 2014). Entitled "Preparing for Discovery: A Rational Approach to the Impact of Finding Microbial, Complex, or Intelligent Life Beyond Earth," the Symposium was not the usual astrobiology meeting where technical aspects were discussed in minute detail. Rather it was a meeting about the *humanistic* aspects of astrobiology, particularly preparing for finding life, and the potential impact if we do. It was billed as a *rational* approach, because it was designed to be a systematic and scholarly (though hardly comprehensive) attempt at tackling the problem, making use of knowledge from a wide range of disciplines. In this spirit the program featured not only scientists, but also philosophers, theologians, historians, and anthropologists. Atypically, the discussion was intended to address not only the impact of the search for extraterrestrial intelligence (SETI), but also microbial and complex life. Indeed, many astrobiologists believe microbial life will be discovered first, certainly if it comes as a NASA discovery, since NASA's astrobiology focus at present is on microbes. Still, the discovery of even microbial life beyond Earth would be perhaps the greatest discovery in the history of science.

Some have asked why now: why not wait to discuss the societal impact of extraterrestrial life until it is actually found? Military planners have an answer for this – waiting for a problem to arise blindly in some part of the world would be considered dereliction of duty, if only because different scenarios require the deployment of different resources, sometimes quickly. Contingency plans are essential, whether an event is likely or not. And it is always better to think ahead, to have time to consider options in a thoughtful way rather than to react in the passion of the moment. Scientists also have an answer, which is why the Human Genome Project has from its beginning sponsored a robust program on the ethical, legal, and social implications of its work. The bottom line is that it is always better to be prepared for events, with the goal of minimizing risks to humanity. Protocols may not work perfectly (as evidenced in the 2014 ebola outbreak), but they work better than nothing at all. And while we can debate how likely any extraterrestrial life discovery scenario might be, in the last few decades the discovery of thousands of planets, some Earth-sized and in the habitable zone of their parent stars, have made the discovery of life beyond Earth more likely.

Given these developments in astrobiology, it is time that we look seriously and systematically at the problem of astrobiology and society. The stakes are high. More than 50 years ago the US National Academy of Sciences compared the impact of astrobiology to the impact of Copernicus and Darwin and concluded, "The scientific question at stake in exobiology is, in the opinion of many, the most exciting, challenging, and profound issue, not only of this century but of the whole naturalistic movement that has characterized the history of western thought for three hundred years. What is at stake is the chance to gain a new perspective on man's place in nature, a new level of discussion on the meaning and nature of life" (National Academy of Sciences, 1962). Ten years later science fiction pioneer and visionary Arthur C. Clarke wrote that, "The idea that *we* are the only intelligent creatures in a cosmos of a hundred million galaxies is so preposterous that there are very few astronomers today who would take it seriously. It is safest to assume, therefore, that *They* are out there and to consider the manner in which this fact may impinge upon human society" (Clarke, 1972). Even for those who consider the discovery of extraterrestrial life a low-probability event, the potentially high impact makes our endeavor prudent, if not essential.

A small interdisciplinary research group, largely under the auspices of the NASA Astrobiology Institute, has been addressing the issues of astrobiology and society over the last few years (Race *et al.*, 2012). It is a sign of astrobiological optimism that several other groups and individuals have also recently taken up the subject of the impact of discovering life (Bertka, 2009; Dick, 2000; Harrison, 1997; Impey *et al.*, 2013; Michaud, 2007; Vakoch, 2013). This volume is intended as a contribution to that effort, spurred on by the interest of NASA and the Library of Congress, as well as almost daily discoveries bearing on the subject. The volume begins in Part I by looking at frameworks for approaching the problems of discovery and impact. We are immediately faced with the problem of how we can transcend anthropocentrism when we talk about foundational concepts like life and intelligence, culture and civilization, and technology and communication. These problems are addressed in Part II. Part III tackles the potential philosophical, theological, moral, and cultural impacts of finding extraterrestrial life, while Part IV tackles the more practical aspects of preparing for discovery – or non-discovery. The questions we ask throughout the volume are foundational, examining the very roots of some of humanity's most cherished concepts. In the end they reflect on an age-old question that never loses relevance: what does it mean to be human? I maintain that even in the event that life is not discovered beyond Earth, the questions addressed in this volume will have been worthwhile because astrobiology forces us to look at ourselves from this foundational extraterrestrial perspective.

This Symposium was held during my tenure as the Baruch S. Blumberg NASA/Library of Congress Chair in Astrobiology, located in the Library's John

W. Kluge Center in the magnificent surroundings of the Thomas Jefferson Building, just across the street from the United States Capitol. I want to thank Congressman Lamar Smith, Chairman of the House Science Committee, whose remarks opened the Symposium; Mary Voytek, Director of the NASA Astrobiology program at NASA Headquarters; and Carl Pilcher and Ed Goolish, Director and Acting Director of the NASA Astrobiology Institute, for their support throughout the year. I also thank the staff of the John W. Kluge Center, especially its two directors during my tenure, Carolyn Brown and Jane McAuliffe, as well as JoAnne Kitching, Jason Steinhauer, and Danielle Turello for their important contributions to the symposium. They provided the congenial and resource-rich environment in which this volume was conceived and implemented. In addition at Cambridge University Press I wish to thank my editor Vince Higgs, as well as Karyn Bailey, Cassi Roberts, Rachel Cox, Jonathan Ratcliffe, and Zoë Lewin.

This book is dedicated to two pioneers: John Billingham, who led a series of workshops on this subject 25 years ago as the head of the NASA SETI program (Billingham *et al.*, 1999), and Baruch S. Blumberg, the 1976 Nobelist in medicine and founding director of the NASA Astrobiology Institute (Pilcher, 2015). Many of us remember Barry fondly for his passionate interest in the subject, not only for the science, but also for the societal aspects discussed here. He always liked to think of astrobiology as exploration in the tradition of Lewis and Clark. This volume should be considered in that light as well – exploration, pushing the envelope of knowledge into uncharted territory, wherever it may lead.

Steven J. Dick
Washington, DC
March, 2015

References

Bertka, C. M., ed. 2010. *Exploring the Origin, Extent, and Future of Life: Philosophical, Ethical and Theological Perspectives.* Cambridge: Cambridge University Press.

Billingham, J., R. Heyns, D. Milne, *et al.*, eds. 1999. *Social Implications of the Detection of an Extraterrestrial Civilization.* Mountain View, CA: SETI Press.

Clarke, A. C. 1972. *Report on Planet Three.* New York, NY: New American Library, p. 90.

Des Marais, D., D. Nuth, L. J. Allamandola, *et al.* 2008. The NASA Astrobiology Roadmap. *Astrobiology*, 8, 715–730.

Dick, S. J. 2000. *Many Worlds: The New Universe, Extraterrestrial Life and the Theological Implications*. Philadelphia, PA: Templeton Press.

Dick, S. J. 2012. Critical issues in the history, philosophy, and sociology of astrobiology, *Astrobiology*, 12, 906–927.

Dick, S. J. and J. E. Strick, 2004. *The Living Universe: NASA and the Development of Astrobiology*. New Brunswick: Rutgers University Press.

Harrison, A. 1997. *After Contact: The Human Response to Extraterrestrial Life*. New York, NY: Plenum.

Impey, C., A. H. Spitz, and W. Stoeger, eds. 2013. *Encountering Life in the Universe: Ethical Foundations and Social Implications of Astrobiology*. Tucson, AZ: University of Arizona Press.

Library of Congress 2014. Astrobiology symposium, video online at https://astrobiology.nasa.gov/seminars/featured-seminar-channels/special-seminars/2014/9/18/nasalibrary-of-congress-astrobiology-symposium/

Michaud, M. 2007. *Contact with Alien Civilizations: Our Hopes and Fears about Encountering Extraterrestrials*. New York, NY: Springer.

National Academy of Sciences. 1962. *A Review of Space Research*. Washington, DC: National Academy Press, pp. 9–2 and 9–3.

Pilcher, C. 2015. Explorer, Nobel Laureate, astrobiologist: things you never knew about Barry Blumberg, *Astrobiology*, 15, 1–14.

Race, M., K. Denning, C. M. Bertka, *et al.* 2012. Astrobiology and society: building an interdisciplinary research community. *Astrobiology*, 12, 958–965.

Sullivan, W.T, III and J. A. Baross, eds. 2007. *Planets and Life: The Emerging Science of Astrobiology*. Cambridge: Cambridge University Press.

United States Congress. 2013. House Committee on Science, Space, and Technology, Hearings on "Astrobiology: Search for Biosignatures in our Solar System and Beyond," December 4, 2013, http://science.house.gov/hearing/full-committee-hearing-astrobiology-search-biosignatures-our-solar-system-and-beyond (accessed December 17, 2014).

United States Congress. 2014. House Committee on Science, Space, and Technology, Hearings on "Astrobiology and the Search for Life in the Universe," May 21, 2014, http://science.house.gov/hearing/full-committee-hearing-astrobiology-and-search-life-universe (accessed December 17, 2014).

Vakoch, D., ed. 2013. *Astrobiology, History and Society: Life Beyond Earth and the Impact of Discovery*. Heidelberg: Springer.

Part I Motivations and approaches

How do we frame the problems of discovery and impact?

Introduction

For the most part in this volume we assume that life exists beyond Earth and ask what the implications are if a discovery is made. In other words, we begin where most scientific discussions of astrobiology end. Before we head down that path, however, it is prudent to ask why we should believe such life exists. A large literature exists on this subject, ranging from the optimistic (e.g. Davies, 2010; Shklovskii and Sagan, 1966) to the skeptical (Gonzalez and Richards, 2004; Ward and Brownlee, 2000). It is not the purpose of this section to adjudicate between the optimists and pessimists, only to see why studying the societal implications of finding life beyond Earth is a valid endeavor.

The first two chapters of Part I summarize the arguments of the optimists from the point of view of both science and philosophy. Seth Shostak, a radio astronomer and Director of the Center for SETI Research at the SETI Institute, discusses the three broad empirical approaches to the search for life: direct exploration by spacecraft, biosignatures in planetary atmospheres, and the search for signals of artificial origin. Should one of these searches prove successful, he believes the societal reaction might be less dramatic than often assumed. Others in this volume beg to differ. Iris Fry, a philosopher who has written extensively on the history of the origins of life controversy (Fry, 2000), examines our deep philosophical presuppositions in the search for life – the Copernican assumption that the Earth is not special, and the Darwinian assumption that life emerged and evolved on Earth by natural processes and might do so wherever biogenic conditions prevail. While these presuppositions are not proven, she argues that astrobiologists are continually testing them, and there are grounds for being optimistic that their assumptions are valid. She contrasts this with the presuppositions of the Intelligent Design movement, some of which implicitly or explicitly drive opposition to the search for life (as in Gonzalez and Richards, 2004). Those assumptions, she argues, are not testable. In other words, some presuppositions are better than others. This does not mean we are lacking good arguments against the existence of life beyond Earth, only that valid grounds exist to proceed with the search and to study potential societal implications.

While the first two chapters frame the likelihood of discovery of life beyond Earth, Chapters 3 and 4 frame the problem of the impact of discovery. In Chapter 3, I examine three approaches from the point of view of human experience: history, discovery, and analogy, laying out a variety of discovery scenarios. History offers lessons from the reaction to cases where life beyond Earth was thought to have been discovered; the nature of discovery teaches us that the event will be an extended affair; and analogy offers cautious but important lessons from the point of view of culture contacts, scientific advances, and changing worldviews. While we certainly cannot predict societal impact for any given scenario, these three approaches arguably can serve as solid guidelines to encounters with life. Taking a very different approach, the philosopher Clément Vidal greatly elaborates possible scenarios and argues that the discovery of extraterrestrial life will either be in the form of microbial life or intelligent life that does not communicate because it is inferior or superior to us. Vidal goes on to elaborate a multi-dimensional impact model from several perspectives, including those of the extraterrestrials. He argues that the extended nature of any discovery means that media and public interest may wane, also making it less impactful than one might predict. A smooth impact, he argues, is what we should seek in any case through proper preparation.

Taken together, these scientific, philosophical, and historical considerations of the problems of discovery and impact set the stage for the remainder of the volume.

References

Davies, P. 2010. *The Eerie Silence: Renewing Our Search for Alien Intelligence.* Boston, MA: Houghton, Mifflin, Harcourt.

Fry, I. 2000. *The Emergence of Life on Earth: A Historical and Scientific Overview.* New Brunswick, NJ: Rutgers University Press.

Gonzalez, G. and Richards, J. 2004. *The Privileged Planet: How our Place in the Cosmos is Designed for Discovery.* Washington, DC: Regnery Publishing.

Shklovskii, J. and Sagan, C. 1966. *Intelligent Life in the Universe.* San Francicso, CA: Holden-Day.

Ward, P. and Brownlee, D. 2000. *Rare Earth: Why Complex Life Is Uncommon in the Universe.* New York, NY: Springer-Verlag.

1 Current approaches to finding life beyond Earth, and what happens if we do

SETH SHOSTAK

Three broad approaches exist in the search for extraterrestrial biology: (1) discover life in the Solar System by direct exploration; (2) find chemical signatures for biology in the atmospheres of exoplanets; or (3) detect signals (radio or optical) transmitted by intelligent beings elsewhere. In this chapter I describe each of these approaches, and then elaborate the multiple ways that we might learn of technologically competent civilizations. I also discuss why society's immediate reaction to the discovery of extraterrestrial intelligence would be less dramatic than often assumed. In all three cases the search for life beyond Earth is the ultimate remote sensing project. With few exceptions (such as sample return missions) this is exploration at a distance. While some reconnaissance is done by spacecraft, the majority of the effort consists of sifting through information brought to us in a storm of photons, either optical or radio.

Introduction

The idea of extraterrestrial biology is hardly new, with written speculation on the subject dating back two millennia and more (Dick, 1982). The first scientific searches are more recent, beginning with Johannes Kepler who, observing the Moon in detail through an early telescope, thought he recognized features carved by rivers. These, he reasoned, were sure signs of biology. Kepler also believed that craters were the surface manifestations of underground cities constructed to protect the citizenry from the relentless sunshine of the two-week lunar day (Dick 1982, 75–77; Basalla 2006, 21).

These pioneering observations were plagued by naïve, anthropocentric assumptions and a lack of information on the true environments on these worlds. Such bugaboos continued to affect attempts to find cosmic company for centuries, extending to the enthusiastic study of Mars by astronomer Percival Lowell. In a series of books, lectures, and articles extending from 1894 until his death in 1916, Lowell proclaimed the existence of a vast, hydraulic civilization on the Red Planet (Crowe 1986; Dick 1996). Just as

Kepler had done, he appealed to morphological evidence – straight-line features that he interpreted as canals – to back up these assertions. Lowell's claims were spurious, although one could argue that the falsity of his discoveries was due more to poor observation than poor interpretation (the trap that had snared Kepler). If the linear features described by Lowell actually existed, they would have been compelling evidence for intelligent beings.

Our knowledge of possible cosmic habitats and their habitability has grown substantially since these early efforts. We've mapped most of our Solar System in detail and have found thousands of planets around other stars. Scrutiny of these worlds has likewise increased: the last half-century has seen the beginnings of radio and optical SETI (the search for extraterrestrial intelligence), robotic exploration of Mars, and spacecraft reconnaissance of moons around the giant planets. In addition, astronomical research has shown that planets are commonplace, and habitable worlds may be plentiful.

A recent analysis of data from NASA's Kepler mission indicates that roughly a fifth of all stars host at least one habitable world (Petigura 2013, 19273). We've also learned of the probable existence of massive liquid reservoirs on five nearby moons, and the likelihood of underground aquifers on Mars. Additionally, the discovery of terrestrial extremophiles able to survive conditions that a few decades ago might have seemed too daunting for life suggests that many worlds – even those with environments rather different from Earth – could be inhabited (Schulze-Makuch, Chapter 5, this volume). For all these reasons, the search for extraterrestrial life – always a subject of interest to the public – has become popular with the research community as well.

The three-way horse race

The various strategies being pursued in the hunt for extraterrestrial biology naturally fall into the three broad categories enumerated above (Shostak 2012). Of the three, direct exploration of the Solar System is the most costly, and requires sophisticated robotic spacecraft and rovers, and eventually manned expeditions. The spectroscopic search for biomarkers in the atmospheres of exoplanets or their moons is dependent upon telescopes that are mostly still unconstructed. The third strategy, SETI, is limited in scope due to very minimal funding. We consider each of these strategies in greater detail below.

Direct reconnaissance of the solar system

Of all the nearby worlds that have tempted scientists with the promise of extraterrestrial biology, none has been more seductive than Mars. Lowell's canals were chimeras, but despite careful research in the early twentieth century proving that the Martian surface was dry, cold, and layered by an atmosphere only 1 percent that of Earth's, many scientists were still convinced that the Red Planet wasn't a dead planet. Acting on that optimism, NASA sent two Viking landers to Mars in 1975. The craft bore sophisticated instruments designed to detect both macroscopic and microbial life, and were launched with both fanfare and high hopes. Carl Sagan, who helped design and manage the mission, ventured that, "The possibility of life, even large forms of life, is by no means out of the question."

The landers provided, at best, ambiguous results. They sampled and sniffed the Red Planet's dusty dirt, looking for microbial metabolism. They didn't find it, although one member of the experimental team maintains to this day that they did (Levin and Straat 1977). The Viking biology team consensus was that the barsoomian landscape is sterile, and is kept that way by stinging ultraviolet radiation from the Sun and oxidizing compounds in the soil. But the issue was reopened after the Phoenix lander discovered perchlorates on Mars in 2008, possibly causing a false positive from the biology experiments (Navarro-Gonzalez *et al.* 2010). And evidence both morphological and chemical has since suggested that liquid water once pooled and flowed on Mars, perhaps fed by underground aquifers that could still exist. The possibility of life remains, although the evidence could be difficult to reach.

Chastened by the experience with Viking, NASA today is taking a more cautious approach to looking for Red Planet residents, and in particular is using its orbiters and rovers to reconnoiter locations that may have been lakes or rivers billions of years ago. By eschewing extant life in favor of life that may have existed in the past, the agency reckons that it has upped its chances for success. The whole history of biology on Mars – assuming there is one – is made fair game for eventual discovery.

At least three moons of Jupiter (most notably, Europa) as well as two moons of Saturn show promise for extraterrestrial biology. Titan is the only other body in the solar system with liquids on its surface, but that surface is at −179 °C, and the lakes there are reservoirs of liquefied natural gas, not water. But Saturn has a second seductive moon – Enceladus – that also shows strong evidence of subsurface aquifers. Thanks to the periodic kneading of this moon as it orbits its host planet, some of the water is erupted into space, where it makes an attractive target for a flyby space mission to grab and analyze. Europa also squirts small amounts of frozen water into space – water that has managed to find its way through the 15 km of ice that separates this moon's surface from the vast oceans below (Figure 1.1). It too may contain evidence of microscopic biology.

Figure 1.1 The icy surface of Europa, one of the four Galilean satellites of Jupiter, is believed to cover vast oceans, possibly harboring life. This mosaic of images was taken by the Galileo spacecraft in 1995 and 1998, and recently reprocessed. NASA/JPL-Caltech/SETI Institute.

In these cases, geysers greatly reduce the difficulty of searching for life by obviating the need for a lander to drill through a thick, icy carapace. In 2015, NASA will consider proposals defining the science instruments for its intended Clipper mission to Europa, and there is at least some planning for a Discovery-class mission to Enceladus. If life exists in the outer Solar System, that fact might become known within a decade.

Spectroscopic search for biomarkers in extra-solar system bodies

Earth's atmosphere betrays the biology below (Figure 1.2). The one-fifth of our air that is oxygen speaks to the presence of photosynthesis. Atmospheric methane is also a clue to both bacteria and bovines. Similar biosignatures could be present on exoplanets (or even on their moons). What's required to find them is a telescope that can separate the light from these worlds and their host sun; in other words, an instrument that can provide a clean, one-pixel (at least) image of the target.

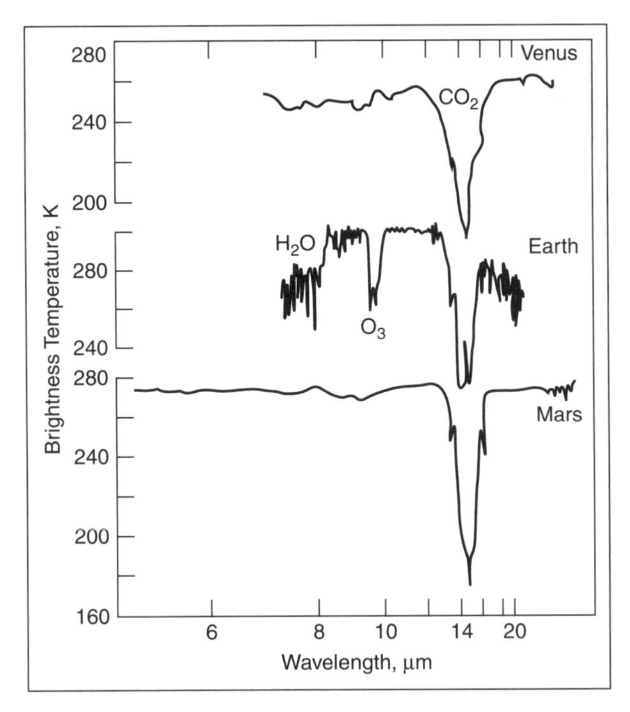

Figure 1.2 Comparison of Mars, Venus, and Earth in water bands, showing the clear presence of water on Earth uniquely. Credit: NASA Workshop, Pale Blue Dot.

In 2002, NASA planned to build two space-based telescopes, known collectively as the Terrestrial Planet Finder, to accomplish this. One was a very large, conventional mirror telescope that would operate at visible wavelengths and use an occulting disk to block out the intense light of stars. The second was an array of infrared telescopes. By a trick of wave interference these instruments could resolve out the star, making it largely disappear and leaving only planets in the visual field. The Planet Finder telescopes were scheduled to scrutinize a small number of nearby star systems, and spectroscopically analyze the reflected light from any planets that were found. Unfortunately, in 2011, the space agency labeled this project as "canceled." Four years earlier, a similar European space-based interferometer called Darwin was also axed.

NASA's James Webb infrared telescope will be able to pick up some of the slack left by these uncompleted endeavors. Scheduled for launch in 2018, this space-based instrument will be able to image the larger planets around the smallest stars, those known as red dwarfs and white dwarfs. With its on-board spectrometer, it can search for water vapor, oxygen, and carbon dioxide in the atmospheres of these planets. In the right proportions, these three ingredients are thought to be a reliable signature of biology.

In this way, even microscopic life – alien analogs of the blue-green algae (cyanobacteria) that infest many environments on Earth – could be detected at light-years distance by means of their waste products. Methane, the smelly exhaust gas from microbes and many ungulates, would also serve as a tip-off for biology. Intriguingly, it's conceivable that such a device could detect a spectral feature of chlorophyll called the "red edge," although that would require considerably more signal-to-noise than finding oxygen or water vapor (Brandt and Spiegel 2014).

As exciting as these prospects are, the Webb telescope has only limited ability to image exoplanets. It may be that only a few dozen will be within its purview. But this number might be greatly increased – perhaps to hundreds or thousands – if a plan to recycle an earlier idea from the canceled Planet Finder is followed, namely to put a "star blocker" in front of a new space-based telescope. An occulting disk of several tens of meters in size would be positioned in the sight line of a relatively inexpensive telescope, covering up the star and enhancing the visibility of orbiting planets. However, while simple in design, this scheme is still tricky: the star blocker and the telescope have to remain aligned even when separated by tens of thousands of kilometers. Nonetheless, this sort of instrument could be operational within two decades.

Searches for intelligence

Arguably, the most speculative of the three approaches to discovering life beyond Earth are experiments designed to find the artifacts of a technically accomplished society, known as SETI. Not everyone is sanguine that, even if life is commonplace, intelligent life is also plentiful (Mayr 1995). Earth's history shows that intelligence can be a late (and so far, brief) evolutionary development that, in our case, occurred halfway through the lifetime of a star.

Nonetheless, a search for technically accomplished beings is clearly interesting, as it would provide a second example of what is unique about our own species: namely, the ability to reason, to understand nature, and turn its laws to our benefit. Intelligence elsewhere would be of considerable philosophical interest, far beyond the discovery of a new species of fauna on this planet, or microbes on another.

Modern SETI can be traced to a famous experiment by Frank Drake in 1960, using an 85-foot antenna at the National Radio Astronomy Observatory in Green Bank, West Virginia (Drake 1961). His pioneering Project Ozma set the stage for much of the SETI effort of the past half-century and featured (1) a search for narrow-band signals in the microwave region of the spectrum, and more specifically in the vicinity of the 1420-MHz line of neutral hydrogen, arguably a universal "hailing frequency," and (2) the targeting of nearby, Sun-like stars, on the assumption that they would be the most likely neighborhoods for planets with complex life.

The observational choices made by Drake have been somewhat altered in the light of new technology and new astronomical knowledge, in particular the following. (1) The discovery of thousands of planets around other stars (exoplanets) has strongly suggested that even stars quite different from the Sun may shelter habitable worlds. We now know that M-dwarf stars (also known as red dwarfs) comprise three-fourths of all stars, and may host life-friendly planets in great abundance. (2) Improvements in both radio technology and digital electronics have broadened the range of frequencies that can be quickly searched for signals. We no longer have to restrict observations to the 1420-MHz band.

Currently, the SETI institute uses a configuration of small antennas known as the Allen Telescope Array, situated in northern California, to examine promising star systems and special locations (such as the galactic center) for signals. Unlike previous campaigns that relied on antennas that were shared with radio astronomers doing conventional research, this array can be used approximately 12 hours a day for SETI observing (Figure 1.3).

Figure 1.3 The Allen Telescope Array. Seth Shostak/SETI institute.

The other large radio SETI group in the United States is at the University of California, Berkeley. Their long-running Project SERENDIP uses the very large (1,000-foot diameter) antenna at Arecibo, Puerto Rico in a commensal mode. By piggybacking on this antenna, the Berkeley group gets nearly non-stop use of the antenna, but the tradeoff is that they have no control of where it is aimed. However, over the course of several years this random scrutiny covers roughly one-third of the sky. A small fraction of the Berkeley data is made available for processing on home computers using the popular screen saver, SETI@home.

At the moment, there is only one full-time radio SETI experiment outside the United States, conducted by a small group at the Medicina Observatory of the University of Bologna, in Italy.

The SETI effort is still small, but new instruments are sure to spawn new efforts. The European Low Frequency Array (LOFAR) is a recently commissioned radio telescope able to examine the sky effectively at frequencies far below the microwave bands normally used for SETI. Its evident advantage for such searches – the fact that it operates in the same wavelength regions as our own radio and television services – is also a complication. There is bountiful terrestrial interference at these wavelengths that must be sorted out before any signals detected by LOFAR can be ascribed to extraterrestrial activity.

A new, very large radio telescope able to work at high frequencies will also be coming on-line sometime around 2025. Known as the Square Kilometer Array, it will have ten times the sensitivity of the Arecibo radio telescope and concurrent use for SETI seems like an obvious project for this largely European radio astronomy initiative.

In addition to the long-running search for radio signals, there are also limited efforts to search for brief laser light pulses from space, an experiment known as optical SETI. Both the University of California at Berkeley and Harvard University field such efforts. Given the relative ease with which light can be beamed, it may be that optical signaling might be used by advanced societies to serially "ping" large numbers of stellar systems that might harbor intelligence, to draw their attention to lower power signals with message content (Shostak 2011).

While no one knows how prevalent signal-generating civilizations might be, conservative estimates of the prevalence of extraterrestrial intelligence suggest that finding a signal demands observation of a million star systems or more. While that is a thousand times the search space investigated so far, it might still be a tractable project for the near future, given the relentlessly increasing power of digital electronics. It is hardly hyperbolic to suggest that scientists could very well discover extraterrestrial intelligence within two decades' time or less, assuming that they can find the resources necessary to conduct the search.

Non-conventional searches for intelligence

In addition to the schemes described above, frequently characterized as "conventional SETI," there are many novel approaches for detecting technically accomplished societies. Foremost among these are those that eschew electromagnetic signaling in favor of other information carriers. Gravity waves have been repeatedly proffered, although the motivation – that somehow this would allow instantaneous communication – are not justified by physics theory, which declares that these waves would move at the speed of light and no faster. In addition, gravity wave detectors are big, because the waves themselves are weak.

Another approach is to send bursts of neutrinos. Neutrino signaling would have the advantage that our receivers could monitor the entire sky simultaneously (even the part of the sky on the other side of the planet, as the particles themselves would pass right through with little hindrance). However, the neutrinos are expensive in terms of energy, and the detection efficiency – at least in the experience of efforts so far – is abysmally low. In any case, neutrino detectors, such as the University of Wisconsin's Ice Cube project (University of Wisconsin-Madison 2014), are already wielded by physicists, and in that sense this sort of SETI experiment is already underway.

In 2004, two computer scientists suggested that to garner the highest average bit rate between star systems, one should simply inscribe information on a substrate and physically transport it to a destination (Rose and Wright 2004). In other words, just loading up a spacecraft with books, or better yet electronic media, could convey enormous numbers of bits to a target at a higher rate than sending them on a radio (or light) beam. While this sort of signaling might make sense once you know what the target is, if that's not the case then this is an expensive approach. It seems, given the limited visibility of *Homo sapiens* from light-years away, we shouldn't expect an alien craft stuffed with Klingon thumb drives to arrive in our neighborhood.

Considering that our species has only recently begun to make our presence known to nearby star systems (since the invention of radar), it's unlikely that any society would be relentlessly targeting our world with a high-powered signal. This suggests that transmissions aimed our way would be more in the nature of "pings," as noted above, designed to get our attention – and sent as part of a large reconnaissance project on the part of the initiators. With this logic, signaling schemes that sent very few bits of information, but did so conspicuously, might make sense. The aliens may only wish to tell us where on the sky to search for other, less obvious, signals.

There have been a few pinging schemes suggested in addition to the optical pulses described earlier. One possibility is a specially shaped light blocker in orbit around the home star of an advanced society. As examples, these could take the shape of a giant triangle or a series of massive shutters (Arnold 2013). These would be perceived by anyone doing transit searches for planets, and – with sufficient signal-to-noise in the receiving equipment – could be readily distinguished from a natural transiting world.

Another technique that advanced societies might employ to signal others is to "tickle" a Cepheid variable star with high-energy radiation, and thereby change its pulse period (Learned *et al.* 2008). Cepheids can be easily seen at intergalactic distances, and this scheme would allow a society to announce its presence even from millions of light-years away. Detection requires nothing more than the daily monitoring of the apparent brightness of Cepheids, something that's routinely done as part of astronomical research.

For decades, there has been speculation about the detection of massive space-based modulators that could insert messages into the strong natural emissions of pulsars, quasars, or interstellar masers. Despite early reports that such modulations had been found (Sholomitsky 1965), there is no reason to suspect that variations in the signals from such astronomical objects are due to anything but natural processes.

In addition to the variety of intentional signals, a second category is inadvertent "signals." The oldest method for detecting intelligence at cosmic distances is simply to look for large-scale artifacts. The canals on Mars are a premiere example of this, but modern approaches include looking for an excess of infrared radiation (waste heat) from star systems or even entire galaxies. The former would betray the presence of Dyson swarms (Carrigan 2009), while the latter might comprise the energy detritus of very advanced, Kardashev Type III civilizations (Kardashev 1964), able to tap the energy of a galaxy.

A recent suggestion for finding technological life (as opposed to biology in general) has been to analyze exoplanets spectroscopically to detect atmospheric pollutants caused by incautious industrial activity on the part of the extraterrestrials. While this works in principle (and has the possible advantage of being able to discover societies that might be long gone), it is a greater technical challenge than the schemes described earlier to find chemical biomarkers in the air of other worlds.

Other approaches include a search for probes in our solar system, placed by other beings for reconnaissance (a self-centered idea, but then again humanity is generally of the opinion that it is the most interesting thing in this part of the universe, or possibly in all of it), looking for artifacts either in the Lagrange points or on the Moon, and the unconventional idea of appealing to alien intelligence on our own internet (Tough 2011).

As can be readily surmised from the above brief inventory, there is no lack of creativity in imagining ways in which we might prove that advanced beings populate space. While none has yet provided compelling arguments for this hypothesis, most of these ideas can either be experimentally explored today, or with the technologies of the near future. These will certainly broaden the search space of existing SETI searches, as well as increasing the speed at which we reconnoiter the sky. In addition, techniques for verifying the extraterrestrial nature of "one-off" signals can better address the possibility that Earth is simply on a list of possibly interesting worlds, and that beings elsewhere are merely trying to attract notice or a response (Shostak 2011).

Reaction to the detection of life beyond Earth

Many members of the public feel that the discovery of extraterrestrial life would be of such momentous consequence that there would be both reason and effort to conceal it. This probably derives from the widespread opinion that governments know about alien visitation (the UFO phenomenon) and are hiding important evidence of extraterrestrials frolicking in our air space. In addition, the non-stop appearance of aliens, mostly hostile, in film and television reinforces the idea that finding extraterrestrial life would be disruptive and possibly dangerous.

There is little historical reason to agree with such dark imaginings. Many people were inclined to believe Percival Lowell's claims of ditch-digging Martians, in orbit a mere 50 million kilometers from Earth, but this had no effect on government policies or public deportment. This indifference to Martian habitation was on view again in 1996, when NASA scientists claimed to have found the remains of Red Planet microbes in a meteorite (McKay *et al.* 1996). While a compelling story that dominated headlines for days, there was neither disquiet nor distraction from the other news of the day. People wanted to know more; they were not rioting in the thoroughfares.

It seems more than likely that the announcement of finding life elsewhere – including a signal that would be proof of intelligent life – would be received with interest rather than fear in most quarters. After all, the public has been conditioned by more than a century of science fiction to expect biology beyond Earth, and roughly one-third of them think that it's already in our skies and occasionally our bedrooms (LiveScience 2012). Even so, there's precious little disquiet.

But if there is little motivation (and no ability, as false alarms have demonstrated) to cover up proof of extraterrestrials, it is hard to dispute that it would be a major discovery. The drama that would accompany the finding of a signal has been taken seriously by the SETI community itself, resulting in the drafting of "protocols" intended to guide the reaction of the researchers should they pick up a transmission (see Michaud, Chapter 18 in this volume). These suggestions, which have no force of law, are straightforward and thoroughly innocuous. They are: (1) carefully verify that the signal is both artificial and extraterrestrial; (2) inform governments, the media, and the public; and (3) refrain from any efforts to transmit a reply without international consultation.

The very existence of these protocols is remarkable. It is unusual in the extreme for explorers, whether they be of the conventional or scientific sort, to fashion documents that are intended to circumscribe their behavior in the event that they make a discovery. In general, science does not manage its finds in this way. Moreover, experience with false alarms (Shostak 2009, 1–19) shows that SETI protocols, while well intended, can easily founder on the exigencies of real-world conditions. For technical reasons, verifying a signal will take days, and in all that time the media will be clamoring for more information. Given that secrecy in SETI has never been implemented or widely desired, the news of the "interesting" signal will be impossible to keep secret. The discovery, whether or not it turns out to be confirmed, will be reported long before the researchers themselves are ready to announce it. The protocols will quickly be overtaken by events.

There has been an effort to investigate some of the more provocative aspects of the public's reaction to finding life, including the response of organized religion. Brother Guy Consolmagno, an astronomer at the Vatican Observatory, has opined that most mainstream theologies would be able to simply incorporate such news into existing theology. If your religion has survived millennia – if it can handle Copernicus, Galileo, and even Darwin – then ET should eventually prove palatable, he maintains (Consolmagno 2007, and Chapter 15 of this volume).

The reaction might be otherwise for non-mainstream believers. In particular, research psychiatrist Paul Lavrakas and his collaborators suggest that fundamentalists, who derive their faith from scripture – an extrinsic source – would note that since extraterrestrial life is not described in the Bible, evidence for its existence should be actively questioned (Rosenbaum *et al.* 1980). The discovery of a signal would likely be called a hoax.

Nonetheless, the general short-term reaction to the discovery of extraterrestrial life – based on historic analogs that are admittedly imperfect – would be dominated by excitement, interest, and a desire to know more. Long-term consequences are far less predictable, and depend on such unknowns as the nature of the life found (consider the difference between discovering microbes under Europa's icy shell and uncovering an alien probe in orbit near Earth) and what we might learn of it.

To think we have much insight into how this discovery might affect the future course of humanity is akin to Europeans of the fifteenth century believing that they knew what the ramifications of discovering an entirely new continent 5,000 kilometers to their west would be. Nevertheless, as this volume demonstrates, serious attempts are now being made to at least lay out impact scenarios in advance of what would be one of the greatest discoveries in the history of science.

References

Arnold, L. 2013. Transmitting signals over interstellar distances: three approaches compared in the context of the Drake equation. *International Journal of Astrobiology*, 12:212–17.

Basalla, G. 2006. *Civilized Life in the Universe: Scientists on Intelligent Extraterrestrials.* New York, NY: Oxford University Press.

Brandt, T. and Spiegel, D. 2014. Prospects for detecting oxygen, water and chlorophyll on an exo-Earth. *Proceedings of the National Academy of Sciences*, 111:13278–83.

Carrigan, R. 2009. IRAS-based whole-sky upper limit on Dyson spheres. *Astrophysical Journal*, 698:2075.

Consolmagno, G. 2007. Personal communication, January 29.

Crowe, M. 1986. *The Extraterrestrial Life Debate, 1750–1900. The Idea of a Plurality of Worlds from Kant to Lowell.* Cambridge: Cambridge University Press.

Dick, S. J. 1982. *Plurality of Worlds: The Origins of the Extraterrestrial Life Debate from Democritus to Kant.* Cambridge: Cambridge University Press.

Dick, S. J. 1996. *The Biological Universe: The Twentieth Century Extraterrestrial Life Debate and the Limits of Science.* Cambridge: Cambridge University Press.

Drake, F. D. 1961. Project Ozma. *Physics Today*, 14:140.

Kardashev, N. 1964. Transmission of information by extraterrestrial civilizations. *Soviet Astronomy*, 8:217.

Learned, J., Kudritzki, R.-P., Pakvasa, S., and Zee, A. 2008. The Cepheid Galactic Internet. arXiv:08090.0339v2

Levin, G. and Straat, P. 1977. Life on Mars? The Viking Labeled Release Experiment. *Biosystems*, 9:165–74.

Livescience. 2012. One-third of Americans believe in UFOs, survey says. Accessed October 19, 2014. http://www.foxnews.com/scitech/2012/06/28/one-third-americans-believe-in-ufos-survey-says/.

Mayr, E. 1995. Can SETI succeed? Not likely. *Bioastronomy News* 7.

McKay, D., Gibson Jr., E. K. Thomas-Keprta, K. L., *et al.* 1996. Search for past life on Mars: possible relic biogenic activity in Martian meteorite ALH84001. *Science*, 273: 924–930.

Navarro-Gonzalez, R. Vargas, E., de la Rosa J., Raga, A. C, McKay, C. P. 2010. Reanalysis of the Viking results suggests perchlorate and organics at mid-latitudes on Mars. *Journal of Geophysical Research*, 115, E12010.

Petigura, E., Howard, A. and Marcy, G. 2013, Prevalence of Earth-size planets orbiting Sun-like stars. *Proceedings of the National Academy of Sciences*, 110:19273–78.

Rose, C. and Wright, G. 2004. Inscribed matter as an energy efficient means of communication with an extraterrestrial civilization. *Nature*, 431:47–9.

Rosenbaum, D., Maier, R., and Lavrakas, P. 1980. Belief in extraterrestrial life: a challenge to Christian doctrine and fundamentalists? *Journal of UFO Studies*, 2, 47–57.

Sholomitsky, G. B. 1965. "Variability of the Radio Source CTA-102." *Soviet AJ*, 9:516.

Shostak, S. 2009. *Confessions of an Alien Hunter*. Washington, DC: National Geographic.

Shostak, S. 2011. "Short-pulse SETI," *Acta Astronautica*, 68:362–65.

Shostak, S. 2012. "How to Find Extraterrestrial Life," *Huffington Post*, July 5. Accessed October 20, 2014. http://www.huffingtonpost.com/seth-shostak/the-threeway-horse-race_b_1647047.html.

Tough, A. 2011. "Invitation to ETI." Accessed October 19, 2014. http://ieti.org/.

University of Wisconsin-Madison. 2014. "*Ice Cube*." Accessed October 10, 2014. http://icecube.wisc.edu/.

2 The philosophy of astrobiology

The Copernican and Darwinian philosophical presuppositions

IRIS FRY

Contrary to common wisdom, science is not exclusively defined by its methods and theories, nor is it constituted only by substantiated empirical hypotheses. Rather, philosophical presuppositions are also a crucial part of the scientific endeavor. Astrobiology, like any scientific field that seeks to learn and understand nature, rests on such philosophical presuppositions. We do not experience nature as a clean slate, as the tabula rasa upheld by the seventeenth–eighteenth-century Empiricists. Philosophical presuppositions guiding science are general, universal claims about nature that transcend limited experience. For example, the notion that natural laws necessarily hold not only on our planet or in our galaxy but in the universe at large cannot be proved or disproved empirically. Nevertheless, it is on the basis of the universal applicability of natural laws that astrobiological research is conducted. Likewise, any other branch of the natural sciences could not function and advance without this principle. Furthermore, philosophical presuppositions express general guiding evaluations of reality that by definition are not open to observation or experience. The claim that nature was created and designed by an intelligent designer or the denial of this claim cannot be empirically settled. Yet, it is the notion that natural processes depend on natural causes and not on supernatural purposes which guides science.

Although the status of philosophical assumptions in science clearly differs from that of theoretical-empirical claims, these two elements are deeply connected. The interaction between the theoretical-empirical and the philosophical becomes apparent when science is examined historically. I argue that this interaction, shaped to a large extent by social and cultural factors, has resulted in the last few centuries in the establishment of the evolutionary naturalistic worldview.[1] The major defining feature of this worldview is the rejection of

[1] Focusing on the contribution of social, political, and cultural elements to the growth of the evolutionary worldview, many historians of science deny an autonomous role to philosophical conceptions. Clearly, philosophical ideas are not "disembodied" and are not detached from their "carriers," at the personal and social levels. Yet, in my view, both the independent contribution of philosophical conceptions and of various social factors should be taken into account in our discussion.

supernatural teleology as necessary for the scientific study and understanding of nature.

Philosophical presuppositions of astrobiology

The natural sciences of today function within the framework of the naturalistic worldview. It is the robustness of this framework which provides validity also to branches of science that are still at the stage of establishing their fundamental data, notably the study of the origin of life on Earth and astrobiology.[2] It has been claimed that the problem of the origin of life, yet unsolved, is the "soft underbelly of evolutionary biology" (Scott 1996). The contention that astrobiology is a field that has yet to establish that its subject matter exists, first voiced in the early 1960s (Simpson 1964), still finds echoes today (Bada 2005; Lazcano and Hand 2012, 160). In addition to the explosive expansion since the 1960s of empirical data relevant to astrobiology (Ćirković 2014), the scientific validity of a field, as argued here, is not determined exclusively by its empirical results but also by its established philosophical underpinning.

We still have no clue whether there is life outside Earth – microbial, multicellular, or intelligent – either in our Solar System or beyond. A positive answer to this question depends on a combination of two factors: first, a habitable planet, i.e. an extraterrestrial planet with physical conditions conductive to life; and second, given such conditions, the actual emergence and evolution of life. The consensual scientific position regarding the physical conditions for life elsewhere is underlined by the Copernican presupposition: Earth is not unique; Earth-like environments enabling life might occur on other planets and moons (see Seager 2013). Recent data attesting to the proliferation of planetary systems in the galaxy and the discovery of an Earth-sized planet in the habitable zone of its star (Quintana *et al.* 2014) give credence to this conception.

There is still no consensus as to the mechanism of the emergence of life on Earth and such a process has not yet been simulated in the laboratory. Yet, the philosophical Darwinian presupposition that terrestrial life emerged and evolved naturally and might do so elsewhere is not in doubt and is commonly guiding research in the field. Recent theoretical and empirical advances in the study of the problem (Powner *et al.* 2009; Adamala and Szostak 2013; Engelhart *et al.* 2013) further serve as a stimulus for research. The origin of life itself is viewed by researchers as an evolutionary process, dependent on the

[2] In fact, astrobiology is conceived today by its practitioners as including the study of the origin of life on Earth (see, Dick and Strick, 2004, 1, 205).

prebiotic emergence of an evolvable physico-chemical infrastructure that could have undergone primitive reproduction, variation, and selection.[3] These processes are predicted to have led to the first living systems (Fry 2011).

The historical development of the Copernican–Darwinian conception

Innumerable studies have been devoted to the Copernican Revolution as well as to the wider framework of the Scientific Revolution, and to each of the major participants in this long historical drama. The literature on changing attitudes to the possibility of life in the universe (the debate on the plurality of worlds) during this period is extensive, especially thanks to the historians Steven Dick and Michael Crowe (Dick 1982, 1996, 1998; Crowe 1999).[4] These historical studies pointed to the crucial role played by both analogical and teleological reasoning in justifying the belief in the plurality of worlds. It is this expansive literature that should be consulted for numerous examples of the complex interaction between scientific achievements and philosophical ideas beginning from Copernicus himself (Figure 2.1) and culminating in the Copernican conception that underlies astrobiology.

In comparison to the historical development of Copernicanism and its scientific significance, the role of the Darwinian viewpoint in shaping astrobiological ideas is less often realized and thus will be examined here in more detail.[5] The establishment of the Newtonian mechanical universe toward the end of the eighteenth century made Copernican astronomy for the first time physically and cosmologically plausible (Kuhn 1985[1957], 261). The scientific tenets of this astronomy, especially the relationship between Earth and other bodies in the universe, are expressed in the Copernican presupposition that underlies today's astrobiology. However, necessary as this presupposition was, it was not sufficient to make the study of extraterrestrial life in the twentieth century scientifically valid. For this to happen, the teleological – anthropocentric – theological reasoning predominant in the traditional

[3] There are several, often competing, hypotheses regarding the nature of such infrastructure (see Fry 2011).

[4] References to good primary sources (among them, by Descartes, Fontenelle, Galileo, Kant, Kepler, Newton) are provided by Dick 1982, 1998 and Crowe 1999. See also Kuhn 1985[1957]; Guthke 1990; Koyré 1994[1957].

[5] The influence of Darwin's theory on supporters of pluralism strengthened the claims for the physical evolution of planets toward habitability and inhabitance (Guthke 1990, 346; Dick 1996, 1998, 18–19; Crowe 1999, 373–374). However, "Darwinian presupposition" as used here refers not to physical evolution or to "cosmic evolution" but to the evolution of life and especially to the evolutionary emergence of life from matter through a natural Darwinian process (see Fry 2011).

Figure 2.1 The heliocentric theory of Copernicus (1473–1543) made the Earth a planet and the planets potential Earths, constituting one of the two major preconceptions that underlie astrobiology. Copernicus scholar Owen Gingerich judges this image to be a modern reworking based on one in Pierre Gassendi's Copernicus biography of 1654.

pluralistic position had to be overcome. This was accomplished only following the rise of the theory of evolution in the nineteenth century and after the establishment of natural selection as its major mechanism in the first half of the twentieth century. Unlike the direct association between Copernican astronomical and cosmological concepts and the ideas supporting life beyond Earth, the influence of Darwinism took a different, less-direct course. The historical development of the Darwinian presupposition matched the complex empirical, theoretical, and philosophical development of Darwinism itself.

The clash between the support of the evolution of species, in particular the evolution of the human species, and the ideas of design and anthropocentrism as related to life in the universe, was evident even before the publication of Darwin's *Origin of Species* in 1859. In his 1853 book, *On the Plurality of Worlds: An Essay*, the renowned philosopher and scientist William Whewell abandoned his long-held support of the idea of plurality. As detailed by Crowe, this change was mostly in reaction to Robert Chambers's 1844 book, *Vestiges of the Natural History of Creation* that promoted a Lamarckian version of evolution, including spontaneous generation, and most significantly the evolution of man.[6] Whewell justly rejected the pluralists' weak analogies between Earth, other planets in the Solar System, and other celestial bodies and their readiness to speculate on flimsy evidence. He relied on new telescopic and spectroscopic astronomical data to make his case for the uniqueness of Earth in the whole universe as the abode of man (Crowe 1999, 286–288). Yet, Whewell's scientifically based criticism was motivated and guided by his wish to promote theological and teleological ideas in support of the argument from design (Crowe 1999, 281). Whewell argued that "the placing of man upon the earth was a supernatural event, an exception to the laws of nature." He thus could not accept Chambers's claim that "man grew out of monkey" as a "natural event, the result of a law" (Crowe 1999, 251).

Darwin (Figure 2.2) did not criticize Whewell publicly. Yet, a revealing comment in one of Darwin's pre-*Origin* 1838 notebooks ridiculed Whewell's judgment that: "the length of day is adapted to duration of sleep in man!! whole universe so adapted!!! & not man to planets. – Instance of arrogance!!" Crowe, quoting this comment, adds that this was Darwin's "effective reminder" against the program of natural theology "taken up in the Bridgewater Treatises," to which Whewell was a famous contributor (Crowe, 1999, 271). At the same time, Darwin himself often manifested a complex, sometimes ambiguous attitude to teleology, design, and religion commented upon by Darwin's scholars (among many, see Ospovat 1980; Sloan 1985; Kohn 1989; Moore 1991).

In his analysis of the history of Darwinism, historian David Kohn suggested that the development of Darwin's ideas toward the mechanism of natural selection amounted to the secularization of biology by the reformulation of teleology, e.g. by eliminating talk about final causes and replacing "perfect adaptation" by "relative adaptation" (Kohn 1989, 220–221, 229). This development was neither linear, nor unequivocal because it was the result of "a

[6] Notably, Chambers was also a supporter of pluralism (Crowe 1999, 265–355; Guthke 1990, 331–337).

Figure 2.2 The work of Charles Darwin (1809–1882) on natural selection gave rise to a second preconception that underlies astrobiology: the assumption that life emerged and evolved on Earth by natural processes and might do so wherever proper conditions prevail. This image showing Darwin late in life is attributed to the British photographer Julia Margaret Cameron.

dialectical relationship between [Darwin's] growing scientific theory and his conflicting metaphysical allegiances" (Kohn 1989, 222). Among these allegiances were Paley's natural theology, the often contradictory religious influence of members of Darwin's family, and his Cambridge Anglican educational background. Kohn insisted that this list also included Darwin's attraction to materialism and even to atheism. Thus, on the one hand "the early Victorian period was profoundly religious," but on the other, the presence of materialistic, atheistic elements in Darwin's thought "suggests that profound secularizing tendencies operated beneath the surface of early Victorian culture, and came to bear on Darwin's science" (Kohn 1989, 218, 219). Indeed, Kohn did not fail to note that, in a number of ways, Darwin was "a theologically transitional figure" (Kohn 1989, 232).[7]

According to several historians of evolution who emphasized the social construction of Darwin's theory, Darwinism changed its face several times from the publication of the *Origin* until Darwin's death and later. This change was a response to social and cultural processes in Victorian England and to fierce theologically oriented opposition to the theory of evolution (see Moore 1991). Peter Bowler claimed that to accommodate this opposition toward the end of Darwin's life, Darwinism was more teleological and "progressive" compared to its later twentieth-century version (Bowler 1990, 86).[8] For various scientific and philosophical reasons, in the early twentieth century several biologists still suggested purposive explanations of evolution and only beginning in the 1930s was natural selection adopted by scientists as the major evolutionary mechanism (Kellog 1907, 1–7; Provine 1988, 59–62; Bowler 2003, 240–273).

The complexity of the transformation toward a secularized, anti-teleological biology was reflected in a remarkable historical case, relevant to the issue of extraterrestrial life: In 1903, Alfred Russel Wallace, the great evolutionist and co-founder with Darwin of the theory of evolution by natural selection, published a radical anti-pluralist book, *Man's Place in the Universe*. Wallace began by examining various physical factors relevant to inhabitance, such as the distance of a planet from its sun, mass of planet, obliquity of its ecliptic, and an atmosphere of sufficient density (Wallace 1904[1903], 259–262). He presented these data as evidence that the Earth is the only inhabited planet in the universe (certainly, regarding "any forms above the lowest and most rudimentary" (Wallace 1904 [1903], 215). However, the astronomical data used by Wallace, in particular the claim that the Milky Way is almost equivalent in size to the whole finite universe and that our Solar System and hence Earth are located almost at the center of the universe, were called into question already at the time of the publication of his book. These claims were overturned and rejected not long after Wallace's death in 1913 (see Dick 2010, 330–331). Furthermore, Wallace drew teleological and anthropocentric conclusions

[7] For a more detailed description of the interplay between empirical study and philosophical assumptions in Darwin's "conversion to evolution," following his visit to the Galapagos Islands, see, Fry 2012, 667.

[8] Kohn pointed out that the reciprocal relationship between Darwin's science and his "metaphysics of evolution" (1989, 27) was regulated by "Darwin's perception of audience, which I identify with social location" (1998, 22).

from his data, and formulated a religious, though not a Christian, view of the universe and man's place within it.[9]

Wallace argued that "in order to produce a world that should be precisely adapted in every detail for the orderly development of organic life culminating in man" a whole universe may have been required (Wallace 1904 [1903], 256–257). He acknowledged that the majority of men of science will " ... explain this conclusion as due to a fortunate coincidence and will argue that if the course of the evolution of the universe has been a little different, there might have been many life-bearing planets or 'none at all'" (Wallace 1904 [1903], 263). His own explanation, on the other hand, was congruent with the "religious view": "[M]ind is essentially superior to matter and distinct from it ... life, consciousness, mind ... must be mind-products". Wallace added that those supporting this view "will see no difficulty in going a little further, and believing that *the universe was actually brought into existence for this very purpose*" (Wallace 1904 [1903], 264, emphasis added). Wallace's anthropocentric beliefs led him to claim that a universe with all its planets inhabited "would imply that man is an animal and nothing more, is of no importance in the universe, needed no great preparations for this advent" (Wallace 1904 [1903], 266).[10]

Wallace's teleological and anthropocentric views can be associated with the change in his attitude toward the evolution of man in the late 1860s.[11] Coming to believe that "savages" (and, by implication, prehistoric humans) share the same mental capacities as Europeans and realizing that much of this potential has no value for men in primitive societies, Wallace concluded that the human mind (and also other uniquely human physical features) were not the product of natural selection but were shaped by higher spiritual forces and were destined to achieve expression only in a civilized state (Wallace 1869, 1870; see Desmond and Moore 1991, 569–570; Bowler 2003, 215–216). Notably,

[9] In his "Notes Added to the Second Edition of Contributions to the Theory of Natural Selection" (1871), Wallace distinguished between a supernatural God or Deity and a superior intelligence (see Dick 2010, 337).

[10] A very similar assertion was made more recently by astronomer Ben Zuckerman who said that if the Milky Way abounds in all kinds of bizarre forms of life, "then life is but a commonplace extension of cosmic evolution ... and we human beings are insignificant – mere cosmic insects" (Zuckerman 1995, xii). A reverse of this logic, but sharing the same presupposition that the Universe should have some meaning, was expressed by Paul Davies: "Only ... if nature has an ingeniously built-in bias towards life and mind, would we expect to see anything like the development ... that has occurred on Earth repeated on other planets" (Davies 1999, 272; see Fry 2000, 279–282).

[11] Wallace's scholars debate whether his scientific, philosophical, and social ideas from the late 1860s till his death reflect a significant change in his thought or whether there is only "one Wallace" from the beginning of his career (see essays in Smith and Beccaloni 2010).

Wallace's skepticism toward the role of natural selection coincided with his growing interest in spiritualism.[12]

The "Wallace case" is a remarkable example of the interaction between scientific data, philosophical conceptions, and cultural–social influences.[13] Wallace alleged that he based his argument for the uniqueness of Earth "wholly on the facts and principles accumulated by modern science" (Wallace 1904 [1903], 263) and derived his philosophical conclusions from his scientific argument. Clearly, this was not the case. Contemporary astronomers reviewing his 1903 book noted that he "seems . . . to have unconsciously got his facts distorted" (H. H. Turner, professor of astronomy at Oxford, quoted in Dick 2010, 330). Analyzing Wallace's ideas, scientific and otherwise, his co-discovery of natural selection with Darwin, his socialism, his spiritualism, and much more, is beyond the limits of this paper (for a detailed discussion, see Smith and Beccaloni 2010). Nevertheless, his intellectual development does demonstrate, among other things, the ongoing interaction between philosophical presuppositions and the search and interpretation of empirical facts, and between such facts and philosophical conclusions. Obviously, whether the general conception resulting from this interplay is teleological or naturalistic depends on the content of the philosophical presuppositions involved. The question of the epistemological status of naturalistic versus super-naturalistic presuppositions will be discussed now, especially in response to recent contentions about their symmetrical status (see, for example, Nagel 2008, 197).

The "Rare Earth" hypothesis

The historical debate on the plurality of worlds was devoted to the question of human-like, intelligent life in the universe. By the mid-eighteenth century, signifying the victory of the Copernican revolution, a universe teaming with intelligent life was almost unanimously endorsed (Dick 1982, 188). Anti-pluralists, e.g. Whewell and Wallace, also focused on the status of the human

[12] Spiritualism became highly popular in Victorian England in the 1860s, and was associated often with democratic, socialistic tendencies as a replacement to traditional religion (Desmond and Moore 1991, 537–8). In a review of Peter Raby's biography of Wallace (Raby 2001), it is argued that Wallace's spiritualism, accompanying his socialism and support of many radical social causes, "was an attempt to save both the social realm from economic laissez-faire and the human mind from [deterministic] biology" (Anker 2002, 414).

[13] There is a proliferation of publications pertaining to the theological aspects of astrobiology, a subject beyond the scope of this paper (see, for example, Dick 2001). Not unrelated, and also not discussed here, is the growing interest in the last few decades among historians and philosophers of science in rejecting the traditional conflict-between-science-and-religion position (see Fry 2012). Wallace, the great nineteenth-century scientist and his deep religious views feature in these discussions (a representative example is Fichman 2001).

species in the universe. The Darwinian revolution helped to undermine traditional anthropocentrism by establishing that rather than being the goal of evolution, the human species, like other species, evolved as a branch of the evolutionary tree. Beginning in the second half of the twentieth century the search for organic compounds and for microorganisms, alive or fossilized, on other planets in the Solar System became a serious subject of investigation. Most research efforts today within the Solar System are invested in this direction. More generally, major lines of astrobiological research focus on solving the problem of the origin of life on Earth and searching for habitable Earth-like planets in the galaxy. Attempts to find signs of extraterrestrial intelligent life via the SETI project are also part of astrobiology (Dick and Strick 2004, 131–154).

On this background, and taking into account the guiding Copernican–Darwinian framework, the book, *Rare Earth: Why Complex Life is Uncommon in the Universe* (2000), aroused much interest. Its authors, paleontologist Peter Ward and astronomer Donald Brownlee, argued that the evolution of complex, multicellular life, including intelligent life, on other planets in the universe is extremely improbable. They based their thesis on the examination of physical factors that indicate the uniqueness of Earth as a possible abode for complex life and on biological factors demonstrating the slim chances for reproduction elsewhere of the long and improbable evolution of complex life on Earth. Critics described Ward's and Brownlee's thesis as serious and stimulating but not necessarily convincing (see Krauss 2000; McKay 2000; Darling 2001; Kasting 2001). Geoscientist James Kasting examined very carefully each of Ward's and Brownlee's physical arguments for the rarity of the Earth. His conclusion was that "while the authors' stances on various issues are well-argued, alternative positions often are equally viable" (Kasting 2001, 118).

Steven Dick compared Ward's and Brownlee's examination and conclusions to Wallace's early-twentieth-century position (Dick 2010, 330). Though the similarity in their physical analysis is notable, the wide philosophical gap that separates the 1903 and 2000 arguments reflects the different worldviews of Wallace on the one hand and Ward and Brownlee on the other. Whereas Wallace did not hesitate to account for the presumed unique existence of man on Earth by purposeful design, Ward and Brownlee suggested that the evolution of complex life and intelligent life on Earth was a result of a rare coincidence of many physical factors reflecting specific astronomical and geophysical processes in the history of our Solar System and of Earth.

In an interview conducted with Ward following the discovery in 2013 by the Kepler space observatory of Earth-sized planets in the habitable zone of their star, Ward indicated that these Earth-sized planets are not necessarily Earth-like. He predicted that "with funding, within 50 years" we will be able to put the Rare Earth hypothesis to the test. This test will depend on being able "to actually image and get spectroscopic analysis of extrasolar earthlike planets" (Dorminey 2013). Thus, the Rare Earth hypothesis is not a philosophical, in-principle rival to the Copernican–Darwinian framework. Rather, the evidence brought by Ward and Brownlee in favor of their hypothesis, convincing or not, is an empirical challenge to the default presupposition that guides all astrobiologists in their work.

The privileged planet: a current argument from Design

Astronomer Guillermo Gonzales and philosopher Jay Richards, both advocates of Intelligent Design (ID) and fellows of the ID Discovery Institute, published in 2004 their book, *The Privileged Planet: How Our Place in the Cosmos is Designed for Discovery*. The authors provided scientific data, similar to those presented in the Rare Earth book, to make their case that the Earth is among the most habitable places for complex life in the universe.[14] In addition to a Ward-and-Brownlee argument, Gonzales and Richards also claimed that the Earth is among the best places, overall, to make a wide range of scientific discoveries in areas as diverse as geology, astronomy, and cosmology (Gonzales and Richards 2005). The fact that, for example, the Earth and Moon harmonious system "produces the best solar eclipses just where there are observers to see them" demonstrates that "Earth – and the universe itself – were designed both for life and for scientific discovery" (Gonzales and Richards 2004).

Although perfect solar eclipses and other examples of the advantages of the Earth's location and physical features (e.g. the transparency of the atmosphere) conductive to exploration and observation of the wonders of nature are a cause for joy, the question is, who cares except us? Gonzales and Richards obviously believe that we were placed on Earth to reveal the glory of God, but could this belief count as empirical evidence? Apparently, Gonzales and Richards cannot imagine astronomers on a far-off planet making discoveries completely unfathomable to us (for a similar point, see Jefferys 2005). Not surprisingly,

[14] Ward and Brownlee gratefully acknowledged that "Gonzales changed many of our views about planets and habitable zones" (Ward and Brownlee 2000, p. x). Gonzales was their colleague at the University of Washington when they were gathering material for their book. As recounted by astronomer David Darling, Ward and Brownlee were however unaware that throughout this time Gonzales was publishing pro-ID articles in various magazines and focusing on "the study of habitability from a design perspective." Gonzales later claimed that he was not open about his beliefs because "of the open hostility to such views" (see Darling 2001, 113–114).

since Ward and Brownlee did not share their philosophical–theological commitment, Gonzales and Richards expressed their disappointment with them: "They [Ward and Brownlee] obviously challenge the letter of the Copernican Principle. But they don't challenge its spirit" (Richards and Gonzales 2004). Indeed, though predicting a "Rare Earth," Ward and Brownlee still presuppose that physical and biological phenomena are the result of natural processes.

Conclusion

To recapitulate: the Copernican presupposition rejects the claim that Earth was uniquely chosen for life and asserts the possible existence of biogenic conditions on other planets. The Darwinian presupposition contends that life emerged and evolved on Earth by natural processes and might do so wherever biogenic conditions prevail. These universal philosophical claims transcend possible experience and therefore can neither be empirically confirmed nor denied. In this sense, and only in this sense, naturalistic and super-naturalistic presuppositions are symmetrical. However, because of their different subject matter – the natural, on the one hand and the supernatural, on the other, they are epistemologically distinct. Specific hypotheses derived from the Copernican argument can and are being continuously tested by astrobiologists. So far, the results of these specific examinations give credence to the Copernican framework. Within this philosophical framework, Rare Earth proponents await the advancement of appropriate technological means to put their claim to the test. As to the Darwinian presupposition, it derives its validity from the strength of the theory of evolution. The theoretical-and-empirical-based hypothesis that the emergence of life itself was part of the evolutionary process enhances the confidence in the natural emergence of life on Earth and possibly on other habitable planets.

In contrast, the Privileged Planet contention arguing for purposeful design of the universe as demonstrated by the uniqueness of Earth cannot, in principle, avail itself to scientific inquiry. This proposal stands, therefore, in stark conflict with the letter and spirit of science.

References

Adamala, K. and J. W. Szostak 2013. "Nonenzymatic Template-directed RNA Synthesis Inside Model Protocells." *Science* 342:1098–1110.

Anker, P. 2002. "Rediscovering Wallace." *MetaScience* 11(3):413–415.

Bada, J. L. 2005. "A Field With a Life of its Own." *Science* 307:46.

Bowler, P. J. 1990. *Charles Darwin: The Man and his Influence*. Cambridge: Cambridge University Press.
Bowler, P. J. 2003. *Evolution: The History of an Idea*. 3rd Edition. Berkeley, CA: University of California Press.
Ćirković, M. M. 2014. "Evolutionary Contingency and SETI Revisited." *Biology & Philosophy* 29:539–557.
Crowe, M. J. 1999. *The Extraterrestrial Life Debate 1750–1900*. Mineola, NY: Dover Publications.
Darling, D. 2001. *Life Everywhere*. New York, NY: Basic Books.
Davies, P. 1999. *The Fifth Miracle*. New York, NY: Simon & Schuster.
Desmond, A. and J. Moore 1991. *Darwin*. New York, NY: Norton.
Dick, S. J. 1982. *Plurality of Worlds*. Cambridge: Cambridge University Press.
Dick, S. J. 1996. *The Biological Universe: The Twentieth Century Extraterrestrial Life Debate and the Limits of Science*. Cambridge: Cambridge University Press.
Dick, S. J. 1998. *Life on Other Worlds*. Cambridge: Cambridge University Press.
Dick, S. J., ed. 2001. *Many Worlds*. Philadelphia, PN: Templeton Foundation Press.
Dick, S. J. 2010. "The Universe of Alfred Russel Wallace." In *Natural Selection and Beyond*, edited by C. H. Smith and G. Beccaloni, Oxford: Oxford University Press, pp. 320–340.
Dick, S. J. and J. E. Strick 2004. *The Living Universe*. New Brunswick, NJ: Rutgers University Press.
Dorminey, B. 2013. "Rare Earth Revisited." Accessed 18 April, 2014. www.forbes.com/sites/brucedorminey/2013/04/21/rare-earth-revisited-anomalously-large-moon-remains-key-to-our-existence/.
Engelhart, A. E., M. W. Powner, and J. W. Szostak 2013. "Functional RNAs Exhibit Tolerance for Non-heritable 2′-5′ Versus 3′-5′ Backbone Heterogeneity." *Nature Chemistry* 5:390–394.
Fichman, M. 2001. "Science in Theistic Context: A Case Study of Alfred Russel Wallace on Human Evolution." *Osiris* 16:227–250. Accessed 6 October 2014. http://www.jstor.org/stable/301987.
Fry, I. 2000. *The Emergence of Life on Earth*. New Brunswick, NJ: Rutgers University Press.
Fry, I. 2011. "The Role of Natural Selection in the Origin of Life." *Origins of Life & Evolution of the Biosphere* 41(1):3–16.
Fry, I. 2012. "Is Science Metaphysically Neutral?" *Studies in the History and Philosophy of the Biological Sciences* 43:665–673.
Gonzales, G. and J. Richards 2004. *The Privileged Planet: How Our Place in the Cosmos is Designed for Discovery*. Washington, DC: Regnery Publishing.

Gonzales, G. and J. Richards 2005. "A Response to Objections by Kyler Kuehn." Accessed July 12, 2014. http://www.discovery.org/f/73.

Guthke, K. S. 1990. *The Last Frontier*. Ithaca: Cornell University Press.

Jefferys, W. 2005. "The Privileged Planet." *Reports of the National Center for Science Education* 25(1–2):47–49.

Kasting, J. F. 2001. "Peter Ward and Donald Brownlee's 'Rare Earth'." *Perspectives in Biology and Medicine* 44(1):117–131.

Kellog, V. L. 1907. *Darwinism Today*. New York, NY: Henry Holt.

Kohn, D. 1989. "Darwin's Ambiguity: The secularization of Biological Meaning." *British Journal for the History of Science* 22:215–239.

Koyré, A. 1994[1957]. *From the Closed World to the Infinite Universe*. Baltimore, MD: Johns Hopkins Press.

Krauss, L. 2000. "Rare Earth." *Physics Today* 53(9):62–63.

Kuhn, T. 1985[1957]. *The Copernican Revolution*. New York, NY: MJF Books.

Lazcano, A. and K. P. Hand 2012. "Astrobiology: Frontier or Fiction." *Nature* 488:160–161.

McKay, C. P. 2000. "All Alone After All?" *Science* 288:625.

Moore, J. R. 1991. "Deconstructing Darwinism." *Journal of the History of Biology* 24(3):353–408.

Nagel, T. 2008. "Public Education and Intelligent Design." *Philosophy and Public Affairs* 36(2):187–205.

Ospovat, D. 1980. "God and Natural Selection: The Darwinian Idea of Design." *Journal of the History of Biology* 13(2):169–194.

Powner, M. W., B. Garland, and J. D. Sutherland 2009. "Synthesis of Activated Pyrimidine Ribonucleotides in Prebiotically Plausible Conditions." *Nature* 459:239–242.

Provine, W. 1988. "Progress in evolution and the meaning of life." *In Evolutionary Progress*, edited by M. Nitecki, Chicago: The University of Chicago Press, pp. 49–74.

Quintana, E. V., T. Barclay, S. N. Raymond, *et al.* 2014. "An Earth-sized Planet in the Habitable Zone of a Cool Star." *Science* 344:277–280.

Raby, P. 2001. *Alfred Russel Wallace: A Life*. Princeton, NJ: Princeton University Press.

Richards, J. and G. Gonzales. 2004. "Are We Alone?" *The American Spectator*. Accessed July 12, 2014. www.discovery.org/a/2143.

Scott, E. C. 1996. "Creationism, ideology, and science." *Annals of the New York Academy of Science*, 775:515–516.

Seager, S. 2013. "Exoplanet Habitability." *Science* 340:577–581.

Simpson, G. G. 1964. "The Nonprevalence of Humanoids." *Science* 143:769–775.

Sloan, P. R. 1985. "The Question of Natural Purpose." In *Evolution and Creation*, edited by E. McMullin, Notre Dame, IN: University of Notre Dame Press, pp. 121–150.

Smith, C. H. and G. Beccaloni, eds. 2010. *Natural Selection and Beyond*. Oxford: Oxford University Press.

Wallace, A. R. 1869. "Sir Charles Lyell on Geological Climates and the Origin of Species." *Quarterly Review* 126:359–394.

Wallace, A. R. 1870. "The Limits of Natural Selection as Applied to man." In *Contributions to the Theory of Natural Selection*, edited by A. R. Wallace, London: Macmillan, pp. 332–371.

Wallace, A. R. 1904[1903]. *Man's Place in the Universe*. 4th Edition. London: Chapman and Hall Limited.

Ward, P. D. and D. Brownlee. 2000. *Rare Earth: Why Complex Life is Uncommon in the Universe*. New York, NY: Copernicus.

Whewell, W. 1855[1853]. *Of The Plurality Of Worlds*. 4th Edition. London: John W. Parker and Son.

Zuckerman, B. 1995. "Preface to the First Edition. 1980." In *Extraterrestrials: Where Are They?* 2nd Edition, edited by B. Zuckerman and M. H. Hart, Cambridge: Cambridge University Press, pp. xi–xii.

3 History, discovery, analogy

Three approaches to the impact of discovering life beyond Earth

STEVEN J. DICK

How do we rationally approach such a "far out" problem as the societal impact of discovering life beyond Earth? In this chapter we examine three possible methods: history, in the form of past reaction to claimed discoveries of life; discovery, including its extended nature and the many discovery scenarios for extraterrestrial life; and analogy, studying those events or changing worldviews in humanity's past that might illuminate the reaction to such a discovery. It would be easy to throw up our hands at the outset and declare that the detection of life beyond Earth will be a unique event, without precedent in human history. Although that may be true, such a position is tantamount to ignoring the problem – a problem in which the US Congress, astrobiologists, social scientists, and the general population have expressed considerable interest. In contrast we conclude that despite the obvious need for caution and the lack of predictive value of our conclusions, history, discovery, and analogy can indeed serve as solid guidelines to cosmic encounters with life.

History

At least six times over the last 200 years Earthlings *thought* life had been detected beyond Earth: the 1835 Moon Hoax/Satire, the canals of Mars controversy (1894–1909), the Orson Welles *War of the Worlds* broadcast (1938), the discovery of pulsars in 1967, the Viking landings on Mars in 1976, and the claim of Martian nanofossils in 1996. The reaction to each of these well-known events could be the subject of a considerable research program; here we focus briefly on three of them as exemplars of their utility to the problem at hand.

The so-called Moon Hoax resulted from a series of six illustrated stories published serially in the *New York Sun* in August, 1835. Readers of the *Sun* for August 25 of that year could hardly miss the headline prominently placed at the top of the front page: "GREAT ASTRONOMICAL DISCOVERIES LATELY MADE BY SIR JOHN HERSCHEL, L.L.D, F.R.S., etc. At the Cape of Good Hope." Though the article, the first of a series of six that concluded on August 31, purportedly came from the "Supplement to the Edinburgh Journal

Figure 3.1 Illustration in the *New York Sun*, 1835, part of a series of articles now known as the Great Moon Hoax. Library of Congress Prints and Photographs Division, Washington, DC.

of Science," readers could hardly be expected to have known that the journal had ceased to exist three years earlier. A few may have heard of John Herschel, who indeed was the son of the famous astronomer William Herschel, had published his *Treatise on Astronomy* in the United States to great acclaim the previous year, and was in fact at the Cape of Good Hope making astronomical observations (Goodman 2008).

According to these articles, Herschel had observed on the Moon large winged creatures, about four feet in height, with faces "of a yellowish flesh-color . . . a slight improvement upon that of the large orang-outang, being more open and intelligent in its expression . . ." (Figure 3.1). Moreover, "these creatures were evidently engaged in conversation: their gesticulation, more particularly the varied action of their hands and arms, appeared impassioned and emphatic. We hence inferred that they were rational beings" (Crowe 2008, 286–287). The articles appeared in the *Sun* anonymously, but within days the author was revealed as Richard Adams Locke (1800–1871), a 34-year-old reporter who had just joined the *Sun* that summer. Within weeks the article was exposed as a hoax, and so it was widely believed to be until a century-and-a-half later, when historian Michael Crowe convincingly argued that Locke was actually writing satire (Crowe 1986, 210–215). What was he satirizing? According to Crowe, no less than advocates of inhabited worlds, of which there were many, especially among German astronomers, but none more enthusiastic than the Scottish astronomer Thomas Dick. Dick once calculated the number of inhabitants on each of the planets in our Solar System, arriving at a figure of more than 21 trillion, not counting those on the Sun! Sincere as he was, such conclusions were ripe for satire, and Locke took the opportunity with aplomb.

Our interest here is not so much in the motivations of the cub reporter Locke, but in the public and scientific reaction to his stories. At the time of their publication the *Sun* was a fledgling tabloid newspaper with a circulation of about 8,000. During the Moon episode its circulation reached 19,000, and remained high thereafter. Moreover, the *Sun* sold 60,000 copies of the story in pamphlet form, as well as lithographs of the lunarians. Nor was interest confined to the United States; French, Italian, Spanish, and German editions of the brochure appeared, and numerous other newspapers reported the story both inside and outside the United States. A detailed study of the series of articles concluded "no other newspaper story of the age was as broadly circulated as Locke's moon series . . . By the time the series had run its course, the *Sun* had become the most widely read newspaper in the world" (Goodman 2008, 12). Obviously the story was profitable for the *Sun* because it had great popular appeal even if not true, a situation that resonates with tabloid journalism today.

But it was not only the general public that fell for the story. Crowe again describes the effect: the *New York Times* pronounced the discoveries "probable and possible," while the *New Yorker* credited them as creating "a new era in astronomy and science generally" (Crowe, 1986, 212–213). Religious journals debated the consequences, Yale was "alive with staunch supporters," including students and professors, and astronomy professors Elias Loomis and Denison Olmsted travelled to New York to unearth more details (though they later denied believing the story). In short, it mattered not that scientists had known for centuries that conditions on the Moon could not support such creatures; it mattered not that no telescope on Earth had the resolution to see such things. And it mattered not that the entire series was the product of a fevered imagination. What mattered was that the upstart *New York Sun* quickly became the most read, and profitable, newspaper in the world. So we arrive at our first rather unsettling lesson from history – in the short term facts won't particularly matter. What the media says will matter, the perception, not the reality.

A century later, on Halloween Eve October 30, 1938, one of the most famous events in radio history occurred. Broadcasting coast-to-coast from Madison Avenue in New York City, the 23-year-old American actor, writer, and producer Orson Welles directed and narrated "The War of the Worlds" as part of the CBS series "The Mercury Theatre on the Air." Based on H. G. Wells' 1898 novel by the same name in which Martians invaded London, Welles transferred the action from London to Grover's Mill, New Jersey, about an hour's drive south of New York City. The 62-minute broadcast was presented as a series of news bulletins, which (despite warnings at the beginning of the show) many took to be real. It is the reaction to this event that makes it an important part of radio history – and of special interest for studying the reaction to the possible discovery of extraterrestrial intelligence in its most extreme form of direct contact (Gosling 2009).

There is no doubt that the reaction was considerable. The *New York Times'* front page headline the next day blared, "Radio Listeners in Panic, Taking War Drama as Fact," adding in the subtitle, "Many Flee Homes to Escape 'Gas Raid From Mars' – Phone Calls Swamp Police at Broadcast of Wells Fantasy." It went on to say that, "A wave of mass hysteria seized thousands of radio listeners between 8:15 and 9:30 o'clock last night when a broadcast of a dramatization of H. G. Wells's fantasy 'The War of the Worlds' led thousands to believe that an interplanetary conflict had started with invading Martians spreading wide death and destruction in New Jersey and New York." Many other newspapers carried similar accounts. A study published in 1940 by respected Princeton University professor and public opinion researcher Hadley Cantril, entitled *The Invasion from Mars: A Study in the Psychology of Panic*, confirmed the idea of a widespread "panic" reaction, and the idea was propagated in both scholarly and popular culture over the next 70 years.

This idea of mass panic from the Welles broadcast, however, has now been thoroughly debunked. On the 60th anniversary of the broadcast in 1998, sociologist Robert Bartholomew reviewed the criticisms of the panic scenario, which included problems with Cantril's estimates of the actual number of people affected, as well as erroneous reports in the media that nevertheless went on to become part of popular culture. He concluded that perhaps tens of thousands rather than millions of people were panicked, and his lesson learned was that the mass media significantly influenced public perception of the event (Bartholomew 1998). It turns out newspaper articles were mostly based on anecdotal reports, and scholars have concluded these stories represented an irresistible opportunity for the print media to rebuke radio, an upstart rival source of news and advertising, as unreliable and untrustworthy. In the process, however, they perpetuated a myth that is very difficult to excise. On the 75th anniversary of the broadcast in 2013, the American Public Broadcasting Service (PBS) was still perpetuating the myth of the War of the Worlds panic in a broadcast in its "American Experience" series. Two professors of communication quickly criticized PBS for perpetuating the myth, which some scholars seriously doubted already in the 1980s (Bainbridge 1987; Pooley and Socolow 2013).

All this notwithstanding, in numbers small or large, there is no doubt that the War of the Worlds broadcast had its effect, both on the night of the broadcast and in contemporary popular culture. And there is no doubt it could happen again on scales large and small. In fact, it did happen again, several times, in a rebroadcast in Chile in 1944, one in Quito, Ecuador in 1949, and on several occasions since. The latter resulted in another front page *New York Times* headline for February 14, 1949: "Mars Raiders Caused Quito Panic; Mob Burns Radio Plant, Kills 15" (Gosling, 2009). Small-scale panics related to anniversary broadcasts of War of the Worlds still occasionally occur, as do other mass panics involving erroneous reports of environmental contaminants, nuclear accidents, and other cultural concerns. Bartholomew's lesson is that only a small portion of the population needs to act on erroneous information over a short period to create large-scale disruptions to society. Sociological study of the causes and effects of collective behavior are thus critically important to the study of the potential societal impact of discovering extraterrestrial life.

A third life detection episode, which occurred 60 years later and centered around a Mars rock dubbed ALH84001, is most interesting because it is a case involving real science. In 1996, NASA scientists announced that organic molecules had been found in a meteorite that was blown off Mars 16 million years ago, had landed in the Antarctic 13,000 years ago, was found there by a meteorite-collecting team funded by the National Science Foundation and the Smithsonian Institution in 1984, and had been recognized as Martian only a few years earlier. Years of exhaustive study had led the researchers to their momentous conclusions, including the claim of organic molecules (already a step beyond the Viking lander results), mineral "carbonate globules" of possible biological origin; evidence of tiny magnetic minerals that on Earth are secreted by certain bacteria; and finally, pictures of strange, hauntingly worm-like structures that they argued might be microfossils. In short, scientists suggested, the Mars rock indicated life had existed on Mars sometime in the planet's distant past, when Mars was warmer and wetter (Dick and Strick 2004, 179–201).

This was not exactly the Martian civilization of Lowell, Wells, and Welles, but compared to the ambiguous results of Viking 20 years before, the claim that a Martian meteorite had landed on Earth bearing evidence of past life on that fabled planet was little short of miraculous. The very possibility of life on Mars, even past life, set the world afire, igniting media hype, public imagination, scholarly discussion, and scientific curiosity alike. Press conferences were held, President Clinton spoke from the South Lawn of the White House, Vice President Gore convened a space summit, Congressional hearings were held, numerous claims and counter-claims were published in scientific journals, and theologians opined on the meaning of it all. The media went wild. Entire books have revealed the history and intrigue surrounding this event (Sawyer 2006). Suffice it to say that Mars rocks are real, but the scientific consensus is that the fossils are not real. In the end the case of the Mars rock offers the most robust example in the modern era of what might happen following the claim of life beyond Earth – even if the claim is only fossils or microbes, much less intelligence. The reaction of government institutions, the media, and the public will occur side by side with the reaction among scientists, who will subject the discovery to withering criticism. The media will play an important role in both reporting and sensationalizing any claims; with the internet and other social media, news of the discovery of life beyond Earth would spread like wildfire in the twenty-first century. But, as the Mars rock demonstrates, the final conclusion will be far from immediate; rather it will consist of an extended process characteristic of all discoveries.

Discovery

The second approach to our problem of the impact of finding life beyond Earth is the nature of the discovery itself. Recent studies of astronomical discovery (Dick 2013) demonstrate that it is never just a matter of pointing a telescope, observing, and making a discovery. There is no such thing as immediate discovery in astronomy, or, I would venture to say, in all of science. The anatomy of discovery always involves at a minimum the three stages of detection, interpretation, and understanding (Figure 3.2). For example, when in 1610 Galileo turned his telescope toward Saturn and detected protuberances on either side, he had not discovered Saturn's rings. He had no idea what they were. The interpretation as rings came only 40 years later with Christiaan Huygens. And even a rudimentary understanding of the nature of the rings began only when James Clerk Maxwell in the mid-nineteenth century demonstrated how they could exist dynamically, not as a solid body, but as a collection of objects orbiting Saturn. The same extended nature of discovery is true of all other astronomical discoveries, and we infer the same will be true of the discovery of extraterrestrial life, in whatever form. A signal, artifact, or

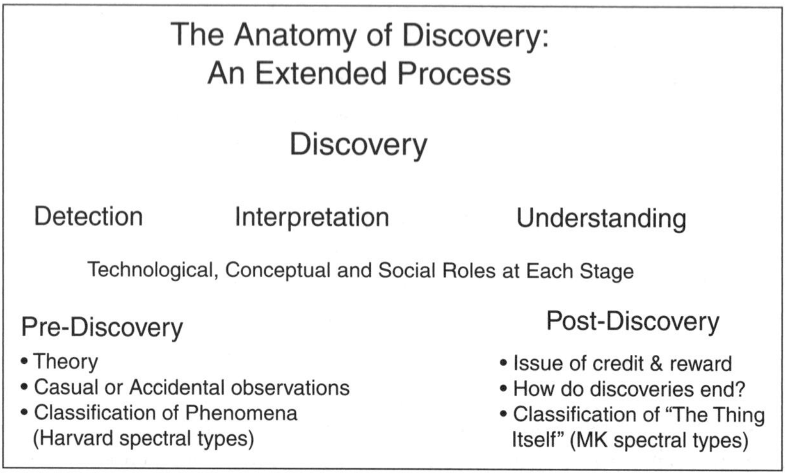

The Anatomy of Discovery:
An Extended Process

Discovery

Detection Interpretation Understanding

Technological, Conceptual and Social Roles at Each Stage

Pre-Discovery

- Theory
- Casual or Accidental observations
- Classification of Phenomena (Harvard spectral types)

Post-Discovery

- Issue of credit & reward
- How do discoveries end?
- Classification of "The Thing Itself" (MK spectral types)

Figure 3.2 Astronomical discovery is always an extended process, and the discovery of extraterrestrial life will be no different. From Steven J. Dick, *Discovery and Classification in Astronomy* (Cambridge University Press, 2013).

anomaly will be detected, there will be years or decades of interpretation (as in the Mars rock), and full understanding will take even longer. These are all essential points if we are going to talk intelligently about the impact of the discovery.

Tantalizingly, the history of discovery also indicates that there is very often a pre-discovery phase, during which the true nature of an object, signal, or phenomenon goes unrecognized or unreported, or during which only theory indicates the phenomenon *should* exist. This is true not only for pre-telescopic astronomy and pre-microscopic biology, but also after these technologies have reached advanced stages. We are perhaps not surprised that Galileo did not recognize his observations of the protuberances of Saturn as rings, or that nebulous objects in the sky were not known to be external galaxies until the twentieth century. But the same is often true today, as objects are not immediately recognized for what they truly are, or are theorized long before they are found (think black holes). Needless to say, the subject of life beyond Earth is still in a pre-discovery phase. Whether anything has been observed that turns out to be an indication of life, or life itself, remains to be seen.

If there is a pre-discovery phase it makes sense that there is also a post-discovery phase, the phase of most immediate relevance to our problem because it deals with reaction and impact. A basic but essential insight about the post-discovery phase is that the reaction will be very much scenario-dependent. Perhaps the simplest way of parsing possible scenarios is according to the physical nature and degree of complexity of the life encountered, ranging from microbial to intelligent. Microbes might be discovered on a planetary surface or subsurface like Mars, in the ocean of a planetary satellite like Europa or in water spouts emanating from such an ocean, as biosignatures in a planetary atmosphere beyond our Solar System, or as fossils or micro-fossils similar to those claimed in the famous Mars rock. The discovery of intelligence offers more hair-raising possibilities, depending on the mode of contact, whether direct or indirect; the moral nature of the life encountered, ranging from inferior to superior (loaded terms to be sure); the altruistic spectrum ranging from competitive to collaborative, and so on. One might further characterize intelligent encounters according to energies harnessed, technology utilized, or communicative ability (Vidal 2014, 203–204). Finally, if the mode of contact is by electromagnetic signal, the nature of that signal is surely paramount when contemplating impact: a clearly artificial "dial tone" raises one set of questions, an undeciphered signal bearing information raises another, and a signal with a deciphered message is on another scale entirely. Each of these scenarios will certainly affect the nature of our encounter with alien life and our reaction to it.

We can summarize discovery scenarios in Table 3.1, a matrix showing direct and indirect encounters in terrestrial and extraterrestrial environments. In other words, we could encounter life *directly* here on Earth or in space, and we could find life *indirectly* here on Earth or in space. Indirect extraterrestrial contact – labeled encounter type 3 in Table 3.1 – is where most of the action is found today, in the form of spacecraft observations, the detection of biosignatures on exoplanets, and SETI programs. Direct encounters of types 1 (terrestrial) and 2 (extraterrestrial) are the most unlikely to occur in the near future, because they imply alien spaceflight to Earth or human spaceflight beyond the Solar System. But UFOs as extraterrestrial spaceships remain a logical possibility, dependent on evidence. A type 3 indirect encounter is the most likely, or at least the most anticipated, scenario, since remote detection of microbes beyond Earth by robotic space exploration is one of the main goals of NASA's astrobiology program, the search for biosignatures is an active research goal, and several SETI programs are in progress. A type 4 encounter raises interesting possibilities not often discussed, including the discovery of an alien artifact on Earth or its vicinity. In its static form Table 3.1 lays out scenarios, but in its dynamic form, proceeding clockwise from encounter types 1 to 4, it arguably depicts impacts from their strongest to their weakest.

Each of these scenarios have triggered a large number of science fiction novels, beginning substantially with H. G. Wells' *War of the Worlds* more than a century ago, accelerating in the second half of the twentieth century with the

Table 3.1 Discovery scenarios for contact with extraterrestrial life

	TERRESTRIAL	EXTRATERRESTRIAL
DIRECT	**Encounter type 1** Accidental contamination by sample or astronaut return Panspermia or interplanetary matter transfer (ALH84001) Alien space exploration UFOs	**Encounter type 2** Human space exploration
INDIRECT	**Encounter type 4** Shadow alien biosphere Unknown alien microbes Earth orbit or vicinity Artifact on Earth or vicinity Artifact in space	**Encounter type 3** Robotic space exploration (Viking, Europa probes, etc.) Biosignatures SETI – radio and electromagnetic spectrum

rise of the Space Age, and continuing unabated today, in novels and movies such as Arthur C. Clarke's *2001: A Space Odyssey*, Carl Sagan's *Contact*, and Stanislaw Lem's *Solaris*. The best of science fiction literature lays out thoughtful scenarios about discovery and possible impacts. The literature in this "alien encounter" or "first contact" genre is very large especially in the case of extraterrestrial intelligence, which is often seen as offering the most drama. Nonetheless, a few science fiction scenarios have been written about the discovery of microbial life, ranging from Michael Crichton's *The Andromeda Strain* (1969) to Dirk Schulze Makuch's *Alien Encounter* (2013) and Nicholas Kanas's *The New Martians* (2013). And new scenarios are still being constructed, as in David Brin's novel *Existence* (2012), which deals with the discovery of alien artifacts in the vicinity of Earth (a type 4 encounter).

Much more could be said about the nature of discovery and discovery scenarios for extraterrestrial life. But the essential point is that the reaction will be extended and will very much depend on the scenario.

Analogy

Our third approach is analogy, and here I must first insist that analogy is not just some wishy-washy form of reasoning deployed in desperation when other arguments are lacking. As cognitive scientist Douglas Hofstadter (of *Godel, Escher, Bach* fame) has written "One should not think of analogy-making as a special variety of reasoning (as in the dull and uninspiring phrase 'analogical reasoning and problem-solving,' a long-standing cliché in the cognitive-science world), for that is to do analogy a terrible disservice . . . To me analogy is anything but a bitty blip—rather, it's the very blue that fills the whole sky of cognition—analogy is everything, or very nearly so, in my view" (Hofstadter 2001, 499). Hofstadter went on to elaborate this claim in a very large book (Hofstadter and Sander 2013). Numerous other scholars have come to the same conclusion, and analogy is now rigorously analyzed and routinely employed in a variety of fields as a robust method of argument (Holyoak and Thagard 1995; Bartha 2010; Launius 2014). In fact analogies are essential in science, and astrobiology itself would not exist were it not for analogical reasoning. On the other hand, analogies do have to be used with caution. As anthropologist Kathryn Denning has written "the problem with analogies is that they are highly persuasive, inherently limited, and easily overextended" (Denning 2014). Nevertheless, if the natural sciences can use analogs to good effect, so can the social sciences and humanities, with caution.

The kinds of analogies we might use with respect to the impact of finding life beyond Earth very much depend on the discovery scenario. These scenarios range from microbes to the multiple ways we might make first contact with extraterrestrial intelligence.

The microbe analogy

What analogies might best represent the reaction to the discovery of extraterrestrial microbes, whether alive, fossilized, or indicated in a more ambiguous biosignature from a planetary surface or atmosphere? Whatever else it may be, the discovery of microbes beyond Earth in any form potentially represents a revolution in biology, depending on the nature of the microbes found. Taking this as our starting point we can therefore ask what discoveries in the history of biology might approximate such extraterrestrial discoveries. More specifically, assuming that such a discovery would change our view of life, what discoveries in biology have changed biologists' view of life? A number of candidates come to mind, including Darwinian evolution by natural selection, the role of DNA in genetics, the discovery of extremophile microorganisms, among others. But perhaps no biological discovery closer approximates the potential fallout than the discovery of terrestrial microbes themselves, representing a new world of life on Earth, what Lynn Margulis and others have called the "microcosmos" (Margulis and Sagan 1986).

It was only in the 1660s and 1670s that two remarkable geniuses, Robert Hooke and Antony van Leeuwenhoek, discovered this new microbial world. And while Hooke's *Micrographia* was a sensation for a while, the surprising and striking fact is that "scant research was done in the study of microbial life during the next century" (Gest 2004; Sapp 2009, 3–7). Progress in microscopy was slow, and in any case no one quite knew what to make of the new discoveries. Once microscopes had been considerably improved, however, the impact was swift and profound. But that was not until the nineteenth century, and not until then did microbiology even begin to develop into a science. Today both the positive and negative effects of microbes in medicine and in the modern world – totally unforeseen before the discovery of the microbial world – cannot be overestimated. The same will probably be true of extraterrestrial microbes. Other revolutions in biology may also prove apt analogies, and it is the common lessons from all of them that will illuminate the problem of the impact of discovering microbes.

The culture contact analogy

A second discovery scenario is contact with extraterrestrial intelligence, either direct or indirect. Terrestrial history abounds with direct contacts between cultures, usually with unhappy effects. One often hears of such contacts as analogies to contact with extraterrestrials, usually in the context of the Western Age of Discovery. An exhibit in the Smithsonian's National Museum of the American Indian in Washington D.C, for example, illustrates several dozen such "first contacts" between Europeans and the Americas. One particularly vivid example, often used in the context of the impact of SETI, is Cortes and the Aztecs in Mexico in 1519. That particular contact started out positively, but very quickly degraded (Figure 3.3), with the destruction in 1521 of Tenochtitlán and the eventual decimation of the native population, with the help of the "Columbian exchange," including microbes. But this is only one example, and it is very dangerous to draw conclusions from a sample of one.

In fact, outside the tradition of Western expansionism, the less well-known great voyages of the fifteenth-century Ming China treasure fleets 50 years before Columbus offer a very different scenario. In contrast to the case of Cortes and the Aztecs, Chinese culture contacts resulted in new embassies in the Near East, deification (to this day) of Zheng He and his cult in southeast Asia, and the publication of works enlarging Chinese knowledge of the oceans and landmasses visited (Dreyer 2007). So my point is that there are other models of culture contact than the destructive ones usually cited, including Jesuit models of culture contact that were not disastrous. Beyond physical culture clashes these contacts illustrate the difficulties among cultures of

Figure 3.3 The conquest of the Aztec capital Tenochtitlán (present-day Mexico City), in 1521. From the Conquest of México series, Mexico, second half of the seventeenth century. Oil on canvas. Jay I. Kislak Collection Rare Book and Special Collections Division, Library of Congress (26.2).

communication and even grasping certain concepts such as the "soul" and "immortality" (Kuznicki 2011).

One of the most trenchant criticisms of using terrestrial culture contacts as analogs to extraterrestrial contacts is that, different as terrestrial cultures may be, they all involve *Homo sapiens*, sharing a common ancestry and thus common mental and behavioral patterns in the most general sense. This warning is well taken. There is one terrestrial analog that partially transcends this criticism: the overlap of modern *Homo sapiens* with *Homo neanderthalensis*. We know this happened in part because about 2% of most non-African genomes consists of Neandertal ancestry, in other words, the two species interbred. As one anthropologist wrote, this culture contact would be a particularly good analogy because Neandertals were possessed of an entirely different sensibility than were *Homo sapiens* (Wason 2011, 2014). Unfortunately, we know very little about that interaction, but it is important to keep in mind in deploying analogies that extraterrestrials and their cultures may be nothing like terrestrial ones, with differences far beyond those of *Homo sapiens* and *Homo neanderthalensis*.

What, in the end, is the lesson from culture contacts? Is it that a supposedly "more advanced" society will overpower and supplant a supposedly inferior one? Far from it, as indicated both by the case of the Jesuits among the native Americans and the Chinese voyages. Though destruction, power projection, tribute systems, and acculturation are indeed sobering historical facts, we cannot conclude based on terrestrial analogy that destruction is the probable outcome when an advanced extraterrestrial civilization makes contact with a less advanced one, even if it is physical contact. More illuminating are the communication and cultural interactions, which include both positive and negative aspects. Taking any one interaction as an exemplar, almost any lesson could be learned. Far better to learn from the entire set of contact experiences, or better yet, to learn collective lessons common to all contact experiences. While we cannot draw definitive conclusions based on a small sample of culture contacts, the determination of characteristics common to all contacts is a respectable research program that could pay significant dividends. Nor should we forget European preconceptions of "the other" prior to contact as analogs to our current preconceptions of extraterrestrials. "Imagining the other" has a long history prior to our preoccupation (at least in the West) with extraterrestrials (Axtell 1992, 31ff).

The decipherment/translation analogy

A third analogy is what I call the "decipherment/translation" analogy (Dick 1995; Finney and Bentley 2014). If, as is widely assumed, first contact with intelligence beyond Earth turns out to be indirect, in the form of an electromagnetic signal that is deciphered with significant information transmitted, the flow of information between civilizations across time is a more appropriate analog. Particularly compelling is the transmission of Greek knowledge to the Latin West by way of the Arabs in the twelfth and thirteenth centuries, only one example of what Arnold Toynbee has called in the terrestrial context "encounters between civilizations in time" (Toynbee 1957, 241–260). Others have pointed out that a better analogy might be decipherment of the Mayan glyphs, since we will not know the alien language (Finney and Bentley 2014). The two might profitably be combined into a decipherment/translation analogy, since both are likely to be phases in the same process. A more modern analogy for information flow is Gutenberg and the printing press, or the impact of the internet. And with the latter, so familiar to all of us in the modern world, we begin to grasp the difficulties of determining societal impact when the analog is a world-changing event.

The worldview analogy

Even if an extraterrestrial message is not deciphered, and perhaps even in the case that "only" microbial life is discovered constituting a second Genesis, a change in worldview would likely gradually take place. I would argue that, in the long term, the discovery of microbial or intelligent life beyond Earth might be analogous to grand changes in scientific worldviews, exemplified in the Copernican worldview originated in the sixteenth century, the Darwinian worldview of the nineteenth century, or the Shapley–Hubble worldview of the twentieth century. The Copernican theory eventually gave birth to a new physics, caused wrenching controversy in theology, and made the Earth a planet and the planets potential Earths in ways we are still unraveling today. Gradually, and more broadly, it changed the way humans viewed themselves and their place in the universe. 150 years after the Darwinian revolution, its implications remain very much controversial, especially among a minority segment of the American public. Like the Darwinian theory, the interpretation of an extraterrestrial signal is likely to be ambiguous and debatable, and the diverse reaction to such a signal may therefore be comparable. The Shapley–Hubble worldview of the twentieth century, in which our Solar System was demonstrated to be at the periphery of our Milky Way Galaxy, which was itself only one of billions of galaxies, is also still playing out with the revelations of the Hubble Space Telescope and other spacecraft, which have demonstrated not only our place in space, but also in time, part of a universe 13.8 billion years old. The long-term implications of Shapley–Hubble revelations have not been absorbed into culture; many religions, for example, have not come to terms with our seemingly insignificant place in the 13.8 billion years of cosmic evolution.

The gradual construction of worldviews, and their influence on our thinking, is a deep philosophical problem requiring more research (Vidal 2007, 2014). The impact of scientific worldviews have been studied extensively (Smith 1982; Blumenberg 1987; Bowler 1989), and cannot help but illuminate the problem of the impact of discovering extraterrestrial life.

In summary, I argue that analogy, used cautiously, can be a very useful tool for our problem. Nevertheless, what I term the "Goldilocks Principle of Analogy" should always be invoked: "Analogy must not be so general as to be meaningless, nor so specific as to be misleading. The middle 'Goldilocks' ground is where analogies may serve as useful guideposts, generating scenarios and setting limits." Just as scientists have profitably employed analogies throughout the history of science, just as historians have investigated analogies to the impact of spaceflight and other human endeavors, and at a time when cognitive scientists have come to see analogy as the "fire and fuel of thinking," so may we cautiously deploy analogy in order to illuminate the impact of discovering extraterrestrial life, even as fundamentally novel aspects may also exist. The human mind, the basis of all behavior, remains the constant factor in our capacity to react to novel events.

Conclusion

History, discovery, and analogy offer a grounding in human experience as approaches to the problem of the impact of discovering life beyond Earth. The lessons of history are indeed ambiguous, and we must be careful to emphasize that we cannot predict impact, only lay out scenarios and guidelines based on human experience. And while the reaction to the discovery of extraterrestrial life may be unique in some respects, the alternative to anticipating societal impact is to give up.

Some may ask why we should study the impact of discovering life beyond Earth now. Why not wait until the day arrives, if ever, when we discover extraterrestrial life? One might as well ask, to use a recent heartwrenching example, why not wait until ebola arrives to do anything about it? Planning for an ebola outbreak had a rather low priority – until it happened. And although the plans that were in place proved inadequate in some cases, the outcome was much better than if there had been no preparation at all. As in events ranging from pathogenic pandemics to military contingencies and near-Earth asteroids, we need to think ahead about the impact of scientific discoveries on society, preferably before the discoveries occur. That was the strategy behind the very robust and successful ethical, legal, and social impact program of the Human Genome Project. In short, it is always better to plan ahead in the hope that benefits can be maximized and bad effects minimized, both for the short term and the long term. In doing so we need to bring together many disciplines, not just from the natural sciences, but also from the social sciences, humanities, philosophy, and theology. And even if no life is found, taking an extraterrestrial perspective on age-old problems can only broaden our human horizons.

References

Axtell, James. 1992. *Beyond 1492: Encounters in Colonial North America.* Oxford: Oxford University Press.

Bainbridge, William S. 1987. "Collective Behavior and Social Movements," in Rodney Stark, ed., *Sociology*. Belmont, CA: Wadsworth.

Bartha, Paul. 2010. *By Parallel Reasoning: The Construction and Evaluation of Analogical Arguments.* New York, NY: Oxford University Press.

Bartholomew, Robert. 1998. "The Martian Panic Sixty Years Later: *What Have We Learned?*," *Skeptical Inquirer*, Nov–Dec, pp. 40–43.

Blumenberg, Hans. 1987. *The Genesis of the Copernican World.* Cambridge, MA: MIT Press (trans. R. M. Wallace).

Bowler, Peter. 1989. *Evolution: The History of an Idea.* Berkeley, CA: University of California Press.

Crowe, Michael J. 1986. *The Extraterrestrial Life Debate, 1750–1900: The Idea of a Plurality of Worlds from Kant to Lowell.* Cambridge: Cambridge University Press.

Crowe, Michael J. 2008. *The Extraterrestrial Life Debate, Antiquity to 1915.* Notre Dame, IN: University of Notre Dame Press, pp. 272–296.

Denning, Kathryn. 2014. "Learning to Read: Interstellar Message Decipherment from Archaeological and Anthropological Perspectives," in Douglas Vakoch, ed., *Archaeology, Anthropology and Interstellar Communication.* Washington, DC: NASA, p. 100.

Dick, Steven J. 1995. "Consequences of Success in SETI: Lessons from the History of Science," in Seth Shostak, ed., *Progress in the Search for Extraterrestrial Life.* San Francisco, CA: Astronomical Society of the Pacific, pp. 521–532.

Dick, Steven J. 2013. *Discovery and Classification in Astronomy: Controversy and Consensus.* Cambridge: Cambridge University Press.

Dick, Steven J. and James E. Strick. 2004. *The Living Universe: NASA and the Development of Astrobiology.* New Brunswick, NJ: Rutgers University Press.

Dreyer, Edward L. 2007. *Zheng He: China and the Oceans in the Early Ming Dynasty, 1405–1433.* New York, NY: Pearson Longman.

Finney, Ben and Jerry Bentley. 2014. "A tale of Two Analogues: Learning at a Distance from the Ancient Greeks and Maya and the Problem of Deciphering Extraterrestrial Radio Transmissions," in Douglas Vakoch, ed., *Archaeology, Anthropology and Interstellar Communication.* Washington, DC: NASA, pp. 65–78.

Gest, Howard. 2004. "The Discovery Of Microorganisms by Robert Hooke and Antoni van Leeuwenhoek, Fellows of the Royal Society," *Notes and Records of the Royal Society of London*, 58, 2, 187–201.

Goodman, Matthew. 2008. *The Sun and the Moon: The Remarkable True Account of Hoaxers, Showmen, Dueling Journalists, and Lunar Man-Bats in Nineteenth-Century New York.* New York, NY: Basic Books.

Gosling, John. 2009. *Waging the War of the Worlds: A History of the 1938 Radio Broadcast and Resulting Panic.* Jefferson, NC: McFarland & Company.

Hofstadter, Douglas. 2001. "Epilogue: Analogy as the Core of Cognition," in Dedre Gentner, Keith J. Holyoak and Boicho N. Kokinov, eds., *The Analogical Mind: Perspectives from Cognitive Science*. Cambridge, MA: MIT Press, pp. 499–538.

Hofstadter, Douglas and Emmanuel Sander. 2013. *Surfaces and Essences: Analogy as the Fuel and Fire of Thinking*. New York, NY: Basic Books.

Holyoak, Keith J. and Paul Thagard. 1995. *Mental Leaps: Analogy in Creative Thought*. Cambridge, MA: MIT Press.

Kuznicki, Jason T. 2011. "The Inscrutable Names of God," in Douglas Vakoch, ed., *Civilizations Beyond Earth: Extraterrestrial Life and Society*. New York, NY: Berghahn, pp. 202–213.

Launius, Roger. 2014. "Power of Analogies for Advancing Space Scientific Knowledge," *Astropolitics: The International Journal of Space Politics & Policy*, 12, 127–131, part of a special issue on analogy.

Margulis, Lynn and Dorion Sagan. 1986. *Microcosmos: Four Billion Years of Microbial Evolution*. Berkeley, CA: University of California Press.

Pooley, Jefferson D. and Michael J. Socolow. 2013. "War of the Words: *The Invasion from Mars* and its Legacy for Mass Communication Scholarship," in Joy Elizabeth Hayes, Kathleen Battles, and Wendy Hilton-Morrow, eds., *War of the Worlds to Social Media: Mediated Communication in Times of Crisis*. New York, NY: Peter Lang.

Sapp, Jan. 2009. *The New Foundations of Evolution*. Oxford: Oxford University Press.

Sawyer, Kathy. 2006. *The Rock from Mars: A Detective Story on Two Planets*. New York, NY: Random House.

Smith, Robert. 1982. *The Expanding Universe: Astronomy's "Great Debate."* Cambridge: Cambridge University Press.

Toynbee, Arnold. 1957. *A Study of History*. London: Oxford University Press (abridgement by D. C. Somervell, vol. 2).

Vidal, Clément. 2007. "An Enduring Philosophical Agenda. Worldview Construction as a Philosophical Method," online at http://cogprints.org/6048/, (accessed May 12, 2014).

Vidal, Clément. 2014. *The Beginning and the End: The Meaning of Life in a Cosmological Perspective*. New York, NY: Springer.

Wason, Paul K. 2011. "Encountering Alternative Intelligences: Cognitive Archaeology and SETI," in Douglas Vakoch, ed., *Civilizations Beyond Earth: Extraterrestrial Life and Society*. New York, NY: Berghahn, pp. 43–59.

Wason, Paul K. 2014. "Inferring Intelligence: Prehistoric and Extraterrestrial," in Douglas Vakoch, ed., *Archaeology, Anthropology and Interstellar Communication*. Washington, DC: NASA, pp. 113–129.

4 A multidimensional impact model for the discovery of extraterrestrial life

CLÉMENT VIDAL

Most science fiction books and movies depict extraterrestrials (ETs) that are similar to us in many ways. They are at our scale, have eyes, limbs, and body symmetries. But what if they don't look like us? What if they are so different that no communication is possible? How would it impact our worldviews to find non-communicative ETs? I first argue that we will most likely find microbial life or what are known as Kardashev Type II stellar civilizations, but nothing in-between, and that any extraterrestrials we find will not communicate, for the simple reason that they would likely be either immensely inferior or immensely superior to us. Then, I show that the discovery of ET life will most likely be very slow, taking years or decades. Finally, to prepare for discovery, I propose a multidimensional impact model. Twenty-six dimensions are introduced, illustrated with spider diagrams, which cover both what extraterrestrials might look like, and how humans may react.

Why extraterrestrials will not communicate

The principle of mediocrity is fundamental in astrobiology. It says that, "we should assume ourselves to be typical in any class that we belong to, unless there is some evidence to the contrary" (Vilenkin 2011). Applied to our *position in space* in the universe, it means that our Solar System, our galaxy, and possibly our universe – if there is a multiverse – are typical. They are not central or special in any way. This insight is well known and well assimilated, and is also known as the Copernican principle. However, what if we apply it to our *position in time*?

If we map our position in time according to the Kardashev (1964) scale (Figure 4.1), we can see that we are in an extremely short transition phase from technical impotence to technical omnipotence. Indeed, the exponential growth of our energy consumption is very recent on evolutionary time scales, but will still be limited by the total energetic output of the Sun.

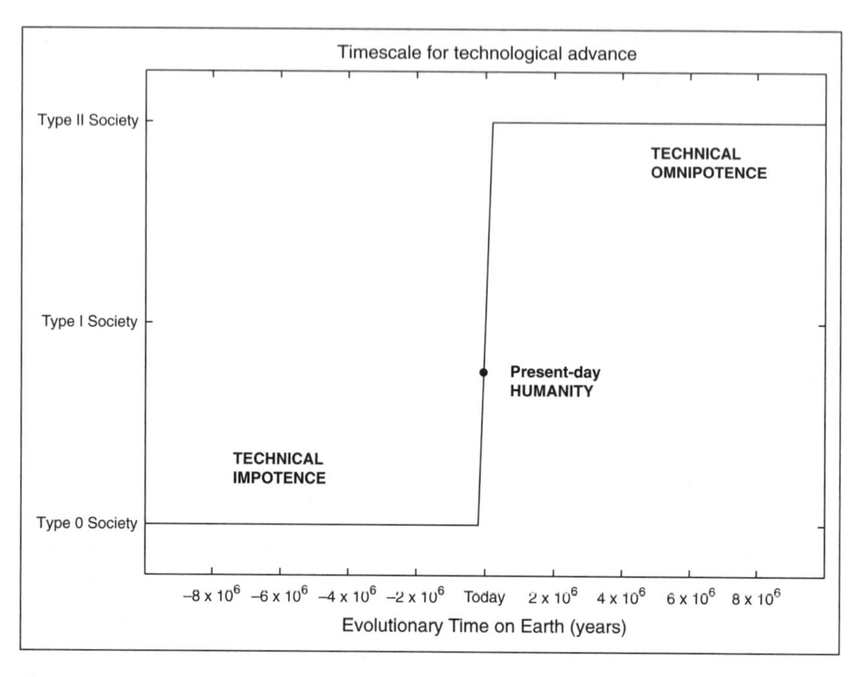

Figure 4.1 From technical impotence to technical omnipotence. We are most likely to find microbial (Type 0) or stellar (Type II) extraterrestrials, but nothing in-between. From *Xenology: An Introduction to the Scientific Study of Extraterrestrial Life, Intelligence, and Civilization*, 1979, sec. 25.2.1, by Robert A. Freitas, Jr. By permission of Robert A. Freitas Jr.

If our position in time is typical, it means that we have most chances to find either microbial life (left side of the step function), or Kardashev's Type II stellar civilizations (right side of the step function). It is extremely unlikely that we would find a civilization that is also in the short transition phase that we are in.

This difference in levels of development further leads to the argument that ETs will not communicate. If they are microbial, they will have not more to say than their biological organization can reveal (which is still very exciting for astrobiologists). If they are stellar civilizations, there might be ethical barriers to communication, as the prime directive in Star Trek which prohibits advanced civilizations to interfere with the development of lesser advanced ones. Of course, it is not an absolute rule, and there may be alien anthropologists who would want to study and communicate with alien races in the galaxy.

However, the difference in development also brings a motivational barrier to communication. Why don't we spend time whispering to bacteria or plants that $E = mc^2$? It doesn't make sense to even try because it's simply a waste of time. In the same way, from the perspective of an advanced stellar civilization, it would probably make no sense to spend time and efforts to try to communicate to a relatively primitive and transitioning civilization such as our own, unless in an anthropological sense.

There is a tradition of searching for non-communicative ETs, through searching for technological manifestations and artifacts (see, e.g., Dyson 1960; 1966; Freitas Jr. and Valdes 1985; Beech 2008; Bradbury *et al.* 2011; Wright *et al.* 2014; Vidal 2014a, ch. 9). But how do we prove the existence of advanced ETs if we cannot communicate with them?

A slow discovery

The ideal SETI (search for extraterrestrial intelligence) scenario occurs if we find a signal easy to observe and to decode, clearly stating the existence of an ET civilization. The issue of proving the presence of ETs is much more difficult if they do not communicate. In this case, we need to compare predictions and explanations hypothesizing that a phenomenon is purely physical, with predictions and explanations hypothesizing that the same phenomenon is living. The living or non-living model which better predicts and explains would gradually be favored. History shows that claims of non-communicative ET life were ambiguous for long periods of time. For example, the hypothesis of microfossils on the Martian meteorite ALH84001 (McKay *et al.* 1996) have fueled controversies which still are discussed today (see, e.g., Wainwright 2014). Turning to more advanced ET life, take the now amusing idea that there are artificial canals on Mars. It was championed by Percival Lowell in the nineteenth century, but was actually ambiguous for 20 to 70 years (Dick 1996, 78)!

More generally, previous discoveries in astronomy constitute an extended process (Dick 2013 and Chapter 3 in this volume), and scientific revolutions took decades to be fully appreciated (e.g. Kuhn 1957; 1970). There is no reason the discovery of ET life should be different.

Extended discovery and lack of communication do not imply there will be no impact. Of course, I can't exclude totally that there won't be a quick discovery with extraterrestrials communicating in plain English. Given this, it is prudent to develop a wide array of impact scenarios.

A multidimensional impact model

The Rio and London scales

I will now introduce an original and complex impact model for the discovery of ETs. Before that, let us discuss two existing scales to assess the social impact of discovering ETs. The Rio scale attempts "to quantify the social impact" of a discovery of extraterrestrial *intelligence* (ETI), and the London scale has the same purpose for a discovery of extraterrestrial *life* (ETL) (Almár 2011). They have the merit of being simple, but too much so. Let us briefly examine the limitations of these scales.

First, the distinction between the Rio and London scale assumes that the distinction between ETL and ETI will be clear cut, and unambiguous. This is of course not guaranteed, for even on Earth it is a contentious issue to decide where simple life stops and intelligence starts.

Second, these scales have much too few dimensions. Only three dimensions are introduced (class of phenomenon, discovery type, and distance) which, when calculated, provide an output number between 0 and 10, representing the significance of the discovery. Giving a result in one single number is similar to other scales, such as Richter's, which assesses the strength of an earthquake. This unidimensional number could give a first approximation to how important the discovery is, and such a simplification might be useful for the media to communicate with the general public. However, a number between 0 and 10 will quickly prove too simplistic and people will want to know more. By contrast, I shall propose below a richer model, with 26 dimensions.

Third, as Shenkel (in Almár 2001) has noticed, using this approach immediately associates the discovery of ETs with a danger, like an earthquake or an asteroid impacting the Earth. But discovering ETs could as well be the greatest opportunity for humanity to grow and learn.

Fourth, Almár (2001) wrote:

> I would like to emphasize that the announcements of putative discoveries we are discussing as well as the circumstances we are ranking are, or will be, most likely connected with physics and astronomy – not social sciences.

It is extremely surprising, if not self-contradictory, to aim to assess the social impact without considering social sciences! By contrast, Harrison's (1997, 244) book, one of the most comprehensive treatments of the subject matter, argues that "even the most clear and detailed announcement will elicit a variety of responses."

Fifth, the one-number outcome implicitly assumes the impact will be uniform for all humans, cultures, religions, and all domains of knowledge and action. This is consistent with the proposition of ignoring social sciences ... but can we really and safely ignore the complexities of human beings and societies?

In preparing for discovery there are actually two equally challenging issues: to envision what ETs will be like, and to envision the reactions of humans. Indeed, an impact is the meeting point of two components, in this case, ETs and humans.

Dimensions and options

The most objective way to study an impact is to be outside of it. This seemingly simple remark is of fundamental importance. In my research, I often try to take a "meta" perspective, which leads to new insights (see Vidal 2014a). It simply means to apply the concept to itself. What could it mean to take the perspective of "extraterrestrials of extraterrestrials"? It means to imagine not only the impact from our perspective, or from the perspective of ETs involved; we should also imagine how a third ET civilization would analyze our impact with an ET life form or civilization.

The 26 dimensions of impact are summarized in three spider diagrams (Figures 4.2–4.4), and the options are laid out in Table 4.1. To choose the dimensions, I drew inspiration mostly from the Rio scale, and from Harrison's (1997) book.

Although a much more comprehensive discussion of these dimensions and options would be desirable, I will aim below to summarize the most important aspects. Importantly, regarding ETs, I systematically include the "unknown" option in Table 4.1 because in a slow discovery, we are unlikely to know all the aspects of ETs at once.

Envisioning extraterrestrials

The distance can be *unknown, near* or *far*, the nearest being in our body, e.g. if you consider genomic SETI (e.g. Davies 2010; shCherbak and Makukov 2013), or if you speculate that (directed) panspermia brought bacteria or viruses from space (e.g. Hoyle and Wickramasinghe 1990). Other near options are in our Solar System, or at radius of 50 light years from Earth, which would allow a back-and-forth travel or messaging within a human lifetime. The *far* option includes everything beyond 50 light years, from galactic to extragalactic sources. The furthest option is if the fine-tuning of the universe itself would be a signature from an advanced civilization. This could result from a successful "search for extrauniversal intelligence" (see, e.g., Pagels 1989; Gardner 2004; Dick 2008; Vidal 2014a).

Table 4.1 Grouped dimensions and options in preparing the impact of discovering extraterrestrials

Group	Dimension	Options
Extraterrestrials	(1) Distance	Unknown
		Near
		Far
	(2) Complexity	Unknown
		Simple life
		Complex life
	(3) Intent	Unknown
		Neutral
		Benevolent
		Malevolent
		Benevolent and malevolent
	(4) Size	Unknown
		Away from our scale
		Near our scale
	(5) Number	One
		Many
	(6) Living state	Unknown
		Fossil
		Dormant
		Living
	(7) Clarity	Clear
		Ambiguous
	(8) Communicative intent	Unknown
		Null
		Null, but we snoop a message
		Limited
		Full, transparent
	(9) Knowledge of us	Unknown
		Nothing
		Some
		Everything
	(10) Influence on us	Unknown
		None
		Low
		High
Humans – objective	(a) Universalization of knowledge	Minimal
		Maximal
	(b) Speed of discovery	Slow
		Fast

Table 4.1 (cont.)

Group	Dimension	Options
	(c) Data quality	Low
		High
	(d) Source of discovery	Astrobiology
		Astroengineering / SETA
		SETI
		Reinterpretation of existing phenomena
		Response to active SETI
	(e) Our knowledge of them	Nothing except existence
		Some
		Everything
Humans – intersubjective	(f) First impression	Positive
		Negative
	(g) Socio-political context	Stable
		Unstable
	(h) Dissemination of the news	Trustworthy
		Unreliable
	(i) World philosophy	Western
		Indian
		Chinese
	(j) Religion	Anthropocentric
		Non-anthropocentric
Humans – subjective	(k) Preconception of ETs	Near from the truth
		Far from the truth
	(l) Worldview	No change
		Major changes
	(m) Developmental stage	Low
		High
	(n) Hierarchy of needs	Basic needs
		Higher needs
	(o) Age	Younger
		Older
	(p) Personality	Sensitive to the discovery
		Insensitive to the discovery

The nature of ETs can be *unknown, simple*, or *complex*. Generally, the more complex, the higher the impact. The intent can be *unknown, neutral* (e.g. if we discover microbes), *benevolent* (like the *E.T.* in Spielberg's movie), *malevolent* (like almost all other movies). The intent can also be *benevolent and*

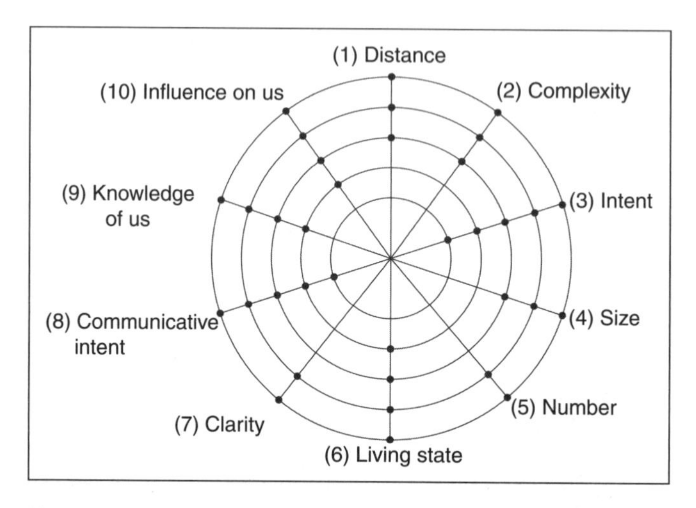

Figure 4.2 Ten dimensions of impact when envisioning extraterrestrials. Each radius represents one dimension, and each dot represents one option. The options are listed in Table 4.1.

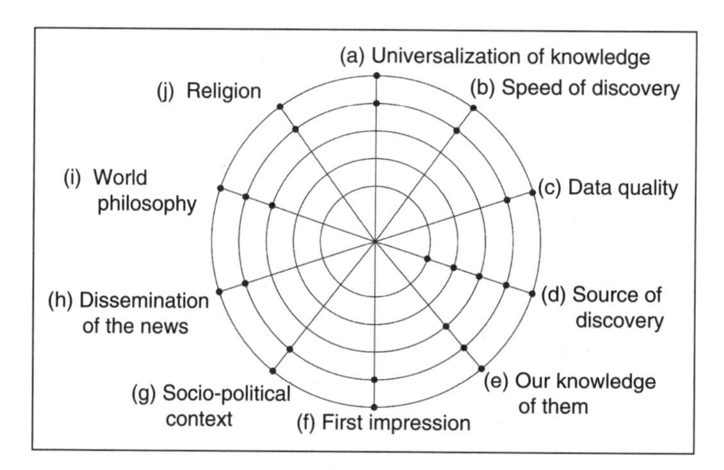

Figure 4.3 Ten dimensions of impact when envisioning humanity. Five dimensions of impact (a–e) concern objective aspects of a putative discovery; and five dimensions (f–j) deal with intersubjective or social aspects. Subjective dimensions are in Figure 4.4, and all the options are listed in Table 4.1.

malevolent, depending on specific contexts and situations. There is no reason that it should be Manichean and uniform. If ten million humans were to colonize a new planet, they would certainly not all be benevolent with its inhabitants. The more malevolent, the higher the impact. Of course, the intent

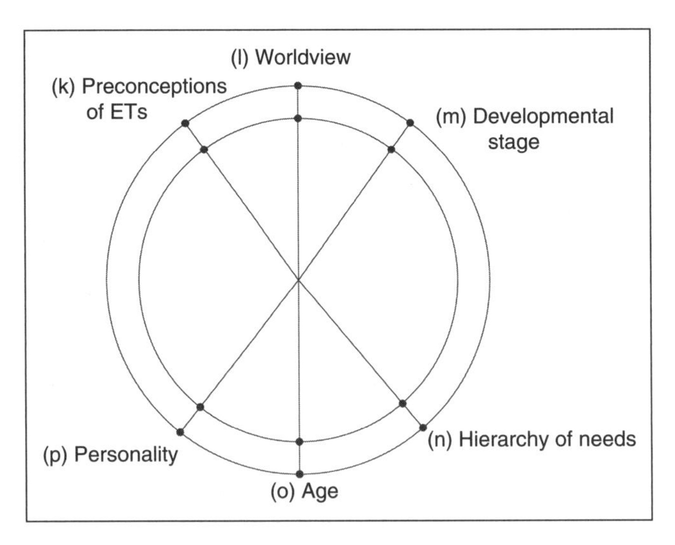

Figure 4.4 Six dimensions of impact regarding the subjective aspects of a discovery of extraterrestrials. The options are listed in Table 4.1.

is largely a subjective dimension. Even the most malevolent ETs killing humans randomly, thus annihilating the human race, could be interpreted by some to be a good thing. Also, a microbe killing the Earth's biosphere might seem malevolent from our point of view, but it is neutral from its point of view.

The size of ETs can be *unknown, away from our scale* or *near our scale*. If the size is away from our scale, the impact is high, because it is disconcerting if we cannot interact directly with an amazingly large galactic Type III civilization on Kardashev's (1964) scale, or with atto-sized ETs living at a very small scale (Type VI on Barrow's 1998, 133 scale; see also Vidal 2014a, ch. 9.2). Both would be equally hard to apprehend. Note that in popular culture and imagination, ETs are almost always at our scale or not far from it, but this is just a prejudice.

The number of ETs can be *one* or *many*. Finding thousands or millions of ET species will be much more distressing or exhilarating than one or two samples.

The living state of ETs could be *unknown, fossil, dormant*, or *living*. The closer to the living, the higher the impact would be. Although a dormant advanced ET could also be quite intriguing or exciting, if not scary.

The dimensions of clarity, communicative intent, knowledge of us, and influence on us presuppose ETI, possibly more advanced than us. If ETI communicates, the message can be *clear* or *ambiguous*. There could be many reasons for a signal or observation to be ambiguous, due to us as observers or them as transmitters, e.g. if ETs broadcast several conflicting messages (see also Harrison 1997, 202).

The communicative-intent dimension is similar to the "class of phenomenon" in the Rio scale. It can be *unknown*, but also *null* (as I argued), or it could be *null, but we snoop a message*. It could also be *limited* in several ways. First, limited for our own good (benevolent), such as the television series *Star Trek*'s Prime Directive; second, it could also be limited to do us harm (malevolent), like in Fred Hoyle's novel *A for Andromeda*. It could also be limited for specific purposes such as beacon navigation, which would be a more neutral message, omnidirectional, not Earth-specific. We could also imagine a *full, transparent* communicative intent, even if this does not guarantee full mutual comprehension. For more on the diversity of signal detection scenarios see, for example, Tarter (1992).

The ETs' knowledge of us could simply be *unknown* to us. It could also amount to *nothing*, or they could know *some* things or even nearly *everything* about us, if they had been observing the Earth for billions of years. To the limit, their knowledge of us could even have turned into an influence upon us. Although it sounds like material for science fiction or ufology, it remains a possibility that ETs have or had a *low* or *high* influence on us (consider, for example, panspermia or directed panspermia). Such an influence could remain *unknown* for a long time or even forever.

Envisioning humanity's reaction: objective dimensions

Let us now turn to the 16 dimensions in Figures 4.3 and 4.4 that are key to understanding the impact on humanity. They are classified in three categories: objective, subjective, and intersubjective categories, which have proven their worth in the history of ideas (Vidal 2012).

The main "objective" or scientific impact of finding ETs will be to literally universalize our knowledge. For now, only physics and some chemistry have been proven to hold beyond the confines of our atmosphere. As Table 4.2 illustrates, all other domains of knowledge remain to be universalized. However, thinking in an astrobiological or cosmological context, authors have already started to universalize many domains of knowledge such as language, sociology, economy, ethics, laws, aesthetics, theology, culture, or eschatology. If we find ETs, the more domains of knowledge become universalized, the greater the impact will be. The *minimal* impact here is that biology will be universalized once we find life beyond Earth. It would already be an immense breakthrough, but at the limit, the universalization of knowledge could be *maximal* (see references in Table 4.2).

Table 4.2 Examples of domains of knowledge which remain to be universalized. Reflecting in an astrobiological or cosmological context, some authors have already started to think in truly universal terms.

Universal	Comments	Example references
Physics	We have it!	Physics and astrophysics textbooks
Biology	Living system theory, universal biology	(Miller 1978; Freitas Jr. 1981; Schulze-Makuch and Irwin 2008)
Language	Anticryptography	(Freudenthal 1960)
Sociology	Astrosociology	(Freitas Jr. 1979)
Economy	Thermoeconomics	(Corning 2005)
Ethics and laws	Thermoethics Thermo-evo-devo ethics Celegistics	(Freitas Jr. 1979) (Vidal 2014a, ch. 10) (Vakoch 2014)
Aesthetics	What is beautiful for an ET?	(Freitas Jr. 1979, sect. 22.5)
Theology	Cosmotheology	(Dick 2000; Peters 2014)
Culture	Cosmos and culture	(Dick and Lupisella 2009)
Eschatology	Facing cosmic doom	(Vidal 2014a; 2014b)

I argued that the speed of discovery will be *slow*, but I may turn out to be wrong. Compare, for example, intercepting a message in plain English versus a cryptic message that takes us 20 years to decode. The most likely scenario is that opinions will oscillate with strong arguments sometimes supporting, sometimes refuting the claim of discovery. The faster the scientific community converges, the higher the impact will be. This dimension is similar to the credibility factor in the Rio scale.

The data quality we gather could be *low* or *high* depending on whether it is steady or transient, complete or incomplete, clear or ambiguous. The clearer, more complete, and steadier, the higher the quality it will be and the more seriously the discovery will be taken.

The source of discovery could come from mainstream *astrobiology*, searching for life in our Solar System, or hunting Earth-like planets. It could also come from *astroengineering*, or *search for extraterrestrial artefacts (SETA)*, or *SETI* as practiced by the SETI institute. But it could also be a *reinterpretation of existing phenomena*. For example, if some UFO reports were to end up convincing the scientific community. Importantly, authors have argued that ETs could already be in our astrophysical data (Shvartsman, cited in Heidmann and Klein 1991, 393; Davies 2010, 124; Vidal 2014a, 205). Should the scientific community be ashamed if ETs were to turn out to already have been in our data? I don't think so, and previous scientific revolutions support this view. Newton's or Darwin's contributions were not in the discovery of new gravitational bodies or new living species. Their contribution was to propose new *theories* to explain and interpret a variety of already known gravitational phenomena and already known living species. Finally, if we were to have a reply to one of the few messages we sent to space, i.e. a *response to active SETI*, the impact would certainly be extremely exciting.

Our knowledge of them could be minimal; for example, if we discovered a cryptic message we would know *nothing except their existence*. We could know *some* more, or even nearly *everything* about them. Note that this dimension needs to be compared with their knowledge of us (9): the more asymmetry in our disfavor, the more distressing it would be for us.

Envisioning humanity's reaction: intersubjective dimensions

Turning to the intersubjective or social group of dimensions, the first impression is crucial, because it could lead to butterfly effects (Harrison 1997, 199; Tarter 1992). Indeed, cycles of interactions, depending on whether they start on good or bad terms, could bifurcate either towards friendly relationships or to fearful ones. Overall, the first impression can thus be *positive* or *negative*.

Whether the socio-political context is *stable* or *unstable* will also play an important role, especially in forming the first impressions and in shaping the first reactions (Harrison 1997). An initial and promising discovery may also lead to a symbolic cold war to secure the discovery, like the race between the Soviets and the Americans to fly to the Moon.

The dissemination of the news can be *trustworthy*, following existing recommendations and protocols. But it could as well become *unreliable*, triggering rumors, disbeliefs, and confusion (Harrison 1997, 206–213).

Academia is strongly influenced by a *Western* world philosophy, but as comparative philosophers have observed (e.g. Bahm 1995; Smith 1957) there are obviously other major world philosophies to take into account, such as the *Indian* and the *Chinese* ones. This is especially important to take into account if we want to study the global impact, considering the large world population which holds a non-western philosophy.

How will various religious groups react? As previous studies have shown (e.g. Vakoch and Lee 2000), the religious affiliation matters less than a belief in *anthropocentrism* or *non-anthropocentrism*. Still, given how important the influence of religions is to billions of people, we must study how religious belonging would affect the impact. Specific cases may be important to monitor, such as the reactions of UFO religions.

Envisioning humanity's reaction: subjective dimensions

In the subjective group (Figure 4.4), the single most important dimension is our preconception of ETs (Connors 1976). Depending on whether we were *near from the truth* or *far from the truth*, reactions will vary immensely. This dimension can thus be assessed differentially, by drawing two versions of the spider diagram in Figure 4.2 (the dimensions 1–10). One version is how we imagine ETs to be, and the second version is how ETs really are, to be drawn at the time of discovery. If the two spider diagrams overlap perfectly, it means that we were spot on, and the discovery will not surprise us. But if we turn out to be wrong in all options of the ten dimensions, it will be a big surprise if not a psychological shock.

The dimension of worldview assesses if the discovery leads to *no change* in worldview, or to *major changes*. Two worldview models can be used for this purpose. First, using the collated model of Koltko-Rivera (2004), which assesses aspects of worldviews such as human nature, will, cognition, behavior, or truth. Again, the impact on worldviews can be assessed differentially, before and after the discovery of ETs. The worldview impact is the difference between your current worldview and the new worldview that finding ETs implies. If your worldview wouldn't change for any ET detection scenario (dimensions 1–10), it means you're truly ready. But double check that you are not in denial, rationalizing and protecting your old worldview. A direct (non-differential) assessment would require confronting the whole landscape of worldviews with the landscape of possible detections of ETs. The other worldview model is the one of Leo Apostel (Aerts *et al.* 1994; Vidal 2008). Here, a worldview is defined as answers to big questions such as: "Where does it all come from?" "Where are we going?" "What is good and what is evil?" "What is true and what is false?" This model allows us to ask: "Does the discovery of ETs change your views on the origin of the universe? On the future? On morality? On the nature of knowledge?" Here we can contrast the London scale with the present approach, taking the example of panspermia. On the London scale, the scenario scores relatively low: 1.4 to 2.8 out of 10 (Almár and Race 2011, 691). However, if proven true, panspermia would require us to re-write our worldview answer to: "Where does it all come from?," regarding the origin of life on Earth. For many people, such a discovery would come as a shock, because it would refute the story of cosmic evolution that most hold today. Note that a disconfirmation of a worldview can be both transformative or catastrophic.

Psychologists have shown that as humans grow, their motor skills, cognitive abilities, moral reasoning, emotional coping mechanisms, and their self-concept develop in stages (e.g. Piaget 1954; Kohlberg 1984; Laske 2008). Different people at different stages of development will certainly react differently to the discovery of ETs. This approach is especially useful to study, because developmental theories aim to relate many psychological dimensions, from feelings and motivation, to ethics and learning strategy (e.g. Graves 1974). People in *lower* stages will react differently than people in *higher* stages. Note that one effect of the discovery could be to trigger a change in developmental stage. Generally, a person grows up in stages, simply because higher stages are more adaptive, allowing one to successfully deal with more and more complex situations. Even if the worldview change is catastrophic at first, hopefully the discovery of ETs could later become transformative and trigger profound psychological changes to higher stages.

A robust psychological theory is Maslow's (1954) hierarchy of needs. Which level or levels of Maslow's hierarchy would the discovery affect? It could affect *basic needs*, or *higher needs*. For example, an ET microbe which would make all our food poisonous would be a disaster for everyone, affecting physiological and safety needs of the whole planet. By contrast, the discovery of a lost civilization, with no impact on daily behavior and functioning would only impact people who wonder about the position of mankind in the universe. Almost all scenarios would not impact basic human needs. In this pragmatic sense, we can argue that most potential discoveries (especially those far from Earth) will have a weak impact on a vast majority of people, because they won't need to change their behavior.

People of *younger* or *older* age will react differently. Younger generations are generally more progressive, more open to new experiences and discoveries. They are likely to be more open and more affected by the discovery, while older generations might either deny the discovery or try to make it fit with their older worldview, instead of changing it radically. Note that the "age" dimension overlaps with the "developmental stage" dimension.

Different personality types will be more or less *sensitive* or *insensitive* to the discovery. Although personality and worldview overlap, little research has been done on this topic (Koltko-Rivera 2004, 44), so it makes sense to consider personality as a dimension on its own. Harrison (1997, 245–248) discussed various personality traits which would affect human reactions. For example, the emotional stability (how well people cope with stress), the rigidity or dogmatism of their worldview (how hard it is to change it), the default affectivity stance (positive and cheerful, versus negativistic and anxious), as well as the self-esteem (low or high). Obviously, those people who are emotionally stable, flexible in their worldview, positively minded, and with a high self-esteem will react much better than those with the opposite attributes.

Discussion

Let us discuss our dimensions and options. Regarding ETs, dimensions 1–3 in Figure 4.2 are the most important parameters. As Harrison (1997, 237) put it: "If we believe that their technology is not much better than ours, that they are safely sequestered in some remote corner of the universe, and that they are basically 'good guys', we will feel less threatened than if they have an overwhelming technical superiority, are already in our neighborhood, and seem unfavorably disposed toward humans."

Regarding the human side, in Figures 4.3 and 4.4 I selected dimensions (a–p) relevant for the question of impact. Assessing the impact on us presupposes that we have a model of humanity. This is of course extremely hard to do, and would require nothing less than a synthesis of social and psychological sciences.

No modeling is neutral and purely objective. Modeling always has a goal (Sterman 2000). What is the goal that we pursue when we try to model impacts of discovering ETs? Different goals will require different models, focusing on different dimensions. My own goal and hope is that these two sets of dimensions will prove useful as a conceptual framework and repository for researchers and policy makers to explore the landscape of impacts related to the discovery of ETs. One could choose to focus on studying beneficial or hazardous impacts, or the impact on a common profile of most people, or some specific categories of people (e.g. decision makers, leaders), or on different human activities (academia, religions, politics, public health, economics, etc.).

Focusing on the impact on humanity may also be too restrictive for a whole class of scenarios where ETL affects not only the human species, but also the whole ecosystem. An example would be a bacterial colony coming from a meteorite and contaminating and disturbing the whole ecosystem.

Mark Neal (2014) recently reviewed the question of ETs in the context of risk management. There are of course many risks we can imagine, but what if the discovery is also a great opportunity to grow and learn? Is there a field of research such as "opportunity management"? If one wants to stick with risk management, then we need to consider the risk of not making the most of the opportunities. For now, we have the opportunity to prepare for discovery, which means to explore an immense landscape of possibilities. In the end, we may learn more now that we are free to explore the landscape of possibilities, rather than after the actual discovery, which would put constraints on the landscape. The astrobiological journey may be more important than the destination.

Does the model have too many dimensions? Multiplying the options, we see that there are 172,800 possibilities (i.e. spider diagrams) for ETs, and 442,368 on the side of humanity. Putting the two together, this totals to more than 76 billion scenarios to study! With Koltko-Rivera (2004, 28), I'd argue that it's a better approach to start with too many dimensions rather than too few:

> Ultimately, multivariate empirical research will reveal which dimensions are actually distinct from one another. Put crudely: "Factor analyze them all – let eigenvalues sort them out." If one starts out with a detailed set of dimensions, research may reveal a simpler structure. Starting out with too small a set of dimensions would make it difficult or impossible to determine whether a more complex approach is needed. Whether tailoring clothing or theory, it is wise to start with more material than the finished product will require.

Further work needs to be done to tailor the model I proposed. As we mentioned, developmental psychology theorists may see patterns at different stages of cognitive, emotional, and moral development, and thus could reduce the space of human reactions. Another approach would be to reduce the number of relevant scenarios by using decision trees (see Billingham 2002). For example, if bacteria are found on Mars, the dimensions of "communicative intent" and "knowledge of us" become irrelevant.

Preparing for this major discovery, educating people to cosmic evolution and a wide array of possible detections and reactions is the most useful thing we can do now. If people entertain a cosmic culture, they will be much more ready to assimilate the news that we are not alone in the universe.

Conclusion

I began this chapter with the argument that humans are unlikely to meet peers, and more likely to discover either microbial life or Kardashev Type II stellar civilization. Those two encounters will be non-communicative, as such extraterrestrials would be either immensely inferior or immensely superior to us.

Then, I suggested that the discovery will be slow, as arguments and counterarguments will likely be difficult to resolve conclusively, leading to waning public and media interest, and possibly an anti-climactic announcement when a discovery is actually confirmed.

Finally, getting ready for impact means that a smooth absorption should be our goal. This can be achieved by remaining aware of the wide array of possible extraterrestrials, and the resulting human reactions. To this end, I have introduced a new multidimensional impact model covering both what extraterrestrials would be like and the possible socio-psychological reactions of humans.

Anticipating and preparing for every single scenario out of the 76 billion entailed by the model is clearly preposterous (see also Harrison 2007, ch. 9). But it would certainly be valuable to explore this expanse of possibilities in more depth and identify the most hazardous ones for human psyche and society.

Taking even wider perspectives, we should also care beyond humanity, to also preserve the Earth ecosystem. Looking at the other side of the impact, we should also study our impact on possible ecosystems at the scales of the Solar System, the galaxy, or the universe.

I hope this model will provide a useful framework for researchers and policy makers to explore and appreciate the diversity and complexity of possible impacts.

Acknowledgments

I thank Mark Lupisella, John Trapaghan, Steven J. Dick, Cadell Last, and Albert A. Harrison for helpful corrections and discussions.

References

Aerts, D., L. Apostel, B. De Moor, S. Hellemans, *et al.* 1994. *World Views. From Fragmentation to Integration.* Brussels: VUB Press. http://www.vub.ac.be/CLEA/pub/books/worldviews.pdf.

Almár, I. 2001. "How the Rio Scale Should Be Improved." *IAA-01-IAA.9.2.03.* International Academy of Astronautics. http://avsport.org/IAA/abst2001/rio2001.pdf.

Almár, I. 2011. "SETI and Astrobiology: The Rio Scale and the London Scale." *Acta Astronautica* 69 (9–10): 899–904. doi:10.1016/j.actaastro.2011.05.036.

Almár, I. and M. S. Race. 2011. "Discovery of Extra-Terrestrial Life: Assessment by Scales of Its Importance and Associated Risks." *Philosophical Transactions of the Royal Society A: Mathematical, Physical and Engineering Sciences* 369 (1936): 679–92. doi:10.1098/rsta.2010.0227.

Bahm, A. J. 1995. *Comparative Philosophy: Western, Indian, & Chinese Philosophies Compared.* Revised edition. Albuquerque, NM: World Books.

Barrow, J. D. 1998. *Impossibility: The Limits of Science and the Science of Limits.* New York, NY: Oxford University Press.

Beech, M. 2008. *Rejuvenating the Sun and Avoiding Other Global Catastrophes.* Dordrecht: Springer.

Billingham, J. 2002. "Pešek Lecture: SETI and Society – Decision Trees." *Acta Astronautica* 51 (10): 667–72. doi:10.1016/S0094-5765(02)00023-1.

Bradbury, R. J., M. M Ćirković, and G. Dvorsky. 2011. "Dysonian Approach to SETI: A Fruitful Middle Ground?" *Journal of the British Interplanetary Society* 64: 156–65.

Connors, M. M. 1976. *The Role of the Social Scientist in the Search for Extraterrestrial Intelligence.* Moffett Field, CA: NASA Ames Research Center.

Corning, P. A. 2005. *Holistic Darwinism: Synergy, Cybernetics, and the Bioeconomics of Evolution.* Chicago, IL: University of Chicago Press.

Davies, P. C. W. 2010. *The Eerie Silence: Are We Alone in the Universe?* London: Penguin Books.

Dick, S. J. 1996. *The Biological Universe: The Twentieth Century Extraterrestrial Life Debate and the Limits of Science.* Cambridge: Cambridge University Press.

Dick, S. J. 2000. "Cosmotheology: Theological Implications of the New Universe." In *Many Worlds. The New Universe, Extraterrestrial Life and the Theological Implications*, edited by S. J. Dick, Philadelphia and London: Templeton Foundation Press, pp. 191–210.

Dick, S. J. 2008. "Cosmology and Biology." In *Proceedings of the 2008 Conference on the Society of Amateur Radio Astronomers. June 29 – July 2.* National Radio Astronomy Observatory, Green Bank, West Virginia: Amateur Radio Relay League, pp. 1–16. http://evodevouniverse.com/uploads/f/f3/Dick_2008_-_Cosmology_and_Biology.pdf.

Dick, S. J. 2013. *Discovery and Classification in Astronomy: Controversy and Consensus.* Cambridge: Cambridge University Press.

Dick, S. J., and M. L. Lupisella, eds. 2009. *Cosmos and Culture: Cultural Evolution in a Cosmic Context.* Washington, DC: Government Printing Office, NASA SP-2009-4802. http://history.nasa.gov/SP-4802.pdf.

Dyson, F. J. 1960. "Search for Artificial Stellar Sources of Infrared Radiation." *Science* 131 (3414): 1667–68. doi:10.1126/science.131.3414.1667.

Dyson, F. J. 1966. "The Search for Extraterrestrial Technology." In *Perspectives in Modern Physics*, edited by R. E. Marshak, New York: John Wiley & Sons, pp. 641–55.

Freitas Jr., R. A. 1979. *Xenology: An Introduction to the Scientific Study of Extraterrestrial Life, Intelligence, and Civilization.* Xenology Research Institute. http://www.xenology.info/Xeno.htm.

Freitas Jr., R. 1981. "Xenobiology." *Analog Science Fiction/Science Fact* 101: 30–41. http://www.xenology.info/Papers/Xenobiology.htm.

Freitas Jr., R. A. and F. Valdes. 1985. "The Search for Extraterrestrial Artifacts (SETA)." *Acta Astronautica* 12 (12): 1027–34. doi:10.1016/0094-5765(85)90031-1.

Freudenthal, H. 1960. *Lincos: Design of a Language for Cosmic Intercourse.* Amsterdam: North-Holland.

Gardner, J. N. 2004. "The Physical Constants as Biosignature: An Anthropic Retrodiction of the Selfish Biocosm Hypothesis." *International Journal of Astrobiology* 3: 229–36. doi:10.1017/S1473550404002162.

Graves, C. W. 1974. "Human Nature Prepares for a Momentous Leap." *The Futurist* 8 (2): 72–85.

Harrison, A. A. 1997. *After Contact: The Human Response To Extraterrestrial Life.* New York, NY: Plenum Trade.

Harrison, A. A. 2007. *Starstruck: Cosmic Visions in Science, Religion and Folklore.* New York, NY: Berghahn Books.

Heidmann, J. and M. J. Klein, eds. 1991. *Bioastronomy The Search for Extraterrestial Life – The Exploration Broadens.* Berlin: Springer, vol. 390.

Hoyle, F. and N. C. Wickramasinghe. 1990. *Cosmic Life-Force.* New York, NY: Paragon House.

Kardashev, N. S. 1964. "Transmission of Information by Extraterrestrial Civilizations." *Soviet Astronomy* 8 (2): 217–20. http://tinyurl.com/Kardashev1964.

Kohlberg, L. 1984. *The Psychology of Moral Development: The Nature and Validity of Moral Stages.* 1st edn. Harpercollins College Div.

Koltko-Rivera, M. E. 2004. "The Psychology of Worldviews." *Review of General Psychology* 8 (1): 3–58. doi:10.1037/1089-2680.8.1.3.

Kuhn, T. S. 1957. *The Copernican Revolution; Planetary Astronomy in the Development of Western Thought.* Cambridge: Harvard University Press.

Kuhn, T. S. 1970. *The Structure of Scientific Revolutions.* Chicago: University of Chicago Press.

Laske, Otto E. 2008. *Measuring Hidden Dimensions of Human Systems: Foundations of Requisite Organization.* 1st edn. Medford, MA: IDM Press.

Maslow, Abraham H. 1954. *Motivation and Personality.* 1st edn. Harper's Psychological Series. New York, NY: Harper.

McKay, D. S., E. K. Gibson, K. L. Thomas-Keprta, *et al.* 1996. "Search for Past Life on Mars: Possible Relic Biogenic Activity in Martian Meteorite ALH84001." *Science* 273: 924–30. doi:10.1126/science.273.5277.924.

Miller, J. G. 1978. *Living Systems.* New York, NY: McGraw-Hill.

Neal, M. 2014. "Preparing for Extraterrestrial Contact." *Risk Management* 16 (2): 63–87. doi:10.1057/rm.2014.4.

Pagels, H. R. 1989. *The Dreams of Reason*. New York, NY: Bantam.

Peters, T. 2014. "Astrotheology: A Constructive Proposal." *Zygon®* 49 (2): 443–57. doi:10.1111/zygo.12094.

Piaget, J. 1954. *The Construction of Reality in the Child*. New York, NY: Basic Books.

Schulze-Makuch, D. and L. N. Irwin. 2008. *Life in the Universe: Expectations and Constraints*. 2nd edn. Advances in Astrobiology and Biogeophysics. Berlin: Springer.

shCherbak, V. I. and M. A. Makukov. 2013. "The 'Wow! Signal' of the Terrestrial Genetic Code." *Icarus* 224 (1): 228–42. doi:10.1016/j.icarus.2013.02.017.

Smith, H. 1957. "Accents of the World's Philosophies." *Philosophy East and West* 7 (1/2): 7–19.

Sterman, J. D. 2000. *Business Dynamics: Systems Thinking and Modeling for a Complex World*. Boston, MA: Irwin McGraw-Hill.

Tarter, D. E. 1992. "Interpreting and Reporting on a SETI Discovery: We Should Be Prepared." *Space Policy* 8 (2): 137–48. doi:10.1016/0265-9646(92)90037-V.

Vakoch, D. A. and Y.-S. Lee. 2000. "Reactions to Receipt of a Message from Extraterrestrial Intelligence: A Cross-Cultural Empirical Study." *Acta Astronautica* 46 (10–12): 737–44. doi:10.1016/S0094-5765(00)00041-2.

Vakoch, D. A., ed. 2014. *Extraterrestrial Altruism: Evolution and Ethics in the Cosmos*. Berlin: Springer.

Vidal, C. 2008. "What Is a Worldview? Published in Dutch as: 'Wat Is Een Wereldbeeld?'" In *Nieuwheid Denken. De Wetenschappen En Het Creatieve Aspect Van De Werkelijkheid*, edited by Hubert Van Belle and Jan Van der Veken, Leuven: Acco, pp. 71–85. http://cogprints.org/6094/.

Vidal, C. 2012. "Metaphilosophical Criteria for Worldview Comparison." *Metaphilosophy* 43 (3):306–47. doi:10.1111/j.1467-9973.2012.01749.x. http://homepages.vub.ac.be/~clvidal/writings/Vidal-Metaphilosophical-Criteria.pdf.

Vidal, C. 2014a. *The Beginning and the End – The Meaning of Life in a Cosmological Perspective*. Berlin: Springer.

Vidal, C. 2014b. "Cosmological Immortality: How to Eliminate Aging on a Universal Scale." *Current Aging Science*. doi:10.2174/1874609807666140521111107. http://student.vub.ac.be/~clvidal/writings/Vidal-Cosmological-Immortality.pdf

Vilenkin, A. 2011. "The Principle of Mediocrity." *Astronomy & Geophysics* 52 (5): 5.33–35.36. doi:10.1111/j.1468-4004.2011.52533.x.

Wainwright, M. 2014. "A Presumptive Fossilized Bacterial Biofilm Occurring in a Commercially Sourced Mars Meteorite." *Journal of Astrobiology & Outreach* 02 (02). doi:10.4172/2332-2519.1000114.

Wright, J. T., R. L. Griffith, S. Sigurdsson, M. S. Povich, and B. Mullan. 2014. "The Ĝ Infrared Search for Extraterrestrial Civilizations with Large Energy Supplies. II. Framework, Strategy, and First Result." *The Astrophysical Journal* 792 (1): 27. doi:10.1088/0004-637X/792/1/27.

Part II Transcending anthropocentrism

How do we move beyond our own preconceptions of life, intelligence, and culture?

Introduction

One of the seemingly intractable problems in addressing the impact of discovering life beyond Earth is the need to transcend anthropocentrism. How can we move beyond our preconceptions of basic concepts such as life and intelligence, culture and civilization, technology and communication? Unfortunately we cannot "get out of our heads," so to speak, no matter how hard we try. But we can at least attempt to imagine a much broader spectrum of each of these concepts than are known to exist on Earth. Indeed, the best science fiction is sometimes very good at doing this. We can also attempt to escape our anthropocentrism by an empirical approach that emphasizes the diversity of life and intelligence on Earth, and potentially in the broader universe. Both natural and social scientists have begun to address these difficult problems.

In this section scholars from a great variety of backgrounds take up these issues. From his perspective in the biogeosciences Dirk Schulze-Makuch surveys the landscape of actual and possible life. He demonstrates that while the limits to life on Earth are much broader than once thought, other planets may exhibit even broader limits based on conditions specific to their planet, whether in the clouds of Venus, on dusty dry Mars, sulfur-rich Io, hydrocarbon-laden Titan, or the great variety of conditions sure to exist on the multitude of exoplanets now being discovered. All of this is based not on science fiction, but on possible real-life adaption mechanisms for life. In a similar way, neuroscientist Lori Marino examines the landscape of intelligence. Despite the importance of the concept to many fields (Sternberg, 2000; 2002), sophisticated studies in an astrobiological context have been lacking, with few exceptions (Bogonovich 2011). Decrying the lack of empirical work in this crucial area within the astrobiology community, she employs scientific data from terrestrial life to establish an expansive concept of the nature of intelligence. Intelligence, Marino argues, is not a binary property in the sense of having or lacking it, but is a continuous multi-staged and multi-leveled property based on the gradual evolution of life on Earth, beginning with the first neurons. Intelligence, she concludes, is a ubiquitous property of life on Earth; even if no life is found beyond Earth, "we are not alone," in the famous phrase often used in the SETI context. Her analysis examines what this expansive view of intelligence on Earth might presage for intelligence on other planets.

Philosopher of biology Carlos Mariscal takes a more theoretical approach to what is known in the field as the "$N = 1$" problem: how can we talk intelligently about properties of extraterrestrial life when we have only one example of life on Earth? He refers here not to one example of microbial life or intelligent life, but to the fact that all life on Earth is related. While there could in theory have been multiple origins, all known life on Earth is based on DNA. This is the domain of the nascent field of universal biology, a long-sought goal for those interested in the search for life. Mariscal argues that a biological claim can be judged universal only if it includes principles of evolution, but no contingent facts of life on Earth. He concludes that biological generalizations are possible, and that we might in fact learn something about the nature of life beyond Earth by studying life on Earth. This conclusion is consistent with the approaches of Schulze-Makuch and Marino.

From life and intelligence we move to the no less problematic concepts of culture, civilization, and communication. Anthropologist John Traphagan argues that our ideas of extraterrestrial civilizations are ethnocentric in the sense that they are affected by terrestrial, and most often Western, notions of civilization and progress. In the same way that Percival Lowell's claim of canals on Mars was arguably affected by cultural factors, so, Traphagan argues, are our notions of civilizations beyond Earth affected by nineteenth-century views of cultural evolution as leading to increasingly better forms of society and moral behavior. Anyone seriously interested in SETI, he suggests, needs to examine these deep preconceptions, and problematize the nature of extraterrestrial civilizations rather than merely accepting them as similar or even analogous to ours. In a parallel to the search for universal biology, he further suggests that the issue might be moved forward by examining behaviors that could be universal because they are advantageous from an evolutionary viewpoint. This is in part a problem of "evolution and altruism," which surprisingly has seen considerable discussion (Vakoch 2014). More generally, we need a more robust framework for the relationship between our cultural values and those we project onto potential extraterrestrials.

Douglas Vakoch, a psychologist with the official title of Director of Interstellar Message Composition at the SETI Institute, takes up the issue of communication in the event of discovery of intelligence, problematizing it in the same way that other authors in this section approach their respective concepts. Again, the issue reduces to a question of universals. The assumption in the field has been that mathematics and science are universal, and that they might therefore be our best hope for communicating with extraterrestrial others. Vakoch shows how philosophers and others have called this Platonic assumption into question, and argues that other approaches might be better applied to message construction. Since we cannot be sure what characteristics humans and extraterrestrials have in common, he argues, it is far better to take a variety of approaches to message construction, each with distinct sets of assumptions. One of these approaches may allow us to build a conceptual framework for mutual understanding.

References

Bogonovich, M. (2011). "Intelligence's likelihood and evolutionary time frame." *International Journal of Astrobiology*, 10: 113–122.

Sternberg, R. J., editor. 2000. *Handbook of Intelligence*. Cambridge University Press, Cambridge.

Sternberg, R. J. 2002. "The search for criteria: why study the evolution of intelligence." In *The Evolution of Intelligence*, edited by R. J. Sternberg and J. C. Kaufman, Lawrence Erlbaum Associates, Mahwah, NJ, pp. 1–8.

Vakoch. D. A. 2014. *Extraterrestrial Altruism: Evolution and Ethics in the Cosmos*. Springer, Berlin.

5 The landscape of life

DIRK SCHULZE-MAKUCH

Earth is a planet that exhibits an immense biomass and an incredible biodiversity. Yet, the question arises as to whether the diversity as observed on Earth reflects the limits of life or whether life elsewhere in the universe could manifest an even greater diversity. To examine the question further, I review some of the limits of life as observed on Earth and then ask what specific adaptation mechanisms could reasonably occur on other planets and moons to extend the limits of life to environmental conditions usually not found on our planet. As we currently do not have any convincing evidence for the existence of extraterrestrial life, these extensions must remain in the field of scientific speculation. Yet, given the enormous creativity and flexibility exhibited by the organisms we know from our planet, it would be odd if life could not adapt to some of the conditions exhibited on other planets. Thus, my conjecture is that life in the universe would exhibit a much larger variety of forms and functions than life on Earth.

The landscape of life provides important basic information when preparing for the discovery of extraterrestrial life. We are familiar with life on Earth, but life might be so strange on another world that we might not recognize it; especially since there is not even a commonly accepted definition of what life actually is. Thus, it is important to be not too constrained and Earth-centric if we do not want to take the risk to miss it even if an organism is in plain sight. This applies to all life, from microbial to more complex, and also to signals from technologically advanced civilizations. Thus, open-mindedness for this quest is an imperative.

The range of life on Earth

The conditions under which life can persist are incredibly broad; however, the range under which life can originate is likely to be much smaller. Since the origin of life is an unsolved puzzle, I will focus on the persistence of life, particularly under which environmental stresses life exists on Earth. The life most familiar to us, consisting of organisms that live at conditions similar to those we are accustomed to, is generally referred to as mesophilic life. However, during the evolution of life, organisms, particularly microorganisms, conquered nearly all available environmental niches on and within our planetary crust. This is exemplified by the range of temperatures under which organisms can grow, extending from at least –15 °C to about 113 °C. There are also reports that the temperature limit is even higher, and McKay (2014) indicated that this limit may be at about 122 °C due to solubility of lipids in water and protein stability. A critical component of tolerance to high temperatures is that water must be still in its liquid phase, which in principle can be achieved up to water's critical temperature of 374 °C at a pressure of 215 bar. The practical limit due to energetic and biochemical constraints under which life can still metabolize and reproduce is surely much lower, however. The pressure tolerance of life, though, is high and extends to at least 1,100 bar (Stan-Lotter 2007). The pH tolerance of organisms, particularly of bacteria, archaea, and fungi, roughly extends from just below 0 to about 13. Examples of organisms which can live at low pH values are *Ferroplasma sp.* and *Cephalosporium*; examples of organisms that live at high pH values are *Natrobacterium* and several species of protists and rotifers (Baross *et al.* 2007; Schulze-Makuch and Irwin 2008).

Since Earth is a water-rich planet and life adapted to it through its evolution, we can expect that life would be sensitive to a lack of water, and in fact, bacteria, archea, and fungi can only metabolize at water activities down to about 0.6 on a scale of zero to one (Stevenson *et al.* 2014). Adaptation to water with high salt content, however, is quite common, as some halobacteria and archea can grow in 35% NaCl solution (Schulze-Makuch and Irwin 2008). Another limit is imposed by radiation, both ultraviolet (UV) and ionic radiation. Tolerance to radiation varies widely, but some organisms such as *Deinococcus radiodurans* can still grow at doses upward of 10,000 Gy. Dormant stages of life can even tolerate much more extreme environmental conditions. A pointed example are the tardigrades, also known as water bears, microscopic animals that usually live in mosses and lichen. It has been shown that tardigrades have survived temperature stresses as low as –273 °C and as high as + 151 °C, pressure stresses from microgravity to 6,000 bars (for a few days), and ionizing radiation doses up to 5,000 Gy. Special adaptation traits include anhydrobiosis and cryptobiosis (Watanabe 2006; Schulze-Makuch and Seckbach 2013), and during its dormant stage the tardigrade can lower its metabolism down to less than 0.01% of normal and its water content down to less than 1%.

Examples of possible life adaptations

Venus – life in the clouds?

As pointed out in earlier publications (Schulze-Makuch and Irwin 2002a, 2006; Schulze-Makuch *et al.* 2004, 2013b) the lower Venusian atmosphere

Figure 5.1 Venus and its cloud structure revealed by ultraviolet observations of the Pioneer Venus Orbiter in 1979. The dark streaks are believed to be caused by larger, non-spherical particles in the clouds that absorb UV irradiation and may be related to Venusian life. Credit: NASA.

can be considered borderline habitable, based on the following observations: (1) conditions in the clouds at 50 km in altitude are relatively benign, with temperatures of 30–80 °C, pressure of 1 bar, and a pH of about 0; (2) the clouds of Venus are much larger, more continuous, and stable than the clouds on Earth; (3) the atmosphere is in chemical disequilibrium, with H_2 and O_2, and H_2S and SO_2 coexisting; (4) the lower cloud layer contains non-spherical particles comparable in size to microbes on Earth (Figure 5.1); (5) the super-rotation of the atmosphere enhances the potential for photosynthetic reactions; (6) COS is present in the atmosphere, which on Earth is a strong indicator of biological activity; (7) CO is less abundant than expected under Venusian atmospheric conditions and could be oxidized as a reactant in plausible metabolic pathways; (8) the biologically critical elements of carbon, phosphorus, and nitrogen are present; and (9) while water is scarce on Venus, water-vapor concentrations reach several hundreds of parts per million in the lower cloud layer. The environmental conditions in the lower cloud layer thus might be sufficient for life to exist and this has special relevance, because Venus used to be located within the habitable zone early in Solar System history. Moreover, it could have retained liquid surface water for a billion years or even longer (Grinspoon 1997).

Thus, there is the possibility of a natural history on Venus. Microbial life could have evolved under conditions very similar to early Earth or alternatively could have been transported from Earth via asteroid impacts, and then gradually adapted to life in the clouds as the surface became more desiccated and hostile to organisms (Schulze-Makuch and Irwin 2002a). Cycloocta sulfur in the atmosphere could not have only protected it from enhanced UV irradiation in the atmosphere, but also allow microbes to employ photosynthesis reactions with visible wavelengths as used on Earth (Schulze-Makuch *et al.* 2004). Alternatively, microorganisms in the Venusian clouds could also use chemosynthesis by reducing sulfur dioxide to either hydrogen sulfide or carbonyl sulfide. The likelihood of these types of organisms in the lower cloud layer of Venus, and perhaps even the presence of a simple microbial ecosystems, hinges largely on the timing and force of the environmental changes that took place, converting Venus' habitable planetary surface to the desiccated, hot, and inhospitable inferno we observe today (Schulze-Makuch *et al.* 2013b). If these changes occurred rather gradually, life that should have been present on early Venus would have had a chance to adapt via directional selection to live in the Venusian clouds to the present day.

Mars – hygroscopic adaptations?

Today Mars is a cold and dry planet. However, that was not always so. It was a much wetter planet billions of years ago, and any organisms that evolved during its warmer and wetter earlier period may have adapted to its current conditions by developing hygroscopic abilities, to attract water directly from the atmosphere. There are analogs for this approach on Earth as some microorganisms in the Atacama desert use hygroscopic salt crystals to attract water from the atmosphere (Davila *et al.* 2010) and even the thorny devil, a type of lizard, has hygroscopic grooves between the spines of its skin to capture water in its desert habitat. Incorporation of hygroscopic compounds such as hydrogen peroxide or perchlorates into the cell could also convey antifreeze properties and support oxidative metabolism, as well as enhance the ability to counter radiation damage, as the biochemical protective mechanisms against both oxidants and radiation that generates them are complementary. The use of hydrogen peroxide by microorganisms was suggested as an explanation for the Viking life detection experiments (Houtkooper and Schulze-Makuch 2007), possibly indicating the presence of life. Organisms that would use hygroscopity to scavenge water molecules from the atmosphere would be extremely vulnerable to abundant liquid water as used in some of the Viking experiments. There are some microbes that use these compounds in their metabolism. For example, *Dechloromonas aromatica* can reduce perchlorate (Coates and Achenbach 2004) and *Acetobacter peroxidans* reduces hydrogen peroxide with the help of molecular hydrogen to water (Tanenbaum 1956). And although hydrogen peroxide is more known as a sterilizing agent, many microbes (e.g. certain *Streptococcus* and *Lactobacillus sp.* (Eschenbach 1989) and even human cells produce hydrogen peroxide, while some microorganisms utilize it (e.g. *Neisseria sicca, Haemophilus segnis*, Ryan and Kleinberg 1995). A stunning example of the use of hydrogen peroxide is that of the Bombardier beetle (Eisner and Aneshansley 1999). The beetle stores an aqueous solution of hydroquinone and hydrogen peroxide in a reservoir, which can be squeezed into a reaction chamber and mixed with enzyme catalysts resulting in a very fast catalytic reaction that heats the solution to a boiling point, which is then discharged for defensive purposes. On a planet such as Mars, in which the abundance of water diminished with time to give place to a dry and cold desert, the use of hygroscopic compounds by life (if it exists or existed) would certainly be conceivable.

Europa – non-terrestrial energy sources?

Jupiter's moon Europa is likely to harbor a deep ocean beneath its ice crust. Recent indications of some form of plate tectonics (Kattenhorn and Prockter 2014) and emanating water plumes (Roth *et al.*, 2014) underline the possibility that the underlying ocean may be habitable and that it may not be that deep underneath the ice crust. Particularly, if oxygenated ice crust is subducted and resourcing the ocean with oxidants and possibly other nutrients, organisms within the ocean could be utilizing these compounds for metabolic reactions. Although the ice crust may not be as thick in some regions (e.g. the Chaos region), photosynthesis as a metabolic pathway on Europa can most likely be excluded. However, in the deep and dark oceans other energy sources in addition to chemosynthesis may be used with which we are not familiar on Earth as a way to harvest metabolic energy. Examples are osmotic or ionic gradients, magnetic energy, thermal energy, or kinetic energy from convection cells (Schulze-Makuch and Irwin 2002b). Osmotic or ionic gradients, for example, could channel water molecules through membranes that couple the movement to the formation of a high-energy covalent bond. Magnetic fields, via the Lorentz force, could drive hydrogen ions across a one-way channel against their concentration gradient into an internal organelle, where they accumulate to a higher concentration than on the outside of the organelle and from which they are later released again, coupling the return reaction again to the production of a high-energy bond. Muller (1985) and Muller and Schulze-Makuch (2006) show in detail how thermal energy would allow biomembranes to convert heat into electrical energy during temperature cycling. Alternatively, convection cells or tidal currents on Europa's ocean floor or ice ceiling could be another way to sustain life in the absence of light and oxygen (Schulze-Makuch and Irwin 2002b). Organisms containing pili or cilia could adhere to a substrate at the ocean bottom or on the underside of the ice ceiling, where they are exposed to currents of moving water that can bend their cilia and open ion channels, through which ions could flow passively down their concentration gradients. This process could then be coupled to the direct formation of high-energy phosphate bonds (Schulze-Makuch and Irwin 2002b). In principle, Europa could provide enough nutrients to support a modest ecosystem with even some multicellular organisms similar in complexity and size to brine shrimps (Irwin and Schulze-Makuch 2003). Due to the tidal energy (Sohl *et al.* 2014) and also internal heating, we would expect hydrothermal vents similar to the ones we find at mid-oceanic ridges on Earth (e.g. Martin *et al.* 2008). These places are oases of life in an environment otherwise largely bare of biomass and biodiversity.

Io – a violent sulfur-rich world

Jupiter's innermost moon Io is a violent world as it is the most volcanically active planetary body in the Solar System. Adding to it the extremely harsh radiation environment originating from the magnetic field of nearby Jupiter, the lack of a significant atmosphere must result in extreme desiccation on the moon's surface. Thus, Io seems to be one of the last places one would expect life. Nevertheless, we should not be too hasty to come to final conclusions, because dynamic activity, even if violent, can be conquered by microbes on Earth as their colonization of black smokers and hydrothermal vents on the ocean floor demonstrates. However, these niches of life have plenty of water and are not bathed in radiation. Would there be a location on Io where these constraints are fulfilled?

Rocks and sulfur compounds that make up lava flows would provide good shielding from radiation. Therefore, we can expect to find more benign radiation environments on Io already several tens of meters below the surface. The lava flows could not only serve as protection from radiation, but could produce chambers that hold a significant amount of volatiles in the subsurface. Spectral analyses indicate that at least some water is present on Io (Salama *et al.* 1990). Also, we should not forget that Io is one of the Galilean satellites and that

Europa, Ganymede, and Callisto all have a rich endowment of water. Thus, early in Solar System history Io probably had a large amount of water as well. If this is correct, life could have originated on Io under more benign conditions, and then retreated to the subsurface adapting to steep thermal gradients and abundant sulfur compounds.

There is one other possibility for life on Io and that is the use of an alternative solvent. Instead of water, hydrogen sulfide or a hydrogen sulfide–water mixture could be used as a solvent for life (Schulze-Makuch 2010). Hydrogen sulfide should be liquid under some of the subsurface conditions encountered on Io and has been observed in the atmosphere (Nash and Howell 1989; Salama *et al.* 1990; Carlson *et al.* 2007). The main disadvantage of H_2S is that its liquidity range is only 26 °C at 1 bar atmospheric pressure (from −86 °C to −60 °C). Other sulfur compounds cannot be entirely excluded as possible solvents. Although a biochemical scheme for those compounds would be much more difficult to envision (Schulze-Makuch and Irwin 2008), sulfur dioxide remains liquid at temperatures between −75 to −10 °C and sulfuric acid from 10 to 337 °C (given temperatures are for a pressure of 1 bar). Surface temperatures on Io range considerably, with a minimum of about 90 K, to 0 °C in the Loki Patera caldera floor (Lopes-Gautier *et al.* 1999), and up to 227–327 °C in volcanic hot spots. Benner (2002) suggested sulfuric acid as a possible solvent on sulfur-rich Venus. Io is a sulfur-rich planetary body and could be an example of how far organic chemistry and perhaps even life could adapt biochemically to a sulfur-rich environment. Unfortunately, the surface environment is so radiation-intense that no space mission will be able to go to Io and probe its subsurface in the near future.

Titan – hydrocarbons as a solvent for life?

Saturn's largest moon Titan is the most exotic planetary environment in our Solar System (Figure 5.2). On its surface are methane- and ethane-containing lakes, which experience occasional methane downpours from Titan's clouds. The atmosphere has a surface pressure of 1.5 bar and is mainly composed of nitrogen with a significant amount of methane. A methane cycle seems to exist on Titan with similarities to the hydrological cycle on Earth (Figure 5.3). Titan appears to have all the basic requirements for life and according to Baross *et al.* (2007) "the environment of Titan meets the absolute requirements for life, which include thermodynamic disequilibrium, abundant carbon containing molecules and heteroatoms, and a fluid environment." Despite the frigid surface temperatures (−178 °C), there have been suggestions of possible life on Titan that uses metabolic pathways involving reactions with photochemical acetylene, hydrogen, and heavier hydrocarbons (McKay and Smith 2005; Schulze-Makuch and

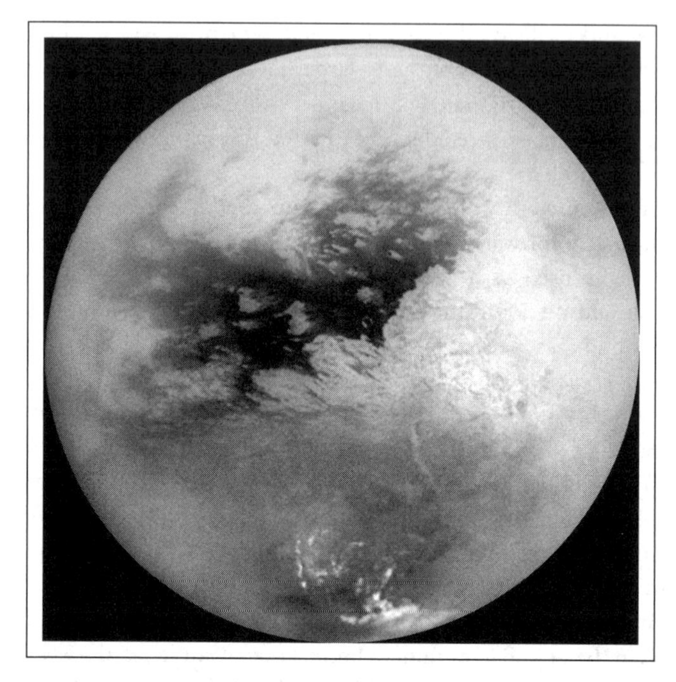

Figure 5.2 A mosaic of images of Saturn's moon Titan taken by the Cassini spacecraft during its first flyby on October 26, 2004. The view is centered on 15 degrees south latitude, 156 degrees west longitude. Bright clouds near the south pole are easily seen; the bright region on the right side of the equatorial region is named Xanadu Regio. Credit: NASA/JPL/Space Science Institute.

Grinspoon 2005). On the other hand, Schulze-Makuch *et al.* (2013a) argued that metal ions may be a missing component that could restrain any type of possible life on Titan. If life exists on Titan, it would surely be "weird" by Earth standards. It may incorporate silane and polysilanes as building blocks in its cellular membranes and could involve huge, very slowly metabolizing cells (Schulze-Makuch and Irwin 2008). Also, instead of chemical redox reactions on which life is based on Earth, powerful radical reactions may be suitable on Titan for various types of metabolisms, such as the reaction of an N_2 radical with a CH_2 radical to cyanamide or the reaction of an N_2 radical with two CH radicals to two hydrocyanic acid molecules (Schulze-Makuch and Grinspoon 2005). Cyanamide in Earth's biochemistry can induce important condensation reactions to form peptides and proteins. Also, given Titan's extremely low availability of unbound oxygen, oxygen molecules might be replaced with nitrogen analogues (Raulin 1998). Far beneath the surface a subsurface ocean is suspected, which might consist of a water–ammonia mixture. Also, the groundwater on

Figure 5.3 Ligeia Mare, the second largest sea on Titan and located near the north polar region. The seas of Titan are filled with liquid ethane and methane. Synthetic aperture radar image mosaic taken by the Cassini spacecraft in 2006–2007. Credit: NASA/JPL.

Titan could be of the same chemical nature. If so, liquid water–ammonia mixtures may be geothermally or tidally driven to the surface, where they mix with liquid methane and ethane in the hydrocarbon lakes. These mixtures then would constitute a rich organic reservoir in which life may exist or perhaps even originate. Recent research of an analog site on Earth, a natural liquid asphalt lake, shows that microdroplets of water in a hydrocarbon matrix can be a suitable habitat of life and contain a diverse ecosystem (Meckenstock *et al.* 2014).

Exoplanets and exotic ideas

The number of detected exoplanets far exceeds 1,000 at the time of this writing. The number of different types of planetary environments is probably staggering, and if life is common in the universe, then so is also the number of organisms and possible adaptations (Irwin *et al.* 2014). Because there are not many constraints on the possible environments that we might encounter on exoplanets due to our very limited remote-sensing capabilities, any speculation of life and possible adaptation mechanisms in these locations borderlines science fiction. However, one of the exotic examples of possible life in an unfamiliar planetary environment is that of a planet that orbits a neutron star. In such a scenario, photosynthesis would likely not develop because by far the majority of the radiation the host star would provide is in the form of magnetic fields. Thus, fluctuating magnetic fields could be the main energy source supporting life – assuming it could exist there. On Earth there are many organisms that are sensitive to magnetic fields (Blakemore, 1982), but it is not used as an energy source for metabolism, probably because it is too low in energy yield compared to visual light and chemical redox reactions. On a planet around a neutron star the situation will be quite different. For example, magnetic energy may be harvested directly via the Lorentz force, which can create a charge separation if the magnetic field lines are oriented perpendicular to the movable charges, like protons and electrons in an elongated cell (Schulze-Makuch and Irwin 2008). If the elongated cell changes its direction with respect to the magnetic field lines, so that they are parallel to the movable charges, then the "stored" energy could be released with the protons and electrons moving toward each other and producing molecular hydrogen. With the prominence of magnetic fields, even the information code could be based on chains of magnets rather than chemical macromolecules (such as DNA or RNA on Earth).

One such fascinating example was introduced by Feinberg and Shapiro (1980). The system would start out with a chain of atomic magnets with their magnetic moments aligned in variable directions. Then, a randomly directed magnet approaching the chain would orient itself with its direction parallel to that of the nearest magnet. If that process were continued for many magnets, the result would be a new chain that duplicates the original chain in the directional arrangement of its magnets. If the magnets in place along the chain retained their alignment and could be protected from any re-magnetization from an exterior magnetic field, then such an informational string could serve as a possible system for the replication and transmission of biological information.

An expansion of the landscape of life?

The proposed possible existence of microbial life on our neighboring planets based on various adaptation mechanisms, including the use of different energy sources and solvents of life, may seem outlandish. Certainly, not all of these will turn out to exist. However, given what we know about chemistry, biology, and physics, these types of adaptations, each of which would increase the landscape of life dramatically, are sensible. Even if such adaptions are not manifested on our neighboring planets and moons, they would likely exist on one of the countless exoplanets. In essence, given the number of stars just in our galaxy, and realizing that many if not most of them are orbited by planets, everything that is chemically, physically, and biologically possible should be manifested somewhere at some time. Thus, if we were to propose that the known diversity of life is all that exists, it would mean that life can only exist under the rather narrow environmental conditions that occur on Earth. Nothing could be more Earth-centric. If we have learnt anything from evolution, then it is that both genotypes and phenotypes of organisms are astonishing plastic, adapting quickly to changing conditions on our planet, some of which have been rather dramatic (such as in the aftermath of some asteroid impacts or other global events).

Whether life can use different energy sources is more controversial. On Earth, only chemistry and light are used as energy sources, but this is likely grounded in the fact that both of these energy sources are common on our planet. Organisms are sensitive to many other types of energy sources such as magnetic fields, osmotic gradients, kinetic energy, pressure gradients etc., and if one of the alternative energy sources was readily available on another world, there does not seem to be any reason why it would not be utilized. The same can be said for an alternative solvent. Water has many properties conducive to life, but also many that make life challenging (Schulze-Makuch and Irwin 2008). Perhaps given the difficulties of organic synthesis in an aqueous medium, it would perhaps not be too outrageous to suggest that life did not originate because of water but rather in spite of it! However, the truth probably lies somewhere in between. Nevertheless, a solvent like methanol, a polar organic molecule, seems to be at least equally suitable to water as a life-sustaining solvent, with its biggest disadvantage being that it is not common on a planetary surface, at least not in our solar system. The proposed possible adaptations for Io, and for a planet in orbit around a neutron star, are surely even more speculative, and are intended as examples of what we know to be compatible with the chemistry and physics of life rather than what we might expect.

Conclusion

We must expect that life on another planet will appear strange and "weird" to us, and it may not be straightforward to recognize it as life. Thus, a collection of possible adaption methods and models of life and how it interacts with its environment, may prove to be instrumental in this quest. The biodiversity we experience from life on Earth can only make us wonder how rich in form and function life in the universe may be.

References

Baross, J. A., S. A. Benner, G. D. Cody, *et al.* 2007. *The Limits of Organic Life in Planetary Systems*. Washington, DC: National Academies Press.

Benner, S. A. 2002. "Weird life: chances vs. necessity (alternative biochemistries)." Presented at "Weird Life" Planning Session for the Committee on the Origins and Evolution of Life. National Research Council, Washington DC.

Blakemore, R. P. 1982. "Magnetotactic bacteria." *Annual Review of Microbiology* 36: 217–238.

Carlson, R. W., J. S. Kargel, S. Doute, L. A. Soderblom, and J. B. Dalton. 2007. "Io's surface composition." In *Io after Galileo: A New View of Jupiter's Volcanic Moon*, edited by R. M. C. Lopez and J. R. Spencer, Chichester, UK: Springer-Praxis, pp. 193–230.

Coates, J. D. and L. A. Achenbach. 2004. "Microbial perchlorate reduction: rocket-fuelled metabolism." *Nature Reviews Microbiology* 2: 569–580.

Davila, A. F., L. G. Duport, R. Melchiorri, *et al.* 2010. "Hygroscopic minerals and the potential for life on Mars." *Astrobiology* 10: 617–628.

Eisner, T. and D. J. Aneshansley. 1999. "Spray aiming in the bombardier beetle: Photographic evidence." *PNAS* 96: 9705–9709.

Eschenbach, D. A., P. R. Davick, B. L. Williams, *et al.* 1989. "Prevalence of hydrogen peroxide-producing *Lactobacillus* species in normal women and women with bacterial vaginosis." *Journal of Clinical Microbiology* 27: 251–256.

Feinberg, G. and R. Shapiro. 1980. *Life beyond Earth: The Intelligent Earthling's Guide to Life in the Universe*. New York, NY: William Morrow and Company, Inc.

Grinspoon, D. H. 1997. *Venus Revealed: A New Look Below the Clouds of Our Mysterious Twin Planet*. Cambridge, MA: Perseus Publishing,

Houtkooper, J. M. and D. Schulze-Makuch. 2007. "A possible biogenic origin for hydrogen peroxide on Mars: the Viking results reinterpreted." *International Journal of Astrobiology* 6:147–152.

Irwin, L. N. and D. Schulze-Makuch. 2003. "Strategy for modeling putative ecosystems on Europa." *Astrobiology* 3: 813–821.

Irwin, L. N., A. Mendez, A. G. Fairén, and D. Schulze-Makuch. 2014. "Assessing the possibility of biological complexity on other worlds, with an estimate of the occurrence of complex life in the Milky Way Galaxy." *Challenges* 5: 159–174.

Kattenhorn, S. and L. M. Prockter. 2014. "Evidence for subduction in the ice shell of Europa." *Nature Geoscience*, doi:10.1038/ngeo2245.

Lopes-Gautier, R., A. S. McEwen, W. B. Smythe, *et al.* 1999. "Active volcanism on Io: global distribution and variations in activity." *Icarus* 140: 243–264.
McKay, C. P. 2014. "Requirements and limits for life in the context of exoplanets." *PNAS* 111: 12,628–12,633.
McKay, C. P. and H. D. Smith. 2005. "Possibilities for methanogenic life in liquid methane on the surface of Titan." *Icarus* 178: 274–276.
Martin, W., J. Baross, D. Kelley, and M. J. Russell. 2008. "Hydrothermal vents and the origin of life." *Nature Reviews Microbiology* 6: 805–814.
Meckenstock, R. U., F. von Netzer, C. Stumpp, *et al.* 2014. "Water inclusions in oil are microhabitats for microbial life." *Science* 345: 673–676.
Muller, A. W. J. 1985. "Thermosynthesis by biomembranes: energy gain from cyclic temperature changes." *Journal of Theoretical Biology* 115: 429–453.
Muller, A. W. J. and D. Schulze-Makuch. 2006. "Thermal energy and the origin of life." *Origin of Life and Evolution of Biospheres* 36: 177–189.
Nash, D. B. and R. R. Howell. 1989. "Hydrogen sulfide on Io: evidence from telescopic and laboratory infrared spectra." *Science* 244: 454–456.
Raulin, F. 1998. "Titan." In *The Molecular Origins of Life*, edited by A. Brack, New York, NY: Cambridge University Press, pp. 365–385.
Roth, L., J. Saur, K. D. Retherford, *et al.* 2014. "Transient water vapor at Europa's south pole." *Science* 343: 171–174.
Ryan, C. S. and I. Kleinberg. 1995. "Bacteria in human mouths involved in the production and utilization of hydrogen peroxide." *Archives of Oral Biology* 40: 753–763.
Salama, F., L. J. Allamandola, F. C. Witteborn, *et al.* 1990. "The 2.5–5.0 micrometer spectra of Io: evidence for H_2S and H_2O frozen in SO_2." *Icarus* 83: 66–82.
Schulze-Makuch, D. 2010. "Io: is life possible between fire and ice?" *Journal of Cosmology* 5: 912–919.
Schulze-Makuch, D. and D. H. Grinspoon. 2005. "Biologically enhanced energy and carbon cycling on Titan?" *Astrobiology* 5: 560–567.
Schulze-Makuch, D., D. H. Grinspoon, O. Abbas, L. N. Irwin, and M. Bullock. 2004. "A sulfur-based UV adaptation strategy for putative phototrophic life in the Venusian atmosphere." *Astrobiology* 4: 11–18.
Schulze-Makuch, D. and L. N. Irwin. 2002a. "Reassessing the possibility of life on Venus: proposal for an astrobiology mission." *Astrobiology* 2: 197–202.
Schulze-Makuch, D. and L. N. Irwin. 2002b. "Energy cycling and hypothetical organisms in Europa's ocean." *Astrobiology* 2, 105–121.
Schulze-Makuch, D. and L. N. Irwin. 2006. "The prospect of alien life in exotic forms on other worlds." *Naturwissenschaften* 93: 155–172.

Schulze-Makuch, D. and L. N. Irwin. 2008. *Life in the Universe: Expectations and Constraints*, 2nd edition, Berlin: Springer.

Schulze-Makuch, D. and J. Seckbach. 2013. "Tardigrades: an example of multicellular extremophiles." In *Polyextremophiles: Life under Multiple Forms of Stress*, edited by J. Seckbach, A. Oren, and H. Stan-Lotter, Dordrecht: Springer, pp. 597–607.

Schulze-Makuch, D., A. G. Fairén, and A. Davila. 2013a. "Locally targeted ecosynthesis: a proactive *in situ* search for extant life on other worlds." *Astrobiology* 13: 774–778.

Schulze-Makuch, D., L. N. Irwin, and A. G. Fairén. 2013b. "Drastic environmental change and its effects on a planetary biosphere." *Icarus* 225: 775–780.

Sohl, F., A. Solomonidou, F. W. Wagner, *et al.* 2014. "Structural and tidal models of Titan and inferences on cryovolcanism." *Journal of Geophysical Research – Planets* 19: 1013–1036.

Stan-Lotter, H. 2007. "Extremophiles, the physicochemical limits of life (growth and survival)." In *Complete Course in Astrobiology*, edited by G. Horneck and P. Rettberg, Weinheim: Wiley-VCH, pp. 121–150.

Stevenson, A., J. Burkhardt, C. S. Cockell, *et al.* 2014. "Multiplication of microbes below 0.690 water activity: implications for terrestrial and extraterrestrial life." *Environmental Microbiology*, doi 10.1111./1462-2920.12598.

Tanenbaum, S. W. 1956. "The metabolism of *Acetobacter peroxidans*: I. Oxidative enzymes." *Biochimica et Biophysica Acta* 21: 335–342.

Watanabe, M. 2006. "Anhydrobiosis in invertebrates." *Applied Entomology and Zoology* 41: 15–31.

6 The landscape of intelligence

LORI MARINO

Introduction: astrobiology and intelligence

The question of how intelligence evolves on different planets is a subject of fervent interest in many scientific and public domains. Yet, it has received little, if any, serious scientific attention in astrobiology. Astrobiology relies on an elegant paradigm: Earth as a natural laboratory. It seeks to investigate how life arose and evolved on this planet and to apply that knowledge to detecting and understanding extraterrestrial life. Thus, the study of the evolution of intelligence fits squarely within the field of astrobiology. But, despite a wealth of accessible data from "mainstream" fields, astrobiology has limited itself to studying the origin and evolution of early life and has not made the connection between these basic processes and intelligence. Why, in its 50-year history, has there been essentially no empirical work within astrobiology on intelligence?

What is intelligence?

Intelligence is, by nature, a fuzzy concept. That is, there are no strict boundaries on it and there is no scientific consensus on its definition (Sternberg 2000). The study of intelligence, therefore, necessitates a strong reliance on "bottom-up" empirical descriptions of a range of phenomena rather than a "top-down" hunt for a precise exemplar. Intelligence is not a binary trait. Rather, it is a multidimensional phenomenon which expresses itself in varying phenotypes and levels of complexity and is interconnected with the entire psychological make-up of any animal. Nevertheless, if we wish to use a working definition of intelligence, then we can refer to intelligence as a level of cognitive complexity, i.e. how an individual acquires, processes, stores, analyzes, and acts upon information and circumstances.

Despite its complexities and fluid boundaries, the phenomenon of intelligence is amenable to empirical scientific investigation just as any other biological property. The absence of the study of intelligence from astrobiology is due to a complex set of historical and psychological roadblocks. One of these may be the mistaken assumption that intelligence is not scientifically tractable. But foremost of these is our species' adherence to the wrong model of life on Earth, one that promotes misconceptions that impede the way forward in the scientific study of intelligence in astrobiology.

The Darwinian revolution

Charles Darwin established that all species on Earth, including humans, descend from common ancestors and, along with Alfred Russel Wallace, identified the mechanism driving biological evolution – natural selection (Darwin 1859, 1871). When the modern evolutionary synthesis emerged in the 1930s to the 1950s a consensus was reached in which natural selection was recognized as the central (but not exclusive) means by which evolution occurs. Darwin's work unified the field of biology and provided a fertile and substantive foundation for studying intelligence. But, well before Darwin another model of life on Earth had already taken hold of the human mind – one that serves a strong desire to be superior to and separate from the other animals, i.e. the *scala naturae.*

Scala naturae

The *scala naturae* is an Aristotelian model of nature from the third century BCE (but which probably goes back much further). According to this model, nature is arranged on a linear scale of progression with inorganic objects such as rocks on the lowest rung of the scale, then plants, then up to "lower" animals (invertebrates), to vertebrates, "higher" mammals such as primates, and, finally, humans, who occupy a separate, superior position above all other animals. Humans are considered the most "advanced" life form on the planet, having a more "perfect" form than the other animals and possessed of unique qualities (Marino 2007). This notion is alive and well today in terms such as "higher versus lower" and "primitive versus advanced" to describe species differences. The *scala naturae* constrains the entire process of thinking about extraterrestrial intelligence because it leads to the assumption that astrobiology has only one example of a unique intelligence on Earth and, therefore, nowhere to go for further investigation.

Sui generis

The *scala naturae* leads to well-known self-perpetuating assumptions, such as *sui generis,* meaning "one of its own kind" and notions of anthropocentrism and directionality – all of which are antithetical to the astrobiology paradigm. The result has been a longstanding argument about contingency versus convergence between those who view extraterrestrial intelligence as highly improbable and those who do not. As far back as 1904, Alfred Russel Wallace expressed strong reservations about extraterrestrial intelligence on the basis that the number of steps it would take to create an intelligence as complex as that of a human is too improbable (Crowe 2008). And over the last several decades many leading scientists have vociferously argued that the emergence of a human-like intelligence is based on a highly improbable set of events (contingencies) that cannot be repeated elsewhere. But these interminable debates miss the point by focusing on human intelligence exclusively and leaving out consideration of the framework of all species in which humans are embedded.

From the Big Bang to the Big Brain?

Scala naturae thinking is also compatible with teleological assumptions about intelligence. Teleology, the appeal to a goal-directed, purposeful progression toward an end point, seems to suggest that all life on Earth has been moving toward the eventual emergence of human intelligence. Adherents point to the fossil record showing an increase in average brain size over the last 200 million years. But this is an illusion that stems from the fact that larger brains are found in more recent species because large brains need time to evolve. When phylogenetic methods are applied, there is no evidence for any progressive linear trends in the evolution of intelligence leading to humans. Life on this planet resembles a branching tree – not an arrow. In short, teleological and anthropocentric notions about intelligence on Earth are not supported by any scientific evidence.

Replacing assumptions with scientific data

Putting assumptions and misconceptions aside, what do we know about the evolution of intelligence on Earth from actual scientific data? Questions about the origin and evolution of intelligence on Earth are more tractable than ever before because of our sophisticated methods of collecting, storing, and analyzing large quantities of data. Although it would be impossible to document each step of this process, the question can be framed as follows: What are the major milestones and characteristics of intelligence on this planet and what do these milestones tell us about the nature of intelligence that may be relevant to astrobiology?

Before beginning, it is important to acknowledge that, with some exceptions, we cannot observe the evolution of life directly and that we must rely on inferential approaches, such as the comparative phylogenetic method. But this method does allow us to reconstruct the traits of extinct species with some degree of confidence, and there are a number of scientific approaches that support informed hypotheses about the evolution of intelligence.

Brain-like functions in unicellular organisms

All organisms on Earth have the same task: to survive and reproduce. Brains exist because of the varied distribution of resources in the environment that affect survival. An immobile organism or one living in a highly predictable, stable environment would not have much need for a brain. But if one moves around then there must be a way to detect and analyze all of the varied inputs one comes in contact with and make adaptive behavioral responses to them. As such, some of the building blocks of nervous systems and brains are found in single-celled organisms. They sense their environment, briefly store the information, integrate it in different channels of input and then act upon it – producing basic adaptive behavior. These brain-like processes work at a molecular level but they reveal a common ancestry with all modern brains. One example is *Escherichia coli*, a bacterium that senses its environment through a dozen different types of protein receptors embedded in its cell wall. Each receptor specializes in a specific kind of information, e.g. toxins, sugars, amino acids, etc. The input is integrated and the result is the "decision" to behave in a certain way, i.e. use its flagella to swim towards or away from the stimulus (Allman 1999). Therefore, the fundamental functional properties of brains such as sensory integration, memory, decision-making, and behavior were likely present in the earliest organisms on Earth.

The property that allows unicellular organisms to detect and respond to their environment is known as "membrane excitability" and involves sensitivity to electrochemical gradients accomplished by the flow of ions into and out of the cell. This basic mechanism was later adapted to specialized cells, i.e. neurons, in metazoans. Evidence for continuity derives from the fact that the membranes of modern neurons operate by the same electrochemical principles (voltage-gated ion movement across a membrane) as in single-celled organisms (Allman 1999). The startling conclusion from these findings is that the basic mechanism for nervous systems was already present approximately one billion years ago, when only unicellular organisms populated the Earth.

Multicellularity and the first neurons

Since its emergence in multicellular organisms 600 million years ago, the neuron has remained the basic unit of information processing in all nervous systems on Earth. A neuron is an electrochemically polarized cell with a characteristic morphology consisting of a body with branching processes containing chemical receptors. When an electrochemical threshold is reached a long axon transmits an electrical pulse to terminal branches, which then release chemicals (neurotransmitters) that are picked up by the next neuron, and so forth. This highly simplified description reveals some important steps in the evolution of nervous systems. First, neurons form fiber tracts that allow information to travel along specific pathways. In the earliest animals with nervous systems, such as jellyfish, transmission is bidirectional. In bilaterians (bilaterally symmetrical animals), neuronal transmission is unidirectional, so information flows only one way in the neuron, thus increasing the specificity of neural transmission. Second, while some animals use non-threshold graded potentials for neural transmission, at some point in evolution the neural pulse, called an action potential, became an all-or-nothing electrical phenomenon, and allowed for a new digital form of neural transmission and integration.

The first organisms to possess neurons were the earliest metazoans – those with radially symmetric bodies such as cnidarians, ctenophores, hydrozoa, etc. – appearing approximately 600 million years ago. Voltage-gated sodium channels are present in jellyfish and all brains. The neurotransmitters coursing through the brains of all modern species have precursors in single-celled organisms and were present in the nerve nets of the earliest metazoans.

Modern versions of these early taxa possess a decentralized nerve net (a loose arrangement of sensory neurons connected to motor neurons by interneurons). In many of these animals there are the very beginnings of the next step in the evolution of nervous systems: centralization, in the form of nerve rings and clusters of neurons called ganglia that allow these animals to learn and exhibit complex behaviors requiring coordination of parts of the body.

Bilateralization, centralization, and cephalization

Bilaterians emerged in the fossil record about 550–600 million years ago as early wormlike creatures with a nerve cord running down the body and enlarged ganglia (early brain) at the head region. The process is called cephalization and an anterior brain connected to a nerve cord became the bauplan of the nervous system thereafter for all organisms with a central nervous system (Arendt *et al.* 2008; Striedter 2005).

All vertebrate brains are segmented into the same parts: forebrain, midbrain, and hindbrain. And all vertebrates possess identical brain structures and functions within those segments. All mammals possess a layered neocortex (Allman 1999) although the way the neocortex is organized or mapped in some species, such as cetaceans, is quite different, in many respects, from other species. The neocortex in all mammals, including cetaceans, is the most evolutionarily recent and pliable part of the brain and, thus, arguably, the vanguard of intelligence, i.e. the source of self-awareness, problem solving, sensory–motor integration, mental representation, etc. And despite differences in neocortical architecture, even cetaceans and primates display a striking degree of convergence in cognitive function (Marino 2002). Likewise, invertebrate and vertebrate brains rely upon the same basic principles of information processing. All of the major vertebrate neurotransmitters, which are ultimately derived from single amino acids, are also found in invertebrate brains (Messenger 1996) and most of them also served as signaling molecules before the first central nervous system evolved (Turlejski 1996). Moreover, embryonic development of nervous systems across species reveals a highly conserved plan, making use of pre-existing genetic plans (Striedter 2005).

There is a striking degree of conservation in the genetic mechanisms underlying nervous system formation across all species. For instance, homeotic genes that segment insect bodies are co-opted to segment parts of the vertebrate central nervous system. Other genes controlling the formation of part of the brain in fruit flies regulate the mammalian neocortex. So even the human brain, which is a very recent evolutionary development, is organized by genes with antecedents going back half a billion years (Allman 1999), showing that ancient conserved mechanisms underlie many of the most complex features of intelligence on Earth.

Brain size

One issue of considerable interest in discourse about intelligence on this planet is brain size. Brain size has to do with the amount of neural tissue available for information processing, under the assumption that "more is better." Measures of brain size range from whole brain mass to the mass of specific structures in the brain to various ratios of neuron types. One measure of brain size or cephalization is the encephalization quotient (EQ). EQ is a measure of a species' average brain size relative to body size taking into account brain–body allometric relations. So the EQ metric allows one to compare directly the relative brain size of a squirrel and an elephant (Figure 6.1). Species with average brain sizes have an EQ of 1. Modern humans have an EQ of 7– i.e. our brains are seven times larger than one would expect for an animal of our body size. Many cetaceans have EQs in the 4–5 range and other primates and elephants have brains 2.5–3 times larger than expected.

Humans like to think that our large relative brain size has allowed our species to achieve a qualitatively different intelligence from other mammals. But brain size is a continuous variable and the human brain is a product of predictable changes in brain anatomy when non-human anthropoid primate brains are scaled up (Sherwood *et al.* 2006). For instance, our degree of cortical folding is exactly as expected for a primate with our brain size (Zilles 1989). Other aspects of our brain, such as relative size of our frontal lobes and the presence of specialized cells for information processing, are all either shared

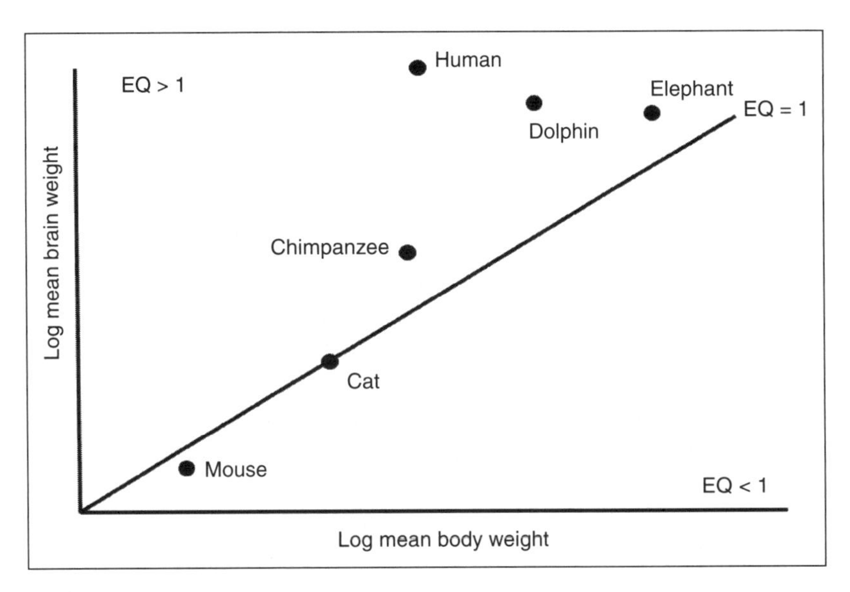

Figure 6.1 Graph showing encephalization levels of animals across a range of body sizes. EQ = 0.5 (mouse), 1.0 (cat), 2.3 (chimpanzee), 7.0 (human) 4.2 (dolphin), 2.3 (elephant). Credit: Lori Marino.

with other species or are expected for a primate of our body size. In summary, decades of neurobiological studies have failed to turn up a single property of the human brain that is qualitatively different from that of other species, i.e. that is not explainable within a common framework of comparative evolution with the rest of the life forms on this planet.

Continuity of mind

As there is continuity of brain and nervous system, so is there continuity of mind. Just as in astronomy, the history of research on animal behavior, comparative psychology, and ethology is a history of decentralization. Increasing understanding of other animals has come with a dwindling of qualitative differences across species and a reframing of intelligence as a continuous character. Basic cognitive processes, such as memory and learning, are found in all animals. Furthermore, even "high-level" cognitive abilities, such as for planning, basic arithmetic, mirror self-recognition, and the possession of a technological culture, are not unique to humans. Some of these shared capacities are reviewed here.

Intelligence in invertebrates

Learning has been confirmed in many invertebrates including roundworms (Rankin 2004), molluscs (Kandel 2001), annelid worms (Friesen and Kristan 2008), arthropods (Morse 2000), and even the echinoderms, which lack a central nervous system (Shulgina 2006). The basic neural mechanisms of learning are the same from molluscs to mammals (Kandel 2009). Of considerable interest to the astrobiology community are the cephalopods (phylum Mollusca), particularly the octopus *Octopus vulgaris*, because of their complex cognitive and behavioral traits. Although octopi brains bear molluscan features, such as numerous interconnected ganglia dispersed throughout the body, they also have a highly elaborated three-lobed architecture, which gives these creatures a vertebrate-like intelligence (Hockner 2006) evincing observational learning (Fiorito and Scotto 1992) and even tool use (Finn *et al.* 2009).

Honey bee and ant colonies possess a collective memory with features shared with the individual memory system of vertebrate brains (Couzin *et al.* 2002); these colonies perform very similarly on a range of psychological tests (Langridge *et al.* 2008; Passino 2010). Importantly, collective or swarm intelligence does not necessarily emerge at the expense of individual intelligence, as individual honey bees have complex cognitive abilities (Giurfa *et al.* 2003) and learning capabilities on a par with those of vertebrates (Bitterman 1996). Individual bees understand the concepts of "sameness" and "difference" (Giurfa *et al.* 2001), can count from one to four (Dacke and Srinivasan 2008), and are able to accurately group visual stimuli into categories (reviewed by Benard *et al.* 2006), to name just a few of their prodigious abilities. Many other invertebrates, such as fruit flies and jumping spiders, possess complex cognitive abilities as well (Greenspan and van Swinderen 2004) and, like octopi, are able to learn from one another, through the use of social information (reviewed by Leadbeater and Chittka 2009).

These findings of cognitive complexity in invertebrates reveal that the basic capacities of the human mind are demonstrable in the minds of beings who we typically view as very different from us and, in the *scala naturae* world, vastly inferior. But scientific findings show that intelligence in many domains is ubiquitous in the animal kingdom and is shared at a very deep evolutionary level across all species.

Vertebrate intelligence: variations on a theme

Setting aside the intelligence of invertebrates, complex intelligence on Earth can be said to have emerged 500 million years ago with vertebrates. All vertebrate brains are variations on the same theme. Likewise, complex cognitive abilities – some found in invertebrates – are shared by all vertebrates. Most differences lie on a continuum of complexity and are shaped by the adaptive needs of each species from pre-existing characteristics. The striking implication of these facts, given the wide scope of astrobiology, is that if we were to find an extraterrestrial intelligence on another planet whose behavior paralleled the complexity of, say, a gobie using long-term memory and visual geometry to navigate back to a home tidal pool, we would know, for all intents and purposes, that a human level of extraterrestrial intelligence almost certainly exists somewhere as well.

A non-exhaustive list of complex capacities found in many non-human vertebrate species includes the following: for planning by chimpanzees and orangutans (Osvath and Osvath 2008) and bottlenose dolphins (McCowan *et al.* 2000); mathematics by chimpanzees (Boysen *et al.* 1993; Tomonaga and Matsuzawa 2000); taking the visual perspective of another by domestic pigs (Nawroth *et al.* 2014), jays (Clayton *et al.* 2007), and dogs (Kaminski *et al.* 2013); empathy in elephants (Plotnik and deWaal 2014); understanding physical causation by chimpanzees and orangutans (Mulcahy and Call 2006); and even intentional deception involving modeling of others' mental states by pigs (Held *et al.* 2000) and chimpanzees (Melis *et al.* 2006).

In many areas of cognition members of other species either do as well, e.g. spatial learning in fishes (Brown 2014), or exceed the capacities of adult humans, e.g. working memory for sequences of numerals by chimpanzees (Inoue and Matsuzawa 2007). In many perceptual domains there are no parallels in our own species, e.g. echolocation and its use in cross-modal mental representation of objects in dolphins (Pack and Herman 1995). All of this is not to deny the remarkable cognitive capacities of humans but, rather, to show that human intelligence is a variation on a highly conserved theme.

The "big five" cognitive domains

There are five cognitive domains in which we have, at one time or another, argued quite vociferously for human exclusivity. These are self-awareness, tool making and use, culture, numerical ability, and symbolic language. But even these impressive capacities are not unique to our species.

Self-awareness

Self-awareness is cognizance of oneself as an individual, i.e. possession of an autobiographical sense of "I." Although all animals must have this capacity at some level, several species have demonstrated a complex human level of self-awareness in experimental studies of mirror self-recognition (MSR), metacognition (reporting on one's own thoughts), and other tests. MSR is the ability to recognize oneself in a mirror and is evinced by using a mirror to investigate parts of one's body. It has been convincingly displayed in all of the great apes (Anderson and Gallup 2011), bottlenose dolphins (Reiss and Marino 2001), Asian elephants (Plotnik *et al.* 2006), and magpies (*Pica pica*) (Prior *et al.* 2008). Bottlenose dolphins are aware of their own body parts and behaviors (reviewed by Herman 2012). Monkeys and dolphins demonstrate abstract forms of self-awareness, such as metacognition (Smith and Washburn 2005). Hence, a complex sense of self is shared with many other species. At this point in time our understanding of the distribution of self-awareness across species on this planet is limited by our own species' ingenuity in developing methods to test such an abstract subjective phenomenon.

Tool use and making

Since the first time Jane Goodall observed wild chimpanzee David Greybeard make and use a tool (a "termite stick") in 1961 we have documented tool making and use in many primates, elephants, birds, dolphins, octopi (see above), and a host of other species. Wild gorillas use branches to gauge the depth of water when crossing streams (Breuer *et al.* 2005). New Caledonian crows create and use stick tools with their beaks to extract insects from logs (Hunt and Gray 2003) and, in the laboratory, use analogical reasoning to use tools for accessing yet more tools (Taylor *et al.* 2007). A group of bottlenose dolphins off the coast of Western Australia uses pieces of sponge wrapped around their rostrum to prevent abrasions when searching for food on the sea floor (Krutzen *et al.* 2005).

Tool making and use is learned socially and through experimentation and involves the creation of novel solutions to various ecological problems. This basic aspect of non-human tool making and use places it on a continuum with human technology and is, thus, highly relevant to hypotheses about the evolution of technological extraterrestrial intelligence. Chimpanzees are not likely to build rockets to Mars any time in the near future but, from a psychological point of view, termite sticks and rockets are born of the same basic capacity to create something new from the environment for the purpose of achieving a specific goal – no matter how simple or complex.

Culture

The basic definition of culture is that of distinctive behavior originating in local populations and passed on through learning from one generation to the next. Culture is the main process by which behavioral innovation manifests itself.

Tool making and use is often referred to as material culture and it exists in a multitude of other species. We now know that the technology of chimpanzees (Boesch 2012), dolphins (Krutzen *et al.* 2005), and New Caledonian crows (Hunt and Gray 2003), for instance, is culturally transmitted. But culture in other behavioral domains also exists. For example, cultural transmission of specific dialects has been documented in orcas (*Orcinus orca*) (Rendell and Whitehead 2001).

Our sophisticated linguistic abilities have allowed human cultures to become extremely complex. But the fact that cultural transmission is shared with many other species on this planet means that we might expect the same in any extraterrestrial organism with a social bend.

Numerical ability

Numerical ability can be categorized into several levels of complexity and abstractness. These include judgments about the relative numbers of objects, subitization (a form of pattern recognition used to rapidly assess differences among small quantities), estimation (of larger quantities), counting and sequencing, and manipulation of quantities (arithmetic). Some of the most complex and abstract numerical abilities are, not surprisingly, found in chimpanzees, who are even capable of mathematical reasoning. They can count or sum up arrays of real objects or Arabic numerals (Beran and Rumbaugh 2001; Boysen and Berntson 1989) and they display the concepts of ordinality and transitivity (the logic that if A = B and B = C, then A = C) when engaged in numerical tasks, demonstrating a real understanding of the ordinal nature of numbers (Boysen *et al.* 1993). Chimpanzees also understand proportions (e.g. 1/2, 3/4, etc.) (Woodruff and Premack 1981). They can use a computer touch screen to count from 0 to 9 in sequence (Inoue and Matsuzawa 2009). Moreover, they have an understanding of the concept of zero, using it appropriately in ordinal context (Biro and Matsuzawa 2001).

But numerical competence is not limited to chimpanzees or great apes. Bottlenose dolphins can learn numerical concepts (Jaakkola *et al.* 2005; Kilian *et al.* 2003) and also appear to have some understanding of the concept of zero (Herman and Forestell 1985). Rhesus macaque monkeys are able to count to nine (Beran 2006). Several species of birds (Garland and Low 2014; Pepperberg 2006;) and other species, such as lemurs (Santos *et al.* 2005), domestic dogs (West and Young 2002), domestic pigs (Held *et al.* 2005), and, as discussed earlier, even bees (Dacke and Srinivasan 2008) have shown numerical competencies, including, in some cases, counting and simple arithmetic.

Symbolic communication

Most natural communication systems on this planet, including human language, derive from common principles of communication and information theory, allowing all organisms to accomplish similar basic communicative feats. But the comprehension and use of symbols is thought to be the defining characteristic of human language, allowing us to express abstract ideas, discuss objects and events which are not present in space and time, and giving us the ability to create an unlimited set of utterances. This quality facilitates the cultural transmission of ideas, as mentioned above.

There have been numerous studies showing that members of other species can acquire a symbolic artificial language, including dolphins who comprehend grammatical sentence structure (reviewed by Herman *et al.* 1993), chimpanzees (reviewed by Rumbaugh *et al.* 2003), African grey parrots (Pepperberg 2009), and many others. Importantly, chimpanzees in captivity use the symbolic elements of language in everyday settings (Lyn *et al.* 2011; Rumbaugh *et al.* 2003). Moreover, the symbolic element that is key to human culture is also found, though to a much more limited degree, in wild chimpanzees. For instance, in one chimpanzee group arbitrary symbolic gestures are used to communicate desire to have sex whereas in another group an entirely different symbolic gesture is used to express the same sentiment (McGrew 2011). The presence of symbolic culture in chimpanzees demonstrates that abstract concepts can be present without human language.

There are several points about communication capacities in other species relevant to astrobiology. First, the natural communication systems of many other species share features of human language. Second, we have vastly incomplete knowledge of the nature of communication in some groups of animals, e.g. cetaceans. Third, members of other species can comprehend and use symbols and human language in appropriate settings and, thus, do possess the capacity to think in symbolic terms. Fourth, chimpanzees use symbolic gestures in cultural settings. Fifth, while symbolic communication has clearly become the forté of the human species, it is not entirely outside the range of mental capacities of many other species. Sixth and most important is the fact that the human ability to communicate with a symbolic language could only have come from shared characteristics with other species.

We are not alone

This review began with the question of why the serious study of intelligence has been ignored by the astrobiology community. It has provided abundant evidence that it is not for lack of empirical data, accessibility, or scientific tractability. Instead, there appear to be other drivers behind this omission having to do with misconceived models of human nature, particularly the *scala naturae*, and the unsupported assumptions they lead to. These notions lead to ways of thinking which block the path to open scientific inquiry about intelligence in astrobiology.

Quite to the contrary, the existing body of scientific knowledge on the evolution of intelligence makes clear several points that support and urge the study of intelligence in astrobiology. These points include the following. There are no scientifically valid reasons for treating the human brain and intelligence as either a unique or singular case. The human brain can only be understood in the same evolutionary-comparative framework as the rest of life on Earth. All nervous systems on Earth are governed by the same electrochemical principles of information processing laid down well over one billion years ago. Once the basic plan for brains evolved, everything that came afterward is a variation on a highly conserved theme. Thus, there is a surprising degree of shared cognition across invertebrates and vertebrates. There are, of course, differences in intelligence and mind across species but they all can be understood in a common framework.

All of this knowledge demonstrates that the central question in astrobiology, "Are we alone?" has already been answered. We are not. Thus, astrobiology should embrace the study of intelligence as a ubiquitous property of life on Earth and one that converts the current impasse to relevant and exciting opportunities for further exploration.

References

Allman, J. M. 1999. *Evolving Brains*. New York, NY: Scientific American Library.

Anderson, J. R. and Gallup, G. G. Jr. (2011). "Which primates recognize themselves in mirrors?" *PLoS Biology* 9(3): e1001024.

Arendt, D., Deans, A., Jekely, G., and Tessmar-Raible, K. 2008. "The evolution of nervous system centralization." *Philosophical Transactions of the Royal Society B* 363: 1523–1528.

Benard, J., Stach, S., and Giurfa, M. 2006. "Categorization of visual stimuli in the honey bee, *Apis mellifera*." *Animal Cognition* 9: 257–270.

Beran, M. J. 2006. "Quantity perception by adult humans (*Homo sapiens*), chimpanzees (*Pan troglodytes*), and rhesus macaques (*Macaca mulatta*) as a function of stimulus organization." *International Journal of Comparative Psychology* 19: 386–397.

Beran, M. J. and Rumbaugh, D. M. 2001. "Constructive enumeration by chimpanzees on a computerized task." *Animal Cognition* 4: 81–89.

Biro, D. and Matsuzawa, T. 2001. "Use of numerical symbols by the chimpanzee (*Pan troglodytes*): cardinals, ordinals and the introduction of zero." *Animal Cognition* 4: 193–199.

Bitterman, M. E. 1996. "Comparative analysis of learning in honey bees." *Learning and Behavior* 24: 123–141.

Boesch, C. 2012. *Wild Cultures*. Cambridge: Cambridge University Press.

Boysen, S. T. and Berntson, G. G. 1989. "Numerical competence in a chimpanzee (*Pan troglodytes*)." *Journal of Comparative Psychology* 103: 23–31.

Boysen, S. T., Berntson, G. G., Shreyer, T. A., and Quigley, K. S. 1993. "Processing of ordinality and transitivity by chimpanzees (*Pan troglodytes*)." *Journal of Comparative Psychology* 107: 208–216.

Breuer, T., Ndoundou-Hockemba, M., and Fishlovk, V. 2005. "First observation of tool use in wild gorillas." *PLoS Biology* 3(11): e380.

Brown, C. 2014. "Fish intelligence, sentience and ethics." *Animal Cognition*, doi 10.1007/s10071-014-0761-0.

Clayton, N. S., Dally, J., and Emery, N. 2007. "Social cognition by food-caching corvids. The Western scrub-jay as a natural psychologist." *Philosophical Transactions of the Royal Society B* 362: 507–522.

Couzin, I. D., Krause, J., James, R., Ruxton, G. D., and Franks, N. 2002. "Collective memory and spatial sorting in animal groups." *Journal of Theoretical Biology* 218: 1–11.

Crowe, M. J. 2008. *The Extraterrestrial Life Debate: Antiquity to 1900: A Source Book*. Notre Dame, IN: University of Notre Dame Press.

Dacke, M. and Srinivasan, M. 2008. "Evidence for counting in insects." *Animal Cognition* 4: 683–689.

Darwin, C. 1871. *The Descent of Man, and Selection in Relation to Sex* (1st edn.). London: John Murray.

Darwin, C. 1859. *On the Origin of Species by Means of Natural Selection, or the Preservation of Favoured Races in the Struggle for Life* (1st edn.). London: John Murray.

Finn, J., Tregenza, T., and Norman, M. D. 2009. "Defensive tool use in a coconut-carrying octopus." *Current Biology* 19: R1069–R1070.

Fiorito, G. and Scotto, P. 1992. "Observational learning in *Octopus vulgaris*." *Science* 256, 545–546.

Friesen, W. O. and Kristan, W. B. 2008. "Leech locomotion: swimming, crawling, and decisions." *Current Opinion in Neurobiology* 17: 704–711.

Garland, A. and Low, J. 2014. "Addition and subtraction in wild New Zealand robins." *Behavioural Processes.* http://www.sciencedirect.com/science/article/pii/S0376635714001909

Giurfa, M., Zhang, S., Jenett, A., Menzel, R., and Mandyam, V. S. 2001. "The concepts of 'sameness' and 'difference' in an insect." *Nature* 410: 930–933.

Giurfa, M., Reisenman, S. M., Gerber, B., and Lachnit, H. 2003. "The effect of cumulative experience on the use of elemental and configural visual discrimination strategies in honey bees." *Behavioral Brain Research* 145: 161–169.

Greenspan, R. J. and van Swinderen, B. 2004. "Cognitive consonance: complex brain functions in the fruit fly and its relatives." *Trends in Neurosciences* 27: 707–711.

Held, S., Baumgartner, J., Kilbride, A., Byrne, R. W., and Mendl, M. 2005. "Foraging behaviour in domestic pigs (*Sus scrofa*): remembering and prioritizing food sites of different value." *Animal Cognition*, 8: 114–121.

Held, S., Mendl, M., Devereux, C., and Byrne, R. W. 2000. "Social tactics of pigs in a competitive foraging task: the 'informed forager' paradigm." *Animal Behaviour* 59, 569–576.

Herman, L. M. 2012. "Body and self in dolphins." *Consciousness and Cognition* 21: 526–545.

Herman, L. M. and Forestell, P. H. 1985. "Reporting presence or absence of named objects by a language-trained dolphin." *Neuroscience and Biobehavioural Reviews* 9: 667–691.

Herman, L. M., Pack, A. A., and Morrel-Samuels, P. 1993. "Representational and conceptual skills of dolphins." In Roitblatt, H. R. *et al.* (eds.) *Language and Communication: Comparative Perspectives.* Hillsdale, NJ: Erlbaum, pp. 273–298.

Hochner, B. 2006. "The octopus: a model for a comparative analysis of the evolution of learning and memory mechanisms." *Biological Bulletin* 210: 308.

Hunt, G. R. and Gray, R. D. 2003. "Diversification and cumulative evolution in New Caledonian crow tool manufacture." *Proceedings of the Royal Society of London B* 270: 867–874.

Inoue, S. and Matsuzawa, T. 2007. "Working memory of numerals in chimpanzees." *Current Biology* 17: R1004–R1005.

Inoue, S. and Matsuzawa, T. 2009. "Acquisition and memory of sequence order in young and adult chimpanzees (*Pan troglodytes*)." *Animal Cognition* 12: S58–S69.

Jaakkola, K., Fellner, W., Erb, L., Rodriguez, M., and Guarino, E. 2005. "Understanding of the concept of numerically 'less' by bottlenose dolphins (*Tursiops truncatus*).” *Journal of Comparative Psychology* 119: 296–303.

Kaminski, J. P., Pitsch, A., and Tomasello, M. 2013. “Dogs steal in the dark.” *Animal Cognition* 16: 385–394.

Kandel, E. R. 2001. “The molecular biology of memory storage: a dialogue between genes and synapses.” *Science* 294: 1030–1038.

Kandel, E. R. 2009. “The biology of memory: a forty year perspective.” *The Journal of Neuroscience* 29: 12748–12756.

Kilian, A., Yaman, S., von Fersen, L., and Gunturkun, O. 2003. “A bottlenose dolphin discriminates visual stimuli differing in numerosity.” *Learning and Behaviour* 31: 133–142.

Krutzen, M., Mann, J., Heithaus, M. R., *et al.* 2005. “Cultural transmission of tool use in bottlenose dolphins.” *Proceedings of the National Academy of Sciences* 102: 8939–8943.

Langridge, E. A., Sendova-Franks, A. B., and Franks, N. R. 2008. “How experienced individuals contribute to an improvement in collective performance in ants.” *Behavioral Ecology and Sociobiology* 62: 447–456.

Leadbeater, E. and Chittka, L. 2009. “Bumble bees learn the value of social cues through experience.” *Biology Letters* 5: 310–312.

Lyn, H., Greenfield, P. M., Savage-Rumbaugh, S., Gillespie-Lynch, K., and Hopkins, W. D. 2011. “Nonhuman primates do declare! A comparison of declarative symbol and gesture use in two children, two bonobos, and a chimpanzee.” *Language and Communication* 31: 63–74.

Marino, L. 2002. “Convergence in complex cognitive abilities in cetaceans and primates.” *Brain, Behavior and Evolution* 59: 21–32.

Marino, L. 2007. “Scala naturae.” In Bekoff, M. (eds.) *The Encyclopedia of Human-Animal Relationships.* Westport, CT: Greenwood Publishing Group, pp. 220–224.

McCowan, B., Marino, L. Vance, E., Walke, L., and Reiss, D. 2000. “Bubble ring play of bottlenose dolphins: implications for cognition.” *Journal of Comparative Psychology* 114: 98–106.

McGrew, W. C. 2011. “*Pan symbolicus.* A cultural primatologist’s viewpoint.” In Henshilwood, C. S. and d’ Errico, F. (eds.) *Homo Symbolicus: The Dawn of Language, Imagination and Spirituality.* Amsterdam: John Benjamins, pp. 1–12.

Melis, A. P., Call, J., and Tomasello, M. 2006. “Chimpanzees (*Pan troglodytes*) conceal visual and auditory information from others.” *Journal of Comparative Psychology* 120: 154–162.

Messenger, J. B. 1996. “Neurotransmitters of cephalopods.” *Invertebrate Neuroscience* 2: 95–114.

Morse, D. H. 2000. “The effect of experience on the hunting success of newly emerged spiderlings.” *Animal Behavior* 60: 827–835.

Mulcahy, N. and Call, J. 2006. “How great apes perform on a modified trap tube task.” *Animal Cognition* 9:193–199.

Nawroth, C., Ebersbach, M., and von Borell, E. 2014. “Juvenile domestic pigs (*Sus scrofa domesticus*) use human-given cues in an object choice task.” *Animal Cognition* 17: 701–713.

Osvath, M. and Osvath, H. 2008. “Chimpanzee (*Pan troglodytes*) and orangutan (*Pongo abelii*) forethought: self-control and pre-experience in the face of future tool-use.” *Animal Cognition* 11: 661–674.

Pack, A. A. and Herman, L. M. 1995. “Sensory integration in the bottlenose dolphin: immediate recognition of complex shapes across the senses of echolocation and vision.” *Journal of the Acoustical Society of America* 98: 722–733.

Passino, K. M. 2010. “Honey bee swarm cognition: decision making performance and adaptation.” *International Journal of Swarm Intelligence* 1(2): 80–97.

Pepperberg, I. M., 2006. “Ordinality and inferential abilities of a grey parrot (*Psittacus erithacus*).” *Journal of Comparative Psychology* 120: 205–216.

Pepperberg, I. M. 2009. *The Alex Studies: Cognitive and Communicative Abilities of Grey Parrots*. Cambridge, MA: Harvard University Press.

Plotnik, J. M. and de Waal, F. B. M. 2014. “Asian elephants (Elephas maximus) reassure others in distress.” *PeerJ* 2: e278.

Plotnik, J. M., de Waal, F. B. M., and Reiss, D. 2006. “Self-recognition in an Asian elephant.” *Proceedings of the National Academy of Sciences* 103: 17052–17057.

Prior, H., Schwarz, A., and Gunturkun, O. 2008. “Mirror-induced behavior in the magpie (*Pica pica*): evidence of self-recognition.” *PLoS Biology* 6(8): e202.

Rankin, C. H. (2004) “Invertebrate learning: what can’t a worm learn?” *Current Biology* 14: R617–R618.

Reiss, D. and Marino, L. 2001. “Self-recognition in the bottlenose dolphin: a case of cognitive convergence.” *Proceedings of the National Academy of Sciences USA* 98: 5937–5942.

Rendell, L. and Whitehead, H. 2001. “Culture in whales and dolphins.” *Behavioural and Brain Sciences* 24: 309–324.

Rumbaugh, D. M., Beran, M. J., and Savage-Rumbaugh, E. S. 2003. “Language.” in D. Maestripieri (ed.) *Primate Psychology*, Cambridge, MA: Harvard University Press, pp. 395–423.

Santos, L. R., Barnes, J. L., and Mahajan, N. 2005. "Expectations about numerical events in four lemur species (*Eulemur fulvus, Eulemur mongoz, Lemur catta* and *Varecia rubra*)." *Animal Cognition* 8: 253–262.

Sherwood, C. C., Stimpson, C. D., Butti, C., *et al.* 2006. "Evolution of increased glia–neuron ratios in the human frontal cortex." *Proceedings of the National Academy of Sciences USA* 103: 13606–13611.

Shulgina, G. I. 2006. "Learning of inhibition of behavior in the sea star, *Asterias rubens.*" *Comparative and Ontogenic Physiology* 42:161–165.

Smith, J. D. and Washburn, D. A. 2005. "Uncertainty monitoring and metacognition by animals." *Current Directions in Psychological Science* 14: 19–24.

Sternberg, R. 2000. *Handbook of Intelligence.* Cambridge: Cambridge University Press.

Striedter, G. F. 2005. *Principles of Brain Evolution.* Sunderland, MA: Sinauer Associates.

Taylor, A., Hunt, G., Holzaider, J. C., and Gray, R. D. 2007. "Spontaneous metatool use by New Caledonian crows." *Current Biology* 17: 1504–1507.

Tomonaga, M., and Matsuzawa, T. (2000). "Sequential responding to Arabic numerals with wild cards by the chimpanzee (Pan troglodytes)." *Animal Cognition* 3: 1–11.

Turlejski, K. 1996. "Evolutionary ancient roles of serotonin: long-lasting regulation of activity and development." *Acta Neurobiologiae Experimentalis* 56: 619–636.

West, R. E. and Young, R. J. 2002. "Do domestic dogs show any evidence of being able to count?" *Animal Cognition* 5: 183–186.

Woodruff, G. and Premack, D. 1981. "Primitive mathematical concepts in the chimpanzee: proportionality and numerosity." *Nature* 293: 568–570.

Zilles, K. 1989. "Gyrification in the cerebral cortex of primates." *Brain, Behavior, and Evolution* 34: 143–150.

7 Universal biology: assessing universality from a single example

CARLOS MARISCAL

Avoiding the $N = 1$ problem about life

Is it possible to know anything about life we have not yet encountered? We know of only one example of life: our own. Given this, many scientists are inclined to doubt that any principles of Earth's biology will generalize to other worlds in which life might exist. Let's call this the "$N = 1$ problem." By comparison, we expect the principles of geometry, mechanics, and chemistry would generalize. Interestingly, each of these fields has predictable consequences when applied to biology. The surface-to-volume property of geometry, for example, limits the size of unassisted cells in a given medium. This effect is real, precise, universal, and predictive. Furthermore, there are basic problems all life must solve if it is to persist, such as resistance to radiation, faithful inheritance, and energy regulation. If these universal problems have a limited set of possible solutions, some common outcomes must consistently emerge.

In this chapter, I discuss the $N = 1$ problem, its implications, and my response (Mariscal 2014). I hold that our current knowledge of biology can justify believing certain generalizations as holding for life anywhere. Life on Earth may be our only example of life, but this is only a reason to be cautious in our approach to life in the universe, not a reason to give up altogether. In my account, a candidate biological generalization is assessed by the assumptions it makes. A claim is accepted only if its justification includes principles of evolution, but no contingent facts of life on Earth.

When biology's sample size became problematic

The discussion of the $N = 1$ problem has a long history. Some Greek and Roman thinkers spoke of infinite worlds, while others, notably Aristotle, ruled it out (Dick 1982, 9–11). In the Western world, many early key figures touched on the issue. Giordano Bruno, Johannes Kepler, and even Immanuel Kant were all optimistic about the existence of many life-bearing worlds. Meanwhile, others, the anti-pluralists, denied the existence of other life, usually from theological principles (Dick 1982, 63–69). Intimately related to the question of other worlds is the question of the origin of life here on

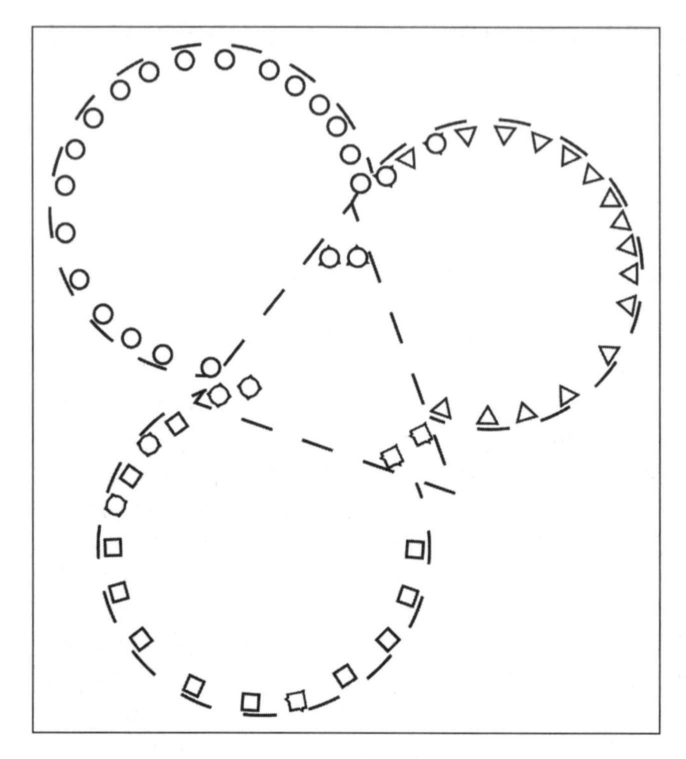

Figure 7.1 In a world in which spontaneous generation is accepted, it might be plausible to view the diversity of life (the triangles, circles, and squares in this figure) on Earth as representative of life in the universe. It might even be expected for enterprising scientists to find predictive generalizations unifying distinct kinds of life (represented as the dashed lines encircling the triangles, circles, and squares). Thus, scientists might see similarities among species and form predictive groupings expected to hold for as-yet-unknown life. Artwork by Carlos Mariscal.

Earth. For nearly two millennia, natural philosophers assumed life regularly originated from non-living material. It is easy to see how such a view of spontaneous generation might seem to justify a universal scope for biology, even independent of evidence of life elsewhere. If life arises spontaneously, then each new example provides a new test case for generalizations claimed to hold for life everywhere (Figure 7.1).

Two of the biggest changes in our conception of life came in the 1800s in the forms of Louis Pasteur and Charles Darwin. With Pasteur, the theory of spontaneous generation was, if not refuted, marginalized (Lahav 1999,

23–29). If life does not form readily, then perhaps our evidence is not as diverse as it might seem – and the origin and existence of species must be explained by appeal to some other process. Another challenge came from the process given by Darwin in the *Origin* (Darwin 1859), which provided a naturalistic explanation of the genesis of species without spontaneous generation. Species descended with modifications from earlier species, which in turn descended from earlier species. This induction, given the difficulty of spontaneous generation, implies all life descended from some last universal common ancestor.[1] Nevertheless, Darwin is non-committal about universal common ancestry (Darwin 1859, 484):

> I believe that animals have descended from at most only four or five progenitors, and plants from an equal or lesser number. Analogy would lead me one step further, namely to the belief that all animals and plants have descended from some one prototype. But analogy may be a deceitful guide. Nevertheless all living things have much in common ... Therefore I should infer from analogy that probably all organic beings which have ever lived on this earth have descended from some one primordial form, into which life was first breathed.

The single origin of all life on Earth naturally leads to the $N=1$ problem. If all organisms had arisen independently via spontaneous generation, biology would be the study of many distinct objects. It would be reasonable for scientists to concern themselves with discovering universal principles governing these distinct phenomena. So, before Darwin, biology could reasonably be considered a universal science on par with physics and chemistry.[2] But given the realization that the sample size of biology is 1, it seems difficult to see how any universality can be justified. See Figure 7.2.

Exploring new ways to address $N = 1$

Not everybody takes the $N = 1$ problem to be insurmountable. In the century and a half since Darwin, many authors have attempted to show the universality of biology in various ways. Herbert Spencer, infamous for the ideas behind social Darwinism, attempted to show evolution was a natural law of universal application (Spencer 1864). Others have maintained we can show aspects of biology are universal from evolutionary principles (Dawkins 1982, 1992), due

[1] Darwin is not the originator of the concept of common descent. British biologist John Ray first introduced the concept (Serafini 1993, 128), but it was Pierre-Louis Moreau de Maupertuis who first postulated *universal* common descent, the view that all life originated from a single organism (Harris 1981, 107). Still, after 1859, arguments for and against common descent tend to trace their roots to Darwin.

[2] Granted, the appearance of a non-mechanistic teleology and the lack of a naturalistic paradigm posed other, major problems for early biologists.

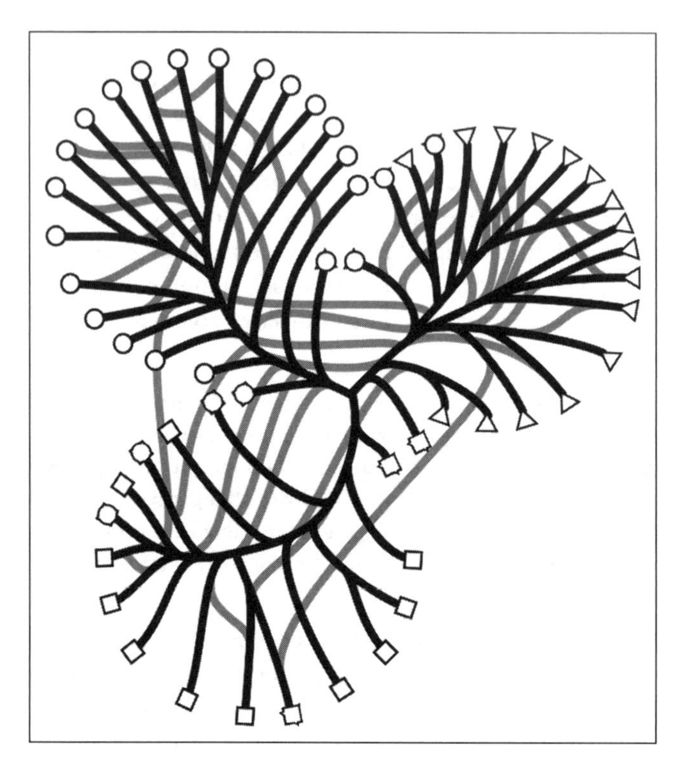

Figure 7.2 We now know all of life on Earth shares common ancestry. In other words, all of our biology forms a single sample. Thus, it appears unjustified to consider life on Earth as representative of life elsewhere. In this case, we can see the explanation for any similarities is due to common descent (black lines connecting the triangles, circles, and squares) or lateral transfer (gray lines) and likely not from underlying universal principles. Artwork by Carlos Mariscal.

to their sheer utility (Dennett 1995), from common patterns or recurrent causal mechanisms (Sterenly and Griffiths 1999), or from an understanding of the physics, complexity, and probability theory that underlie them (Kauffman 2000). Some biologists have pointed to the long history of life on Earth as a series of natural experiments, from which we might be justified in assessing some generalizations as more robust (e.g. McGhee 2013, Powell and Mariscal in press). Still others have attempted to explore universal biology through discovering a second example of life (e.g. Davies and Lineweaver 2005) or creating it (e.g. Langton 1989, Gibson *et al.* 2010).

The most interesting modern approach to the $N = 1$ problem comes from astrobiology. Astrobiology is the scientific investigation designed by NASA in the wake of several discoveries in the mid 1990s.[3] Periodically, key astrobiologists gather to produce "roadmaps" intended to serve as assessments of astrobiology's progress and guides for future directions. Continuing in the most recent roadmaps, the authors propose beginning a new science of "universal biology" (Des Marais *et al.* 2008, 6):

> We must move beyond the circumstances of our own particular origins in order to develop a broader discipline, "Universal Biology." . . . we need to exploit universal laws of physics and chemistry to understand polymer formation, self-organization processes, energy utilization, information transfer, and Darwinian evolution that might lead to the emergence of life in planetary environments other than Earth. Clearly an inventory of molecules must exist that is capable of gaining chemical, structural, and functional complexity and eventually assembling into living systems. This is strongly conditioned on temperature, solvent, energy sources, etc.

Astrobiology's approach is clearly grounded on physical and chemical principles. Given the assumptions that life is based on chemistry and physics, which we accept as universal, the project seems justified. Within these constraints, there is active debate in the astrobiological community as to what specifics – if any– are essential to life. Astrobiologists worry about assuming that life elsewhere would require water, carbon, or other chemistry important to life on Earth. The astrobiology research program is both speculative in questioning our "Earth chauvinism" and constrained by the conditions we expect to exist in the universe. For astrobiologists, possibility is grounded in actuality. Astrobiologists focus on hypotheses testable through astronomical techniques, modeling, Earth analogs, or robotic space travel. With a few stellar exceptions (e.g. Irwin and Schulze-Makuch 2012), astrobiologists disregard large-scale ecological and evolutionary effects such as major transitions, increasing trends in complexity and diversity, or mass extinctions, which could have major implications for life everywhere. In the next section, I will describe an approach that accepts contributions from physics, chemistry, and biochemistry, while allowing for insight from evolutionary and developmental biology, ecology, philosophy, and other disciplines. I hope my account becomes accepted as an expansion of the universal biology project in astrobiology.

[3] The change of direction was spurred by the discovery of suspected water on Mars and Europa, increased research into extremophiles, the discovery of extrasolar planets in 1992, and the Allan Hills 84001 meteorite, which was thought to contain microscopic fossils of Martian bacteria in 1996 (Dick and Strick 2004).

Separating universal biology from non-universal biology

Consider two generalizations, commonly made about life on Earth, but which could apply to all biology in the universe:

(1) The genetic code holds for all life.
(2) Hereditary information is digital not analog. That is, all life will use *some* biochemical code (Maynard Smith 1986, 21; Dawkins 1992, 26).

For the sake of argument, let us grant both are true.[4] What justifies them? By understanding the justification of these generalizations, we will be better able to assess *why* they hold, *how* they hold, and to *what* extent may hold in the future.

Generalization (1) refers the usual set of 20 amino acids coded for by the 64 possible sequences of three nucleotides (codons). Francis Crick (1968) argued (1) is only true because all life on Earth shares common ancestry to the point in history when the genetic code was settled (though Vetsigian *et al.* 2006 have convincingly shown lateral gene transfer is also necessary in this story). Since the genetic code's success at producing amino acids depends on its consistency and stability, any changes would introduce new amino acids into established proteins, many of which would have proved lethal. It is akin to swapping the letters "e" and "r" on a keyboard: *it eaisrs the peobability mrssagrs will br misundrestood.* Crick's historical explanation accounts for both the generality and rigidity of the genetic code. If this is true, it follows that with different starting points or different contingent factors, the genetic code could have been different.[5] So even if we expect (1) to be true everywhere on Earth, we do not expect it would apply to any organisms that lack this shared history. If scientists ever create life in a laboratory or discover alien genetic material on Mars, they would not be surprised if (1) did not hold.

Meanwhile, generalization (2) is based on an argument from evolutionary principles to the conclusion that an analog system of heredity would prove catastrophic. Suppose I make a photocopy of a poem. Then I make a

[4] There are many examples of (1) being violated, but none of these violations are wholesale new examples of genetic codes (Elzanowski *et al.* 2000). And it is not obvious that all genetic information is digital (2). Some epigenetic and environmental methods of inheritance seem perfectly analog. A charitable understanding of (2) would see it as claiming all life in this universe requires *some* digital process, regardless of any analog processes surrounding it.

[5] It has since been argued that 99.97% of codes are less robust than our current code, when accounting for such features as hydrophobicity and production of physicochemical-similar amino acids (Wagner 2005, 25. See Philip and Freeland, 2011 for a review). Nevertheless, there are thousands of equally optimal possible codes and it is possible one of those possible codes would have evolved given a replay of the tape of life. Thus, the case for (1) is stronger than Crick makes it out to be, but his explanation is still apt.

photocopy of the resulting photocopy (and so on). After enough iterations, the latter photocopies will accumulate enough specks and noise as to prove incomprehensible. If faithful inheritance is a universal goal in biology, then analog reproduction is a terrible way to do so (but see Mariscal 2014). If we expect some element of heredity to reproduce in a potentially endless way, then it follows that all biochemical life would use some sort of lossless code.

We can see that (1) depends on contingent events in the early history of life, but (2) is justified by principles that are independent of the foibles of Earth's history. So if (1) is true, it's sensitive to initial conditions in a way unlike (2). In this sense, (1) is less robust and less general than (2). Since the justification for digital codes does not depend on Earth's initial conditions, (2) would be expected to hold wherever life exists. If (2) is true, it has nothing to do with the contingencies of evolution on Earth. An Earth-based explanation wouldn't even make sense.

In the next section, I propose a way of drawing a sharp distinction between generalizations like (1) from ones like (2). By assessing the underlying justifications of purported universal features of biology, we can adequately address the $N = 1$ problem. My proposal is an account of how to discover, understand, and justify generalizations in biology. It is *not* a defense of the generalizations themselves, which must be defended by the relevant kinds of scientists.

On universality given a single sample

I have proposed defining universal biology as "the study of evolutionary generalizations whose justification does not assume contingent facts about Earth's history (and so are expected to apply elsewhere in the universe)" (Mariscal 2014). This section will explain, clarify, justify, and expand that definition.

Consider the *kinds* of justifications biological claims can have. I propose six major divisions based on the kind of necessity each is taken to invoke:

a. claims that are applications of *a priori* concepts such as logic or probability theory;
e. claims based on or derivable from *natural necessity*, such as laws of physics or chemistry;
i. claims based on or derivable from *principles of evolution*, including natural selection and drift;[6]
o. claims that follow from the *initial state* of the universe or planet as a whole;

[6] Some have argued that evolution is derivable from physics and so ***i*** is contained in ***e*** (cf. Rosenberg 2014). In Mariscal 2014, I argue against this view.

u. *historically contingent* claims following from more recent events in the system; and
y. *state-dependent* claims, which occur as a result of other co-occurring factors (whether external or internal).

Roughly, each division sets an envelope by which the subsequent category is constrained. These categories are neither exclusive nor exhaustive and I will only consider claims that make reference to all and only their justifications. A claim can only have as wide a scope as the least general justification it invokes. My definition can now be restated in these terms: universal biology is "the study of generalizations that are (***i***) evolutionary and whose justification does not make reference to (***u***) contingent facts about Earth's history." A generalization must reference (***i***) to count as "biological" in this view.[7] The claim "all organisms face entropy" might be universal, but it is not based on evolutionary principles.

Additionally, if an explanation in biology makes reference to a historical explanation (***u***),[8] it will be excluded. This exclusion demarcates the difference between universal biology, the study of life as it *must* be, and possible biology, which I take to be the study of life as it *could* be. The $N = 1$ problem is a bigger problem for universal biology, which strives to be more conservative in its claims. I disallow mere possibilities in universal biology because possibilities are infinite and I worry that without an explicit limitation, possible biology could be wildly speculative (but see Schulze-Makuch in this volume for interesting, scientific, and fruitful ways to delimit possible biology). Those seeking to understand life *as it must be* should proceed in a cautious manner, if we are to address the concerns raised by the $N = 1$ problem.

This account is broader in scope than that which astrobiology currently investigates. It is able to accommodate any understanding of "polymer formation, self-organization processes, energy utilization, information transfer, and Darwinian evolution," as well as universal elements of ecology and development. A justification-centered approach makes explicit the degree to which we expect these factors to exist elsewhere in the universe. The constraints of my

[7] We do not have a good definition for life and there is reason to suppose that a consensus on a definition for life will prove impossible (Cleland 2011, Machery 2011, Tsokolov 2009, Mariscal 2014). On the other hand, we have a decent grasp of how evolution works and we think it will apply to all naturally arising living systems. There is good reason to suppose the kind of life we will find will be complex and reasonably well adapted to its environment. The concept of "life" adds nothing to the issue that cannot be better accomplished by a concept of adaptive complexity, which is best explored in terms of (***i***) the principles of evolution.

[8] I will only consider claims that make reference to all and only their justifications. I could reference the price of tea in China (***u***) when I explain why I wasn't able to subdivide a prime number of fruits equally, but that detail is inessential in the explanation, which only requires math (***a***) to be explained (see Lange 2012).

account are designed to make it clear in which biological systems our generalizations are expected to apply. A universal generalization that depends on other states in the system (***y***) might no longer apply if that system is disrupted, while one that depends solely on conditions ***a***, ***e***, and ***i*** will likely continue holding true until the living system goes extinct.

More concerns for a universal biology

Discussing a related issue, Carol Cleland draws lessons from the early history of alchemy in which some alchemists classified various liquids as water, including nitro-hydrochloric acid (HNO_3+3HCl) and a solution of nitric acid in water (HNO_3+H_2O) (Cleland 2011). They even referred to these liquids as strong water (*aqua fortis*) and royal water (*aqua regia*), respectively. It makes a certain sort of sense to think of them as unified since all three are clear, liquid, and serve as solvents. But these characteristics are not essential to water; it is the chemical structure of H_2O that is the essential determinant. A study of the necessary and sufficient conditions for the set of water, strong water, and royal water would have been pointless. Similarly, Cleland looks at the debate about definitions of life as premature. She argues the features of life we think are essential, such as replication, metabolism, or Darwinian evolution,[9] might rest on mistaken assumptions and we cannot say anything definitive about life until we discover a second genesis.

Cleland proposes a situation in which we are tasked with uncovering the essential features of mammals based on a single species, such as a zebra (Cleland 2012). A scientist could study the zebra forever without truly understanding mammals. She might disregard mammary glands, for example, because only half of zebras – the females – have mammary glands. Instead, the scientist might infer all mammals are hooved, striped, or walk on four legs. This, she argued, is the plight of biologists doing universal biology. If we can't generalize from zebras– what hope does any such project have?

The objections Cleland raises are interesting and well worth addressing. Not only do we not know the essential features of life, but life itself may also be a spurious category (see Jabr 2013, Mariscal 2014, ch. 4). Still, with respect to

[9] Darwinism has been amended and augmented numerous times since the first publication of the *Origin*. Many scientists now use the term "Darwinian" with some fluidity to refer to some kind of evolution by natural selection. Current biological theory only accepts a portion of Darwin's work. Darwin's notion of heredity was superseded by Mendelian genetics, his acceptance of gradualism was complicated by punctuated equilibrium (Eldredge and Gould 1972), his principle of divergence has been challenged by notions of random-walks (Kimura 1984, McShea and Brandon 2010, Fleming 2013), and the tree of life has been cut down (Doolittle 1999). What's left of Darwinism is unclear.

zebras, plenty of scientists *would* find mammary glands relevant. They would be fascinated to study a subject in which half the population had one set of traits and the rest had a different, but complementary set of traits. Perhaps the scientist might refuse to conjecture such traits were universal, but she might correctly infer such traits would probabilistically apply to half the individuals of any similar population.

The generalization from traits found in *zebra* females to traits found in all *mammalian* females is likely a tenuous one, but this is only problematic if we accept the analogy as representing the situation. In fact, the analogy is suspect. Mammals are a historically contingent evolutionary group, unified solely by a common origin. There is still an open debate in the philosophy of biology as to whether broad categories, like "mammal," are philosophically *real* in any sense that matters to biology (Claridge *et al.* 1997, Mishler 2009). So why should biological generalizations be limited in such a way? It's true we cannot know if Earth-life is representative of life in the universe or an extreme outlier among all possible life worlds, but that is not the issue under investigation. Rather, the question is about whether there is *anything* we could know about life in the universe *whatsoever*. Universal factors can intersect with biology in a number of different ways: geometric constraints, probability theory, physical laws, and so on. A scientist studying zebras might infer some facts about the environment in which they arose: why all four limbs are roughly the same length, that vision is an adaptation, and the role that grass has in producing energy. This scientist would be perfectly justified in inferring all similar creatures facing similar ecological pressures would have corresponding adaptations. Such justifications would be independent of whether these inferences lined up with the kinds of generalizations we initially might have wished a universal biology would provide.

There is a broader concern alluded to in this discussion. What bothers us about the zebra thought experiment is that it suggests our inferences might be unjustified, *no matter how thorough our methodology*. We can always imagine a biological system violating generalizations we think hold true for all of biology. If we can imagine it and evolution is cleverer than we are, then biology in the universe will surely be far stranger than our conceptions of it will allow. Skepticism about any particular claim or about the utility of the project is fair. But appeals to the $N = 1$ problem implicitly assume *any* biological generalization can potentially be understood as a local phenomenon. This cannot be true. There are features of biology for which a purely local explanation is inadequate. The explanation for why there are few large predators relative to the number of primary producers is fully captured by the way energy is passed along the food chain (Colinvaux 1979). If an opponent were to deny the universal scope of this phenomenon, she would have to show how local features of life on Earth – but not evolution or thermodynamics – explain each instance of it. If such an alternative is unpalatable, then, to some extent, biology *must* be universal.

One might be justified in being skeptical about these and other generalizations, but the move toward stopping universal biology is too quick. We could be wrong about how we characterize a biological phenomenon or the justifications we presume underlie a biological phenomenon, of course, but the enterprise is not flawed simply because it may produce claims of which we might be skeptical. Certainly we are going to be wrong about *some* claims of universal biology, but the same holds true for any other branch of inquiry. Universal biology might be in a more difficult situation than many sciences, but it is a difference of degree, not a difference in kind.

Ending thoughts regarding the $N = 1$ problem

The $N = 1$ problem should be viewed as merely an explanation of why we should be cautious about confounding evidence, not the stronger claim that any biological generalizations are at base unjustifiable. By stating the justification of each claim and being conservative in accepting new claims, my approach provides sufficient caution. In my view, if the explanation of a certain biological phenomenon makes no reference to any contingent facts here on Earth, includes causal or constitutive factors, and its conditions are expected to hold in other places in which life might develop, then we are justified in accepting the biological phenomenon as being universal in scope, regardless of our sample size of one. In a sense, I move away from claims like, "robin's eggs must be blue" and toward claims like "eggs must be spheroid."

Some claims will be sufficiently justified by the history of life on Earth. For instance, Earth has suffered several mass extinctions. Shortly after each of these, diversity has increased (Erwin 2001, Krug and Jablonski 2012). With many independent natural experiments showing the same phenomena and a causal explanation (perhaps dealing with niche construction and energy available in the environment), a biological generalization of some sort might be justified. However, for universal biology, it is necessary to specify the underlying assumptions in order to make any justified predictions about other life in the universe. This and other examples from the various biological sciences need to be taken seriously for astrobiology's approach to be maximally effective. Other claims might be justified by new science. A biologist interested in universal biology would find no shortage of candidate biological generalizations to test in the deep ocean or in the microscopic world. Finally, a good number of these claims are not yet testable, allowing for future work and development to render them good science.

Because the kinds of claims made in biology are so varied, it is key to seek a precise statement of a generalization and its assumptions before we can test it, assess it, and, hopefully, accept it. I hope to have proposed a worthy project in astrobiology that explores universal biology from a number of different sciences and methods, including experimenters, modelers, and theorists. Independent of whether we ever create new life or discover life elsewhere, there is a benefit for practicing scientists in knowing whether a biological feature is expected without exception or only given certain preconditions. This is not merely a speculative or hypothetical pursuit – it should have real consequences for practice and theory in the study of life.

Acknowledgments

I would like to thank Robert Brandon, Alex Rosenberg, Steven J. Dick, and Zoë Lewin for help with earlier versions of this paper. I am also grateful to the Katherine Goodman Stern Fellowship and the Natural Sciences and Engineering Research Council of Canada (grant GLDSU/447989) for support of this research.

References

Claridge, M., H. Dawah, and M. Wilson (eds.) 1997. *Species: The Units of Biodiversity*. London: Chapman & Hall.

Cleland, C. E. 2011. "Life without definitions." *Synthese*, 185:125–144.

Cleland, C. E. 2012. "Is a general theory of life possible? Seeking the nature of life in the context of a single example." *Biological Theory*, 7: 368–379.

Colinvaux, P. 1979. *Why Big Fierce Animals are Rare: An Ecologist's Perspective*. Princeton, NJ: Princeton University Press.

Crick, F. H. 1968. "The origin of the genetic code." *Journal of Molecular Biology* 38:367–379.

Darwin, C. 1859. *On the Origin of Species by Means of Natural Selection, or the Preservation of Favoured Races in the Struggle for Life*, first edition. London: Murray.

Davies, P. C. and C. H. Lineweaver. 2005. "Finding a second sample of life on Earth." *Astrobiology* 5:154–163.

Dawkins, R. 1982. "Universal Darwinism." In *Evolution from Molecules to Men*, ed. D. S. Bendall. Cambridge: Cambridge University Press, pp. 403–425.

Dawkins, R. 1992. "Universal biology." *Nature* 360:25–26.

Dennett, D. C. 1995. *Darwin's Dangerous Idea: Evolution and the Meaning of Life*. New York, NY: Simon & Schuster.

Des Marais, D. J., J. A. Nuth III, L. J. Allamandola, *et al.* 2008. "The NASA Astrobiology Roadmap." *Astrobiology* 8:715–730.

Dick, S. J. 1982. *Plurality of Worlds: The Origins of the Extraterrestrial Life Debate from Democritus to Kant*. Cambridge: Cambridge University Press.

Dick, S. J. and J. E. Strick. 2004. *The Living Universe: NASA and the Development of Astrobiology*. New Brunswick, NJ: Rutgers University Press.

Doolittle, W.F. 1999. "Phylogenetic classification and the universal tree." *Science* 284: 2124–2128.

Eldredge, N. and S. J. Gould. 1972. "Punctuated equilibria: an alternative to phyletic gradualism." In *Models in Paleobiology*, ed. T. J. M. Schopf. San Francisco, CA: Freeman, Cooper & Co., pp. 82–115.

Elzanowski, A., J. Ostell, D. Leipe, and V. Soussov. 2000. "The genetic codes." Bethesda, MD: National Center for Biotechnology Information (NCBI). http://www.ncbi.nlm.nih.gov/Taxonomy/Utils/wprintgc.cgi, accessed May 29, 2014.

Erwin, D. H. 2001. "Lessons from the past: biotic recoveries from mass extinctions." *Proceedings of the National Academy of Sciences* 98: 5399–5403.

Fleming, L. 2013. "The notion of limited perfect adaptedness in Darwin's principle of divergence." *Perspectives on Science* 21: 1–22.

Gibson, D. G., J. I. Glass, C. Lartigue, *et al.* 2010. "Creation of a bacterial cell controlled by a chemically synthesized genome." *Science* 329: 52–56.

Harris, C. L. 1981. *Evolution Genesis and Revelations: with Readings from Empedocles to Wilson*. Albany, NY: State University of New York Press.

Irwin, L. N. and D. Schulze-Makuch. 2012. *Cosmic Biology: How Life Could Evolve in Other Worlds*. New York, NY: Springer-Praxis.

Jabr, F. 2013. "Why life does not really exist." *Scientific American*, December 2. http://blogs.scientificamerican.com/brainwaves/2013/12/02/why-life-does-not-really-exist/, accessed: May 29, 2014.

Kauffman, S. 2000. *Investigations*. Oxford: Oxford University Press.

Kimura, M. 1984. *The Neutral Theory of Molecular Evolution*. Cambridge: Cambridge University Press.

Krug, A. Z. and D. Jablonski. 2012. "Long-term origination rates are reset only at mass extinctions." *Geology* 40: 731–734.

Lahav, N. 1999. *Biogenesis: Theories of Life's Origin*. New York, NY: Oxford University Press.

Lange, M. 2012. "What makes a scientific explanation distinctively mathematical?" *British Journal for the Philosophy of Science* 64: 485–511.

Langton, C. G. 1989. "Artificial life." In *Artificial Life* (Santa Fe Institute Studies in the Sciences of Complexity, Proceedings, vol. IV), ed. C. G. Langton. Redwood City, CA: Addison-Wesley, pp. 1–47.

Machery, E. 2011. "Why I stopped worrying about the definition of life . . . and why you should as well." *Synthese*, 185: 145–164.

Mariscal, C. 2014. *Universal biology*. Ph.D. dissertation, Duke University, Durham, NC.

Maynard Smith, J. 1986. *The Problems of Biology*. Oxford: Oxford University Press.

McGhee. G. R. 2013. *Convergent Evolution: Limited Forms Most Beautiful*. Cambridge, MA: MIT Press.

McShea, D. W. and R. N. Brandon. 2010. *Biology's First Law: The Tendency for Diversity and Complexity to Increase in Evolutionary Systems*. Chicago, IL: University of Chicago Press.

Mishler, B. D. 2009. "Species are not uniquely real biological entities." In *Contemporary Debates in Philosophy of Biology*, eds. F. J. Ayala and R. Arp. Hoboken, NJ: Wiley-Blackwell, pp. 110–122.

Philip, G. K. and S. J. Freeland. 2011. "Did evolution select a nonrandom 'alphabet' of amino acids?" *Astrobiology* 11: 235–240.

Powell, R. and C. Mariscal. In press. "Convergence as natural experiment: the 'tape of life' reconsidered." *Journal of the Royal Society Interface*.

Rosenberg, A. 2014. "How physics fakes design." In *Evolutionary Biology: Conceptual, Ethical, and Religious Issues*, eds. R. P. Thompson and D. Walsh. New York, NY: Cambridge University Press, pp. 217–238.

Serafini, A. 1993. *The Epic History of Biology*. New York, NY: Perseus Publishing.

Spencer, H. 1864. *The Principles of Biology*. London: Williams and Norgate.

Sterelny, K. and P. E. Griffiths. 1999. *Sex and Death: An Introduction to Philosophy of Biology*. Chicago, IL: University of Chicago Press.

Tsokolov, S. A. 2009. "Why is the definition of life so elusive? Epistemological considerations." *Astrobiology* 9: 401–412.

Vetsigian, K., C. Woese, and N. Goldenfeld. 2006. "Collective evolution and the genetic code." *The Proceedings for the National Academy of Sciences* 103: 10696–10701.

Wagner, A. 2005. *Robustness and Evolvability in Living Systems*. Princeton, NJ: Princeton University Press.

8 Equating culture, civilization, and moral development in imagining extraterrestrial intelligence: anthropocentric assumptions?

JOHN W. TRAPHAGAN

We shall not cease from exploration. And the end of all our exploring will be to arrive where we started and know the place for the first time.

– T. S. ELIOT

When scientists, philosophers, theologians, and others think about extraterrestrial life, a fundamental roadblock is that the nature of such life is a matter of human imagination. Lack of data beyond Earth limits our capacity to think about extraterrestrial life, even if that capacity is being increasingly stretched by the discovery and study of extremophiles on our planet and exoplanets around other stars. Despite our growing ability to imagine previously unimaginable forms of life, when we think about extraterrestrial life, intelligent or otherwise, Earth remains our only example. And when it comes to thinking about extraterrestrial intelligence, this presents several very significant limitations, since it is quite difficult to imagine what we might share with beings that evolved on a world different from our own (DeVito 2014, 2).

First, the Earth-bound nature of the data on life we have predisposes us to think about extraterrestrial intelligence in human terms. When we talk about alien beings, cultures, or civilizations, it is difficult to move beyond what we know – imagination does not happen in a vacuum; it is based upon our observations of the world around us and the only observations of life, intelligence, and civilization we have are right here on Earth. Second, when we do attempt to think about extraterrestrial intelligence, there is a tendency to ignore the fact that we are building our ideas on comparison with our own world, and often that comparative structure is not grounded in a nuanced and complex understanding of our world and the history and nature of culture and civilization on Earth. Indeed, ideas like culture and civilization need to be problematized when we think about our own world, which makes this doubly important when thinking about what intelligent life might be like on other worlds.

There are two questions I want to explore in this chapter. First, I am interested in how notions of progress and development shape the imaginations of scientists and others when thinking about extraterrestrial intelligence (ETI), particularly when we are contemplating the moral nature of that intelligence. Second, I am interested in exploring the extent to which our imaginings of ETI are shaped by anthropocentric assumptions about intelligence and civilization, and even the extent to which many of our assumptions may be ethnocentric in the sense that they are based upon Western notions of civilization and progress.

Culture and imagination

In his book *Modernity at Large*, anthropologist Arjun Appadurai (1996, 31) argues that in modern societies imagination has moved from the realm of fantasy or escape for the elite and educated, to organized fields of social practice that contribute to thinking about what is possible while also limiting the scope of our imagination in reference to what might be. Imagination, according to Appadurai, is a process through which we create "imaginaries," or constructed and invented social and cultural landscapes that reflect our collective interests, desires, ideas, and goals.

Although I think Appadurai makes an important point here, I don't think this is anything new – cultural production is always at some level about imagination. Culture is a function not only of how we are but how we imagine the world to be or how we want it to be at any given time – it is a product of our aspirations as well as a reflection of our social condition. That said, Appadurai's notion of imaginaries is a helpful tool for thinking about the relationship between science and culture, particularly when it comes to speculative research related to the nature of extraterrestrial life and intelligence, because science does not exist in a cultural vacuum. The questions scientists choose to ask and the interpretations they develop are not simply products of objective observations of the world and neutral interpretations of the data collected. As philosopher of science Thomas Kuhn (1962) has argued, although normal science allows for openness to alternate ways of thinking, it also shapes and limits the ways in which scientists think about the world and the specific questions they select to ask in order to direct their research. Scientific inquiry is largely focused on the articulation of observed phenomena and theoretical frameworks that a given paradigm supplies, rather than the creation of new theories (Traphagan 2015, 12). In short, scientific inquiry operates within the confines of a paradigmatic structure that influences and can even limit the range of questions that are normally asked, and any given paradigm is ultimately shaped at some level by broader social forces that shape culture.

There are, in fact, numerous instances of scientific knowledge being closely tied to a broader cultural imaginary – a good example is the Ptolemaic approach to astronomy which was deeply intertwined with the Christian imaginary until people like Copernicus, Galileo, and Newton came along to reconfigure our understanding of Earth's (and humanity's) place in the cosmos (Traphagan 2015). A more recent example of this can be found in the late-nineteenth-century discussion about the possible presence of a civilization on Mars. In 1877, Italian astronomer Giovanni Schiaparelli observed a network of straight lines he called *canali*, which was unfortunately translated into English as "canals." In Italian, the word *canali* refers to channels, which can either be natural or artificially created; however, in English the term canals has a very strong connotation of being a product of intentional construction. This problem in translation, of course, could have been corrected, but the fact is that many scientists and others interested in astronomy at that time *wanted* canals, and thus some sort of civilization, on Mars. And one of the most vocal and influential of these was Percival Lowell, a Harvard-educated businessman and amateur astronomer who played a key role in fueling speculation that an advanced civilization existed on the red planet (Dick 1998, 31).

We need not go into the errors and limitations in Lowell's observations of Mars that led to his conviction that there was evidence of Martian intelligence. In her superb work *Geographies of Mars: Seeing and Knowing the Red Planet*, Lane (2011) demonstrates that the Martian canals were not simply an example of a stunning error in scientific inquiry, but that this interpretation of Mars as hosting an advanced civilization reflected dominant geopolitical themes associated with Western imperialism and American westward expansion. In particular, Martian canals were a product of the intersection of these themes with an emphasis on mapping, which itself was a central tool used as Western societies moved around the globe and exerted dominance over "alien" societies, as well as setting up hierarchies between the advanced and the primitive much in the way Lowell did in relation to Mars and Earth.

Scientists did not, of course, simply accept Lowell's ideas about Martian civilization, and by the 1920s it had been clearly shown that there was no such civilization to be found. But the imaginary that developed from the ideas of Lowell and others continued to support a public discourse about Martian civilization that continued into the middle part of the twentieth century. Were the canals of Mars a product of poor science? On the one hand, yes, but much more was going on. In fact, Martian canals and the related Martian civilization were cultural products of a prevailing imaginary that brought together techniques of mapping, notions of progress and civilization closely related to Social Darwinism, poor translation, and a set of technical limitations in astronomical observation that existed at the time. These factors also were augmented by a new and growing awareness of the idea that Earth might not be the only place to host life, as seen in the emergence of the genre of science fiction. In short, both the public and the scientific community were primed to imagine an extraterrestrial civilization on Mars, in part due to the fact that writers like Jules Verne and H. G. Wells were expanding the range of imaginary activity and these fictional accounts were being strengthened by the "scientific" writing (much of which was in popular media outlets) of people like Lowell, a proactive and creative individual with the public relation skills needed to sell his vision of extraterrestrial life.

The important point about Lowell's Martian civilization isn't really about errors in scientific method and interpretation or limits in technology; it is about the fact that science, whether natural or social, does not exist in isolation from social and cultural environments. The flow of scientific inquiry, as well as the scientists who conduct research, are embedded in cultural contexts and the ideas and interpretations generated are not simply objective responses to data, they are shaped by the imaginary in which they arise. In other words, the capacity to imagine is limited and shaped by prevailing cultural ideas, values, and scientific paradigms that prevail at a given point in history, as well as by limitations in technology.

Think of it this way; a book like the one you are reading now would not have been possible five hundred, or even two hundred years ago. It certainly would not have been possible when the Ptolemaic/Christian geocentric worldview prevailed – it would be extremely difficult to imagine life on other worlds in such a cultural milieu (Traphagan 2015). A book similar to this might have been possible in the time of Lowell, but the structure of the discussion would have been very different. There would have been much more discussion about whether or not intelligent life could exist elsewhere; for the most part in this volume we are taking for granted – with good reason – the idea that life exists on other planets somewhere in our galaxy and that intelligent life is likely to exist as well. You would not be reading this book if the prevailing imaginary about Earth and the universe did not allow for this kind of thinking right at this particular historical juncture.

Lowell's imagined Martian canals and advanced civilization represent an overt example of how cultural context and scientific inquiry are intertwined, but there are considerably more subtle examples of this kind of interaction, as well. A particularly good example – also salient for the study of extraterrestrial intelligence – can be found in the application of ideas from Darwinian evolutionary theory to the study of human societies in the second half of the nineteenth century.

Anthropology, evolution, and civilization

Anthropologists during the nineteenth century were often concerned with identifying the nature of civilization and the differences between so-called civilized and less civilized (or uncivilized) people. The discussion of civilizations was surrounded by the broader discourse on Darwinian evolution that captured the interest of scholars in many areas of the emerging social science disciplines. Much of this literature attempted to place different forms of human social organization into evolutionary schema that showed the gradual transition from primitive to civilized societies.

This model of cultural evolution, known as Social Darwinism, applies concepts related to the process of natural selection and "survival of the fittest" as found in biological evolution to human social and political structures as well as to individual behavior within groups. According to this model, human societies evolve through progressive stages of development via processes of evolution driven by the same selection principles that shape biological evolution.

Among the more notable of scholars working in the area of cultural evolution in the nineteenth century was anthropologist Lewis Henry Morgan (1818–1881), a pioneer of American anthropology and the study of kinship. Morgan (1877) developed a typology for classifying different stages in the development of human societies that progressed from what he deemed "lower savagery" to "civilization" (see Table 8.1).

It is clear from this model that Morgan, and others of his period, assumed that cultural evolution inherently involves progress and is directional: as societies evolve, they get better. And the pinnacle of this process for Morgan, as for many of his contemporaries, is the "civilization" of the nineteenth-century North Atlantic countries. For Morgan, changes (again improvements) in social organization, as well as moral abilities, are closely equated with technological advancement. Note that all of the measures of evolutionary change in Table 8.1 are technologically based – technological development reflects a level of advancement or development as a society. It is important also to recognize that terms like savagery, barbarism, and civilization are value-laden ideas that are, as they were in the nineteenth century, reflections of moral assumptions. Lower levels of technological development reflect ideas about backwardness, lack of development, and lack of civilization and a basic "brutishness," while higher levels of technological achievement are associated with higher levels of moral and cultural achievement. At the same time, the idea of the "savage" had complex meaning that could indicate a lack of sophistication and moral inferiority, but could also index ideas about the "noble savage" that

Table 8.1 Anthropologist Lewis Henry Morgan's nineteenth-century typology of cultural evolution. This model, which assumes progress from lower to higher forms of society, has greatly affected SETI thinking about extraterrestrial civilizations.

Type of society	Technological features
Lower savagery	Fruit, nut subsistence
Middle savagery	Fish subsistence and fire
Upper savagery	Bow and arrow
Lower barbarism	Pottery
Middle barbarism	Domestication of animals (Old World)
	Cultivation of maize, irrigation, adobe and stone architecture (New World)
Upper barbarism	Iron tools
Civilization	Phonetic alphabet, writing

worked as a trope of the pure native not corrupted by the problems of civilization.

That said, in general, scholars of the nineteenth century viewed cultural evolution as a form of progress that led to increasingly *better* forms of social organization and moral behavior, with European values and forms of social organization as expressing the most evolved form – whether it was in terms of political organization (simple tribal society vs. complex state society) or religious life (polytheism vs. monotheism). We see this idea either overtly or more subtly reflected in the work of many scholars from the late-nineteenth and early-twentieth centuries, including Marx, Durkheim, and Weber, and the idea that cultural evolution is equated with progress has shown up in many theories of social change since, including modernization theory in the 1950s and more recently in globalization theories.

SETI, cultural evolution, and intelligence

My aim here is not to go into a detailed discussion of these ideas, but to emphasize the fact that this model of cultural evolution has deeply influenced the SETI imaginary related to thinking about extraterrestrial civilizations. In a great deal of writing related to SETI, cultural evolution is viewed as progressing toward civilization and civilization is equated with technological and moral progress from lower or earlier stages of development to our current (sometimes referred to as adolescent) stage, to some imagined mature stage associated with ETI. But there has been very limited work in the area of SETI that questions or problematizes what ideas like cultural evolution mean and how moral progress might, or might not, be connected to technological progress.

This raises a more fundamental question important to SETI: what is progress? That seems like a fairly easy one to answer. Dictionary.com defines progress as "a movement toward a goal or to a further or higher stage." Progress involves improvement and growth and, "the development of an individual or society in a direction considered more beneficial than and superior to the previous level." That said, while we may be able to agree on what progress is, it is much more difficult to determine in what ways it actually happens. Technological progress at one level seems obvious – computers are faster today than they were twenty years ago, cell phones do much more, stealth fighters are considerably more capable than the fighters of World War II. When we consider technology, we can see the accumulation of knowledge being expressed in new and more capable technologies. But can we say the same thing for other aspects of change in human groups over time? Does ethnographic and historical evidence support a notion of cultural evolution as a form of progress in the same sense that we see in technological change?

In fact, there is very little evidence that technological improvement is accompanied by ethical or social "improvement." Some authors, such as Steven Pinker (2011) argue that there is clear evidence that human societies have become less violent as compared to the past. He finds, for example, that the percentage of deaths as a result of warfare has dropped with state-level societies as opposed to earlier forms of social organization such as hunter–gatherer groups. He also argues that since the 1400s, there has been a decline in battle- or war-related deaths as a percentage of population. Pinker's data themselves are open to considerable debate about their validity and reliability. For example, he discusses homicide rates in Western Europe and non-state societies such as the !Kung bushmen in Africa and presents the idea that in the 1200s the averages for non-state societies were considerably higher than for contemporary Western Europe, which has shown a steady decline since that time.

This is a spurious comparison for many reasons. First, it seems very unlikely that we have good data for homicide rates anywhere in the 1200s, but certainly not among the so-called non-state societies, and Pinker doesn't provide longitudinal data for the non-state societies, only for Western Europe. Second, non-state societies have existed within the context of interactions with and domination by state-level societies for hundreds of years. How do we measure the numbers of homicides inflicted upon groups like the !Kung or Native American groups by Europeans (and Americans) in the contexts of colonialism and imperialism? And how do we factor in the influence of colonialism and imperialism on the violence that occurred in these non-state level societies? Is Pinker's drop in violence within state-level societies also reflected in the levels of violence Western European and American societies have inflicted upon those non-state societies?

Third, Pinker does not deal well with counter examples. In the case of Japan, for example, after a long period of isolation (from 1600 to the mid 1800s), the country rapidly shifted from a collection of loosely organized feudal domains to a nation-state. What accompanied that was a major increase in belligerent behavior towards neighbors such as Korea and China that culminated in World War II. One might argue that following the war, Japan renounced warfare officially in its constitution (although this was imposed by the United States) as evidence that the Japanese have become less violent. Certainly, rates of violent crime in Japan are quite a bit lower than in many industrial societies. That said, in the past ten years we have seen an increased arming of the Japanese military and growing nationalistic and militaristic ideologies, and suicide rates have climbed significantly since the 1990s (Traphagan 2010).

Finally, measuring the level of violence in this society (as in any society) is in part a matter of defining what constitutes violent action. Pinker makes it clear that societies that have slavery are more violent than those which don't, but what do we do with the kind of forced labor we see in factories during the nineteenth century in the industrial areas like the northern United States? How do we measure and assess the psychological violence (as well as physical violence) associated with police brutality or domestic violence in countries like the United States? And, should we falter and let loose our nuclear arsenals, or some biological weapon that devastates much or all of humanity, it would seem that in that single event we would have eclipsed the violence of our past.

In the end, I think it is difficult to quantify the amount of hatred, violence, or war that has existed in human history or that exists today, which makes arguments for cultural evolution as being equated with civilizing processes quite difficult to support. Much of the discussion about cultural and moral evolution also depends upon what we deem good or moral. If we assume an egalitarian society is morally better – more civilized – than a highly class/stratified society, then our tribal ancestors were morally superior and more civilized than we are, since members of the group generally had equal opportunity to achieve, power was diffuse without clearly defined leaders (it tended to be based upon skills rather than positional authority like heredity), and resources were fairly evenly distributed. Do such societies represent "civilizations" and if so, are they "better" than our civilization?

What we see in the work of thinkers like Pinker is an ongoing application of a concept of cultural evolution that assumes change is directional and progressive. Interestingly, this is quite the opposite of how scientists see evolution in the biological realm. Evolutionary theory simply describes processes of biological change, where evolution is usually defined "as the historical process that leads to the formation and change of biological systems" (Johnson and Lam 2010, 880). There is no directionality or *telos* in that process – it's just change (Traphagan 2015). While this is not an entirely random change, because the types of changes that happen are limited by the environment in which prior changes occur, there is not an end point and we cannot see one type of organism as being "higher" or better than another type. Each organism is adapted appropriate for the environment in which it resides and when that environment changes (which it always is doing) the species either adapts through the process of natural selection or it goes away. Additionally, complexity is not a measure of improvement, it is only related to fitness for a given environment. In short, biological evolution does not have a target – it isn't *advancing* or progressing – and there is no design. As Dawkins (Dawkins 2006, 103) notes, Darwin made it abundantly clear that "natural selection produces an excellent simulacrum of design, mounting prodigious heights of complexity and elegance," but this seeming design does not presuppose that said change actually has an intentional, or even unintentional, direction that might lead to intelligence or anything else.

Evolution and altruism

So we are left with a problem. Even if we can't identify a directionality to cultural or biological evolution, perhaps we can identify certain types of behaviors or values that are of evolutionary benefit for social beings and that, therefore, might represent universal aspects of human behavior as well as of the behavior of any intelligent extraterrestrial beings we might encounter. One behavior – or value – that has been presented this way in SETI research has been altruism. Several SETI scientists have argued that altruism should be expected among intelligent extraterrestrials because in order for their civilization to survive the long periods of time necessary for us to have a good chance of making contact, they must have developed values related to not only self-preservation but to the success and preservation of others (Vakoch 2014). This idea is summed up well by astronomer Frank Drake, who writes (Drake 2001; see also Drake 2014):

Should altruism be expected, perhaps be ubiquitous, even universal? There seems to be an easy answer to this question – yes. As noted by many writers, altruism can be expected in intelligent creatures simply as a Darwinian imperative. Communities of mutually supportive individuals, practicing altruism, will possess greater potential for survival than the same individuals acting alone. Even greater survivability accrues when individuals have the will to endanger their own well being for the good of the community.

I'm not convinced that this is actually an easy question to answer, because altruism is itself rather difficult to define from a universal perspective and the behaviors that we think of as being altruistic are shaped by culture (Traphagan 2012). That said, there is an argument to be made that altruism is in some way biologically encoded in our DNA. Some animals sacrifice themselves for others, particularly kin, a behavior that is discussed in theories of kin selection and inclusive fitness, which represents a measure of how the actions of one individual (the donor of the behavior) influence the fitness of others (the recipient).

There are several examples of this evolutionary strategy evident in the natural world, such as alarm calls among ground squirrels (and other animals) that sometimes alert other members of the same species about danger (Milius 1998). Clearly, such an alert can draw attention to the caller and increase the risk of predation for that individual. This type of behavior is found in a variety of animals. It normally happens when kin are in the immediate vicinity of the caller, suggesting that this is a way to warn kin of impending danger. This increases the reproductive benefit to the recipient of the altruistic act such that it outweighs the potential reproductive cost to the donor – this is known as Hamilton's rule: $rB > C$, where r = the genetic relatedness of the donor to the recipient, B = the additional reproductive benefit gained by the recipient, and C = the reproductive cost to the donor performing the act.

The theory of kin selection certainly is interesting and provides one way to think about the possibility that ETI might be predisposed toward altruistic behavior and, thus, not be a threat to humans. But it is not without its criticism. Alonso (1998) argues that the behaviors kin selection theory tries to explain are not necessarily altruistic – they may, in fact, favor the donor and thus would represent ordinary examples of natural selection or they may simply be byproducts of social structures that have developed in which different individuals perform specific tasks, such as what we see in bee colonies.

In essence, the hang-up comes in whether or not any of these behaviors identified with kin selection actually represent anything we might call altruism, and this depends upon how we define altruism. Should we consider the squirrel who shrieks a warning to be actually engaged in altruistic or self-sacrificial behavior? Doesn't this require intent? And do we want to attribute volition to these types of actions among squirrels or bees or any other non-human organism? Altruism is not an act, it is a value that is tied into notions of self-sacrifice and concern about others. As such, it inherently involves intent on the part of the donor, but intent is extremely difficult to identify without ambiguity when it comes to non-human animals and even when it comes to humans.

Even if we accept that humans are capable of altruistic behavior, the nature of specific acts can be quite ambiguous. Suppose, for a moment, that the Chair of my department asks me to teach an overload due to a colleague suddenly deciding to quit the department two days before classes start. I accept the request, and the Chair shows her appreciation of my willingness to take one for the team by offering her thanks and gratitude. Was mine an altruistic act? Did it involve self-sacrifice? On the one hand, the answer seems to be an unambiguous yes – I have more work to do and don't appear to gain anything from that, although most likely I would be paid extra for the overload. Assuming that, in fact, I did not get paid extra, what do I gain? Well, there could be quite a bit. Perhaps I am interested in becoming the next Chair of the department, which brings monetary advantages as well as a degree of (very limited) power and prestige. Is my action motivated out of a desire to help others with a willingness to sacrifice my time? Or is it the product of a very subtle set of calculations directed at achieving personal gain in terms of cash, status, and power in the long run?

There must be some form of action that is unambiguously altruistic. Suppose I push a complete stranger out of the way of an oncoming bus and get squashed in the process (so much for kin selection). Doesn't this meet the requirements for altruism? The problem here is that we don't actually know the motivation. It may be that the action was motivated by a desire to save another life, but it may also have been motivated by a desire to attain fame even in death or some sort of reward in a presumed afterlife such as entrance into heaven – or some combination of these motivations. Humans are complex animals and rarely have simple motivations behind their actions – but it does seem clear that completely unambiguous altruism would be difficult to detect in any act and is potentially impossible to identify. It may well be that all actions by humans are at some level motivated by selfish interests or self-preservation (remember, guaranteeing a good afterlife is a form of self-preservation), which raises the possibility that altruism is not actually a feature of human behavior, may not be necessary for social beings, and, thus, may not be a feature of the behavior of ETI. Perhaps altruism is nothing more than a value we aspire to, but one that we cannot actually achieve because all of our actions are ultimately rooted in self-preservation and ensuring selfish gain, but some of these actions also have positive social consequences because selfish gain is often tied to recognition by others.

The point that I want to emphasize here is that a great deal more thinking about the nature of altruism and its connection to biology is needed before we make any assumptions about the possible altruistic nature of ETI. Theories like kin selection do not explain altruism toward non-kin, nor from an imagined non-human intelligence toward humans. Furthermore, there is a tendency in SETI research to often tacitly assume a linkage between technological progress and cultural/moral progress, as noted above in the discussion of cultural evolution, which often is conflated with the idea that altruism is some sort of biological imperative of social beings and of civilization. The evidence related to this assumption about Earth-dwelling life remains open to considerable interpretation.

SETI, anthropomorphism, ethnocentrism

Indeed, in SETI literature the concept of civilization often is imagined in terms of ideas about human development – words like infant, adolescent, and mature are frequently used to describe both our own civilization and the civilization of potential interstellar interlocutors. For example, Lemarchand and Lomberg (2009) argue that: ". . . all civilizations should evolve ethically at the same time as they evolve technologically" and that when said civilizations reach their "technological adolescent stage" they necessarily must have some sort of "ethical societal mutation" or fall to extinction. The number of assumptions in this argument is quite impressive. First, the authors assume that we have the ability to determine what a technological adolescent stage is – and they assume that we're it. Since ours is the only data point, it seems spurious to suggest that we are representative of universal trends or can be represented as being at any particular stage of cultural/social development. Second, one is left wondering why it is necessary to have an ethical societal mutation; why do we assume that ours is the only path that might occur as a civilization develops? Perhaps the path to civilization for extraterrestrial society X is the product of values that never involved violence, while society Y is able to continue to exist despite the centrality of violence in its culture, even while that is moderated by a distaste for mass destruction or the civilization spends relatively long periods in equilibrium, has short burst of intense and highly destructive violence, and then rebuilds and re-enters long periods of relative equilibrium. This would be a sort of punctuated equilibrium for cultural evolution (cf. Eldredge and Gould 1972).

Put another way, when we imagine the civilizations of ETI, we tend to anthropomorphize those civilizations by applying our own – often outdated – theories about the nature and development of human civilizations, as well as a very terrestrial definition of civilization itself, to whatever we might find in the cosmos. Imagined extraterrestrial civilizations become a product of imagined and observed aspects of our own history (Figure 8.1).

What these factors combine to generate is a highly anthropomorphized conceptualization of extraterrestrial intelligence and civilization, which is, in

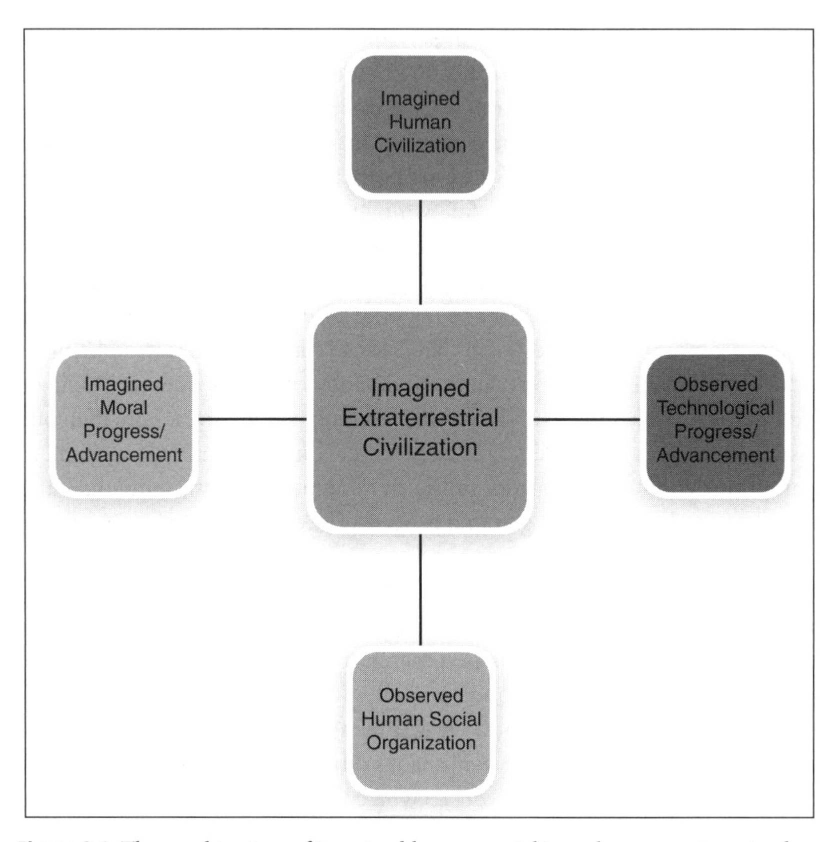

Figure 8.1 The combination of imagined human social/moral progress, imagined human civilization as a product of cultural evolution and often Euro–American in nature, observed technological progress, and observed patterns of human social organization that lead to the construction of extraterrestrial civilizations in the imaginations of SETI scientists and many others.

turn, based upon ethnocentric ideas about human civilization and the nature of progress over time. This influence can be summarized in three key points: (1) our imaginings about ETI are deeply influenced by Social Darwinism and Western notions of both technological and moral progress; (2) our imagination about ETI is based upon a linear concept of time that assumes history is directional and, thus, that the development of *any* civilization will involve progress that is not only technological in nature, but also moral; and (3) these assumptions derive not only from ideas related to Social Darwinism, but also from Christianity.

Indeed, this last point that there are important tacit assumptions about progress in cultural evolution theories – and by extension in theories about the nature of ETI – that derive from Christian ideas about human history and its future is particularly important. Christian-centered assumptions about the nature of civilization and life in general include a sense that both individual lives and time itself is directional – we are born, live, and die and then end up in another place (heaven or hell) following the end of corporeal existence. Humans live once – there is no cycle of life and death and no reincarnation as is found in other religio-philosophical systems such as Buddhism. As a result, the physical, bodily aspect of life ends at death, but the spirit or soul persists, passes through a judgment that leads to some eternal state of being in the form of a resurrected spiritual body. That eternal state may be utopic or involve endless punishment and Christian eschatology involves more than one possible outcome, which are not of equal value, but which may contribute to assumptions about directionality in the development of both individual and collective moral abilities – e. g. moral progress (Friesen, personal communication, 2014).

If we consider the implications of Christianity and a more generally ethnocentric approach to thinking about the nature of ETI, several features of our imagined extraterrestrial friends become clear. ETI is altruistic, unified as a single civilization, exists in some form of utopia (a product of its having evolved to a higher level of moral development), and is a product of a linear development with an end point that reflects our own aspirations about evolution, progress, and the nature of a civilized society. In other words, the imaginary in which ETI is constructed by SETI scientists and others is one that is, in fact, a reflection of ourselves. Our speculations about the nature of ETI is a product of both how we think we are and what we think we'd like to become – and by "we" I really mean Western scholars and others deeply influenced by Western ideas about human nature and society.

In closing, I want to leave the reader with two main points to think about when considering how in the SETI community we have tended to imagine the nature of extraterrestrial civilizations. First, most of these imaginings are laced with assumptions about progress, development, and directional change that are a product of a Western worldview that has been deeply influenced first by Christianity and later by Social Darwinism. Our speculations about the nature of ETI are not neutral, nor necessarily scientific, but are products of assumptions about civilization that come from our own culture and history. This is in many ways unavoidable, because we have no non-human civilization from which to draw data in order to develop ideas about what ETI might be like. However, we do have other civilizations on Earth that have developed very different ways about thinking about human nature, progress, and civilization (such as societies influenced by other religious frameworks like Islam and Buddhism) and incorporation of thinking from these areas might help to stretch our ideas about the possible nature of ETI (see Chapter 19).

Second, which is really an extension of the first point, much of our thinking about ETI does not simply involve anthropomorphism, but is ultimately quite ethnocentric in nature. It is a product of the Western imaginary, largely because the vast majority of scientists working in this area are a product of the Western imaginary, as is science itself. There are very few non-Western scholars working in the area of SETI and this influences the manner in which we think about the nature of alien civilizations. In other words, there is a great need for the inclusion of non-Western scholars in the SETI community to develop more diverse ways of thinking about the nature of an extraterrestrial other and to inform our ideas about communication with that other with perspectives outside of the Western cultural milieu. Perhaps the most important point to be derived from this exploration of culture and SETI is that there is a need within the SETI community to develop a robust framework for reflexive thought about how our own cultural values influence the ways we imagine extraterrestrial others.

References

Alonso, W. J. 1998. "The role of kin selection theory on the explanation of biological altruism: a critical review." *Journal of Comparative Biology*, 3, 1–14.

Appadurai, A. 1996. *Modernity at Large: Cultural Dimensions of Globalization*. Minneapolis, MN: University of Minnesota Press.

Dawkins, R. 2006. *The God Delusion*. New York, NY: Houghton Mifflin.

DeVito, C. 2014. *Science, SETI, and Mathematics*. New York, NY: Berghahn.

Dick, S. J. 1998. *Life on Other Worlds: the 20th-Century Extraterrestrial Life Debate*. Cambridge: Cambridge University Press.

Drake, F. 2001. "Encoding Altruism." *Science and Spirit*, 12, September–October, online at http://archive.seti.org/seti/projects/imc/encoding/altruism.php

Drake, F. 2014. Foreword. In Vakoch, D. A., (ed.), *Extraterrestrial Altruism: Evolution and Ethics in the Cosmos*. Berlin: Springer, pp. vii–x.

Eldredge, N. and S. J. Gould. 1972. "Punctuated Equilibria: An Alternative to Phyletic Gradualism." In Schopf, T. J. M. (ed.), *Models in Paleobiology*. San Francisco: Freeman, Cooper & Co, pp. 82–115.

Johnson, B. R. and S. K. Lam. 2010. "Self-organization, Natural Selection, and Evolution: Cellular Hardware and Genetic Software." *BioScience* 60:879–885.

Kuhn, T. 1962. *The Structure of Scientific Revolutions*. Chicago, IL: University of Chicago Press.

Lane, K. D. M. 2011. *Geographies of Mars*. Chicago, IL: University of Chicago Press.

Lemarchand, G. A. and J. Lomberg. 2009. "Universal Cognitive Maps and The Search for Intelligent Life in the Universe." *Leonardo* 42:396–402.

Milius, S. 1998. "The Science of Eeeeek!" *Science News* 154: 174–175.

Morgan, L. H. 1877. *Ancient Society*. London: MacMillan and Co.

Pinker, S. 2011. *The Better Angels of Our Nature: Why Violence has Declined*. New York, NY: Penguin Books.

Traphagan, J. W. 2010. "Intergenerational Ambivalence, Power, and Perceptions of Elder Suicide in Rural Japan." *Journal of Intergenerational Relationships* 8: 21–37.

Traphagan, J. W. 2012. "Altruism, Pathology, and Culture." In Oakley, B., G. Madhavan, and D. S. Wilson (eds.), *Pathological Altruism*. Oxford: Oxford University Press, pp. 272–288.

Traphagan, J. W. 2015. *Extraterrestrial Intelligence and Human Imagination: SETI at the Intersection of Science, Religion, and Culture*. New York, NY: Springer.

Vakoch, D. A., ed. 2014. *Extraterrestrial Altruism: Evolution and Ethics in the Cosmos*. Berlin: Springer.

9 Communicating with the other

Infinity, geometry, and universal math and science

DOUGLAS A. VAKOCH

If the search for extraterrestrial intelligence (SETI) detects an artificial signal from a distant civilization, our next challenge will be to understand any encoded message, and then to decide what we may want to transmit in reply. The few intentional messages humans have sent into space thus far reflect the assumption that mathematics and science are universal. Any civilization able to build technology capable of interstellar communication must certainly know at least the basics of these areas, it is often argued. How accurate is this Platonic notion that our math and physics tap into universal principles? Might different civilizations have their own versions of math and science that are perfectly adequate for explaining the universe, but that do not directly map onto our notions?

Lingua Cosmica

In the standard approach to constructing interstellar messages that may be comprehensible to an independently evolved intelligence, we start with concepts presumably shared by sender and recipient. If the goal of interstellar communication is to share information not previously known by the recipient, the sender has the additional challenge of identifying a sequence that will lead from shared to unique information. For example, logician Hans Freudenthal began his interstellar language *Lingua Cosmica*, or *Lincos*, with an exposition of mathematical concepts (Freudenthal 1960). He then moved to a discussion of time; then human behavior; and finally notions of space, motion, and mass. In the process, he attempted to convey some of the idiosyncrasies of human life in terms of potentially universal principles of mathematics and science.

A recurrent critique of *Lincos* is that ambiguities at one point in the exposition may make subsequent sections unintelligible. In this chapter, I examine the potential problems raised by Freudenthal's reliance on concepts related to infinity early in his message, and propose an alternative approach that introduces notions of three-dimensional (3D) space and motion as a foundation for discussing human behavior, with the goal of enhancing intelligibility.

Infinity questioned

One of the core assumptions of prior approaches to creating interstellar messages is that extraterrestrial intelligence, by virtue of partially shared technology for interstellar communication, will also have shared understanding of basic mathematical principles. Implicit in this assumption is the belief that mathematics as known and used by humans taps into some ontologically fundamental aspect of reality. Moreover, it is often assumed that any independently evolved rationality that shares certain technological capabilities will identify many of the same mathematical principles that we have. This Platonic notion that numbers and other mathematical concepts somehow exist independently of the human mind is commonly accepted among practicing scientists, although questioned by a number of philosophers of mathematics. Some have argued that mathematics does not simply find preexisting truths, but that the particular form that our mathematics has taken is influenced by social factors (Hersh 1997), our physical embodiments (Lakoff and Núñez 2000), and theological presuppositions (Rotman 1993).

A closer examination of the ontological status of mathematical concepts, especially in light of mathematical prerequisites for various technological capabilities, can help guide the selection of basic mathematical principles used to design interstellar messages. That is, to the extent it is conceivable that extraterrestrial intelligence could create advanced technologies without knowledge of certain mathematical concepts, messages can be designed that either avoid the use of those concepts or instead, explicitly *teach* those concepts, without presupposing prior familiarity with them. In addition, to the extent that compelling reasons can be provided to assume that other concepts are very plausibly known by extraterrestrial intelligence, those concepts can be used as foundations for interstellar messages. In this chapter, I provide examples for both of these cases by considering why it may be wise to avoid putting too much reliance on notions of infinity in interstellar messages, and instead emphasize geometry.

Rotman (1993) has argued that the concept of infinity is based on essentially theological assumptions of a God-like entity that can generate numbers without limit, with accuracy and regularity undiminished by entropy, an action impossible for an embodied counting agent constrained by the material limitations of time and space. A mathematics based on Rotman's alternative, "a non-Euclidean arithmetic," would not include the notion of infinity. The real world consequences of such an alternate foundation for higher mathematics could be quite significant. For example, Rotman (1993, 11) suggests that a physics based on counting that can be done only by physical agents, and not immaterial processes, would yield:

> . . . a description of the physical universe strikingly different from the one currently on offer. In the absence of unlimited subdivision everything numerically measurable – space, time, mass, energy, charge, gravitation, information – would be quantized, there would be no arbitrarily-approximatable-to "points" in space or "instants" of time, all movement would be discrete and ultimately discontinuous, and so on.

Mathematician Carl DeVito (2014, 173) emphasizes the critical role that infinite processes have played in the human development of differential and integral calculus, but he suggests our emphasis on infinity may arise from our reflections on the limited span of our lives: "Perhaps our preoccupation with infinite sets somehow reflects our awareness of the finiteness of our lives. Our desire to know the infinite seems somewhat like a religious impulse, and this branch of mathematics seems a little like theology." Linguist George Lakoff and mathematician Rafael Núñez (2000) offer a contrasting analysis of the origins of our sense of infinity. It is not from the negation of the finite, they say, but from our embodied experience of certain actions: "Of course, in life, hardly anything one does goes on forever. Yet we conceptualize breathing, tapping, and moving as *not having completions*" (Lakoff and Núnez 2000, 156). Would differently embodied extraterrestrial intelligence (ETI), or ETI having a different stance toward their own finitude, have a similar predilection for concepts of infinity?

Psychologist Louis Narens (1997) has argued that extraterrestrial intelligence capable of radio transmissions might not be familiar with notions of infinity, and that some of the very principles that have often been suggested in the SETI literature for interstellar communication might be alien to extraterrestrials. Narens (1997, 563) cautions that "one should be wary of the use of concepts such as 'prime numbers' or a 'binary form of the decimal expansion of π' as a common basis for communication with ETIs." While Narens acknowledges that extraterrestrials with advanced technologies may well need to be familiar with basic principles of arithmetic, these beings may not need the concept of "natural numbers." That is, extraterrestrials may have the ability to use numbers for mathematical purposes without reflecting on more general categories that terrestrial mathematics uses to describe the fundamental nature of these numbers. For example, extraterrestrial mathematicians may never need the notion of infinity to create a science suitably flexible to guide the development of technologies that make possible interstellar communication.

Common geometry

Freudenthal (1960) introduces the notion of infinity relatively early in the first chapter of *Lincos*, which deals with mathematics. If Narens is correct, however, it is not so obvious that extraterrestrial intelligence would find transparent a mathematics based significantly on this concept. Rather than presupposing the universality of the notion of infinity, Narens suggests that concepts from Euclidean geometry are more likely to be universally known by extraterrestrials that engage in interstellar communication. He argues his point in several ways. First, he suggests that the ability to describe the movement of objects in 3D space is a prerequisite for advanced civilizations to launch objects into Earth orbit, a capability likely among astronomically curious extraterrestrials who attempt interstellar communication. Moreover, he argues, any alternate descriptions of the same movements are ultimately inter-translatable, meaning that a human geometry and an extraterrestrial version would be mutually comprehensible. Narens also expects extraterrestrials to have the ability to form mental models of objects that allow them to recognize objects that move; that is, an object that changes position over time nevertheless remains the same object. Even though the perceptual categories used to sense such objects may vary from species to species, Narens expects some form of "object constancy" to be selected in the course of evolution.

We might create a potentially comprehensible framework for conveying 3D messages using various approaches. For instance, we might "layer" two-dimensional (2D) pictograms to form 3D messages, expanding on Drake's (1978) approach of encoding 2D images. Standard cryptologic techniques could be especially helpful for reconstructing the format of such messages, since these techniques do not presuppose a comprehension of the meaning of the message contents (Deavours 1985). As an alternative, we could derive a 3D coordinate system from scientific and mathematical knowledge potentially known by both transmitting and receiving civilizations (DeVito and Oehrle 1990).

Narens (1997) argues that geometrical intuitions may provide a more intelligible foundation for interstellar messages than intuitions about mathematical concepts based on the set of all natural numbers, the latter being presupposed by many previous proposals (e.g. Freudenthal 1960). Following this line of reasoning, early parts of interstellar message sequences could profitably emphasize descriptions of geometrical objects, providing a connection between pictorial and mathematical representations of abstract concepts (Morrison 1963).

We might begin with a group of geometrical objects with clearly defined features. For instance, we might start with the Platonic solids, which can be defined as all of the convex polyhedra that have regular polygon faces, in which all vertices and faces are identical. There are precisely five Platonic solids, each with a different number of faces: four faces (the tetrahedron), six faces (the cube or hexahedron), eight faces (the octahedron), twelve faces (the dodecahedron), and twenty faces (the icosahedron). The Platonic solids can be described either through formulas (e.g. Euler's rule specifies that for each Platonic solid, the number of edges is equivalent to the number of vertices plus the number of faces minus two) or through 3D representations (noting the coordinates of the vertices on the x-, y-, and z-axes). In addition, the coordinates of both the icosahedron and the dodecahedron are partially defined by the golden mean (1.61803 . . .), a proportion that is found repeatedly in human judgments of beauty (Lemarchand and Lomberg 2011). The golden mean is also seen in biological phenomena (e.g. the nautilus shell and branching patterns of trees), physical phenomena (e.g. the hydrogen atom's spectral lines [DeVito and Little 1988]), and a range of mathematical contexts (e.g. the Fibonacci series).

Once we have defined the Platonic solids, we can use them to show relationships between objects. For instance, we can introduce various *prepositions* in a sequence of images that depict, for example, (1) a tetrahedron *between* a cube and an octahedron, (2) a tetrahedron *outside* an octahedron, and (3) a tetrahedron *inside* an octahedron. We could pair 3D representations like these with symbolic, non-pictorial representations of the same relationships, transmitting the two types of representations one after the other. Assuming the recipient is able to make a link between these pictorial representations and the corresponding arbitrary symbols, later communications can make use of the symbolic system's "shorthand."

Representing physical objects

One virtue of emphasizing geometrical concepts from the outset is that the same methods for representing abstract mathematical objects can also be applied to objects in the physical world. Relatively early in an interstellar message sequence, we might profitably encode 3D representations of objects familiar to both ETI and humankind. As one example, we might describe a physical environment commonly known by both receiver and transmitter: an interstellar-sized 3D space defined by the receiver–transmitter axis, with one end of the axis defined by our Sun, and the other by the ETI's home star. When these two points are combined with a cosmically significant third point, such as the center of the galaxy, a common reference plane could be defined. The surrounding 3D space would contain stars known by both receiver and sender, yielding a confirmation that this coordinate system maps onto the macroscopic physical universe.

In addition, geometrical representations could be used to describe scientific constructs at a microscopic scale, such as molecular structure. For instance, atoms of methane molecules (CH_4) are arranged in a tetrahedral configuration, with four hydrogen atoms located at the vertices of a tetrahedron and a carbon atom located at the tetrahedron's center. Such a 3D pictorial representation could also be linked to set theoretic descriptions of the periodic table of elements (DeVito and Oehrle 1990).

After describing physical objects known by both ETI and humans, we could begin to describe objects not familiar to ETI. For example, the general morphology of the human body can be described geometrically. Part–whole relationships could also be conveyed, for instance, by first showing an image of the entire human body, followed by an image showing only one arm, with the arm in the same location within the frame that it appeared in the previous picture of the whole body (Vakoch 1978; Vakoch 2011). Symbols standing for the words "body" and "arm" could be paired with the matching pictures, providing a method to teach ETI our symbolic languages.

Bodies in motion

ETI plausibly has geometric intuitions akin to ours because it is evolutionarily adaptive to have the capacity to recognize the constancy of objects in the physical world, even when these objects move in 3D space (Narens 1997). Such positional translations let us depict objects as dynamic rather than static. We can depict humans, for example, as bodies in motion – either individually or in interaction with one another.

Animated movies use sequences of 3D images to tell stories, and we might use the same approach in interstellar messages. To convey human action and intent in such messages requires a clear exposition of the basics of human movement and gesture. With a sufficiently detailed tutorial, we might construct messages that serve as interstellar morality plays, letting us communicate notions of altruism, for example.

3D animation sequences can describe human movements through the same concepts of object constancy that Narens has suggested may be widely comprehensible by extraterrestrials capable of interstellar communication. Motion-capture technology records actual human movements in 3D space, generating digital data that can be encoded directly into interstellar messages, in the process avoiding many of the ambiguities of interpreting 2D pictures (Vakoch 2000). For depicting complex motions of human beings interacting, motion capture has the advantage over other animation techniques of yielding movements that are more realistic and cost-effective. This ease of gathering large amounts of data will be critical for generating the thousands of scenarios necessary to convey the complexity of human behaviors.

Interstellar messages can be constructed using this type of technology, yielding series of images similar to those in Figures 9.1, 9.2, and 9.3, which were generated through less-efficient animation techniques. In this sequence, designed by the author, one person is shown helping another person, even though the act of helping puts the altruist in danger. Descriptions of the physical environment and biomechanical explanations of the dangers to the human body of falling help provide a context for extraterrestrials to understand the potentially fatal consequences of altruism. In the process, human vulnerability might be communicated within a framework describing the physical and temporal dimensions of human finitude. Interstellar messages incorporating data from motion-capture technologies can also be combined with parallel descriptions of the events encoded in a logic-based interstellar language.

Figure 9.1 Interstellar messages that include 3D depictions of humans interacting in physical environments can set the stage for introducing even complex behaviors reflecting human values. Here one person sees another in danger of falling and must decide whether to help. Credit: Douglas Vakoch.

Figure 9.2 By definition, altruism involves some risk or cost to the person offering help. While it is possible that both the person helped and the altruist will make their way to safety, that is not guaranteed. Credit: Douglas Vakoch.

Figure 9.3 In this case, the altruist pays the ultimate price of self-sacrifice, fatally falling while attempting to save someone else. In a full interstellar message, this sequence depicting altruism would be contrasted with other sequences, emphasizing that acts of altruism can have different outcomes. Credit: Douglas Vakoch.

Universal math and science?

While we might be hopeful that some geometrical concepts would be recognizable to ETI for the reasons outlined by Narens, reflections on the contingency of the development of mathematics and science on Earth lead to a more guarded, if not outright pessimistic, conclusion. Even assuming that ETI and humans have a shared commitment to making increasingly more accurate models of physical reality, it is possible that these models will remain fundamentally incommensurable (Vakoch 1998). "Admittedly there is only one universe, and its laws, as best we can tell, are everywhere the same," notes philosopher Nicholas Rescher (1985, 90). "But the sameness of the object of contemplation does nothing to guarantee the sameness of the ideas about it. It is all too familiar a fact that even where human (and thus *homogeneous*) observers are at issue, different constructions are often placed upon 'the same' occurrences" (Rescher 1985, 90).

It is essential to distinguish between descriptions of the world – whether in natural language or in specialized languages such as mathematics – from the nature of the world as it exists independently of our attempts to understand it. As George Sefler (1982, 70) observes, "The issue is not whether there exists some external physical world, but whether this world comes linguistically prearticulated as we know it, or whether empirical language is in some way a construction of, a shaping of, the stuff of experience according to certain human structures." To the extent that human and extraterrestrial scientists use different languages to explain reality, they may end up with radically different models of the same shared physical substrate.

Peter Barker (1982) has examined the challenges that terrestrial scientists from different times and cultures would have in understanding one another; how much more difficult might it be for terrestrial and extraterrestrial scientists to understand one another, given they have evolved in different environments? "Science can only form a basis for translation if we happen to encounter the aliens at a point in their scientific history parallel to our own," Barker (1982, 82) argues. "It seems perfectly possible to imagine an alien culture proceeding from a different starting point through different scientific revolutions to a set of theories which we have never come across," he continues. "To the extent that our science reflects our culture and history it is no more useful as a starting point for communication than any other aspect of our culture and history" (Barker 1982, 83). Nor can we automatically hope to find a readily intelligible basis for communication by shifting from science to mathematics. "Mathematics, too, is a product of individual cultures," Barker (1982,

83) continues. “Mathematics goes through stages of noncumulative growth which may hinder communication between people with different views.”

Even the goals of extraterrestrial scientists could differ markedly from the goals of their human counterparts, given that the two species evolved on different worlds and have different needs. As Rescher (1985, 85) argues, “We can hardly expect a ‘science’ that reflects our parochial preoccupations to be a universal constant. The science of a different civilization would presumably be closely geared to the particular pattern of their interaction with nature, as funnelled through the particular course of their evolutionary adjustment to their specific environment.” An aquatic intelligence might develop a more sophisticated understanding of hydrodynamics than a land-based intelligence, simply out of greater need.

Analogous arguments about scientific change can be found in philosopher Larry Laudan’s book *Progress and Its Problems*, as he argues that research is assessed not by reference to any absolute criteria for the rationality of scientific theories, but by communities that compare the effectiveness of one research tradition with competing traditions (Laudan 1977). By his analysis, scientific rationality is in part “a function of time and place and context. The kinds of things which count as empirical problems, the sorts of objections that are recognized as conceptual problems, the criteria of intelligibility, the standards of experimental control, the importance or weight assigned to problems, are all a function of the methodological–normative beliefs of a *particular community of thinkers*” (Laudan 1977, 130–131; emphasis added). If we extend Laudan’s argument to extraterrestrial scientific communities, whose theories would have developed in isolation from terrestrial science, it becomes conceivable that extraterrestrial science could differ considerably from terrestrial science.

The prospect of radically distinct extraterrestrial science has significant implications for interspecies communication. On the one hand, it suggests that mutual comprehension may be tremendously challenging. However, on the other hand, it also raises the possibility of learning a great deal from another independently evolved civilization, if we are ever able to establish some foundation for common understanding. In fact, the prospect of a plurality of sciences in the galaxy provides reason to think that even a young civilization on galactic timescales – such as ours – might provide novel insights to even much-longer-lived civilizations. If there are multiple paths towards increasingly sophisticated models of physical reality, each contingent on the idiosyncratic biological and cultural histories of their respective worlds, then the scientific accounts of even nascent civilizations might add fundamentally to the scientific understanding of other civilizations. Rescher (1985, 93) finds a similar view in the work of psychologist William James, insofar as the latter contended that "in the course of cognitive evolution, nature, metaphorically speaking, works towards being known in all its various aspects, evolving beings capable of comprehending it in many different and diversified modes. Here we are dealing with a teleology of diversity – a nature striving to be known in a variety of forms and aspects." Viewed in this way, even much older civilizations in the universe might substantially benefit from learning about science that has been developed by a much younger civilization – even as practiced here on Earth at the beginning of the twenty-first century.

In the end, we can never be certain which of our human intuitions might find a counterpart in ETI. Is it our reflections on infinity, our penchant for describing objects in geometrical terms, or something else? Perhaps it is unrealistic to think that we can ever find one obviously universal foundation for interstellar communication. A more modest goal would be to identify multiple plausible starting points, and to send a series of different messages, each based on its own distinct set of assumptions about what we and ETI have in common. If our goal is to make contact with another world, we have no need to find a truly universal language, applicable for all possible ETI at whatever their level of development. Simply finding *any* conceptual framework that lets us begin to understand the one extraterrestrial civilization we encounter in our first contact – that will be enough.

References

Barker, Peter. 1982. "Omnilinguals." In *Philosophers Look at Science Fiction*, edited by Nicholas D. Smith, 75–85, Chicago, IL: Nelson-Hall.

Deavours, Cipher A. 1985. "Extraterrestrial Communication: A Cryptologic Perspective." In *Extraterrestrials: Science and Alien Intelligence*, edited by Edward Regis Jr., 201–214. Cambridge: Cambridge University Press

DeVito, Carl L. 2014. *Science, SETI, and Mathematics*. New York, NY: Berghahn.

DeVito, Carl L. and W. A. Little. 1988. "Fractal Sets Associated with Functions: The Spectral Lines of Hydrogen." *Physical Review A* 38:6362–6364.

DeVito, Carl L. and Richard T. Oehrle. 1990. "A Language Based on the Fundamental Facts of Science." *Journal of the British Interplanetary Society* 43:561–568.

Drake, Frank. 1978. "The Foundations of the Voyager Record." In *Murmurs of Earth: The Voyager Interstellar Record*, edited by Carl Sagan, 45–70. New York, NY: Random House.

Freudenthal, Hans.1960. *Lincos: Design of a Language for Cosmic Intercourse, Part I*. Amsterdam: North Holland.

Hersh, Reuben. 1997. *What Is Mathematics, Really?* Oxford: Oxford University Press.

Lakoff, George and Rafael E. Núñez. 2000. *Where Mathematics Comes from: How the Embodied Mind Brings Mathematics into Being*. New York, NY: Basic.

Laudan, Larry. 1977. *Progress and Its Problems: Towards a Theory of Scientific Growth*. Berkeley, CA: University of California Press.

Lemarchand, Guillermo A. and Jon Lomberg. 2011. "Communication Among Interstellar Intelligent Species: A Search for Universal Cognitive Maps." In *Communication with Extraterrestrial Intelligence (CETI)*, edited by Douglas A. Vakoch, 371–395. Albany, NY: SUNY Press.

Morrison, Philip. 1963. "Interstellar Communication." In *Interstellar Communication: A Collection of Reprints and Original Contributions*, edited by A. G. W. Cameron, 249–271. New York, NY: W. A. Benjamin.

Narens, Louis. 1997. "Surmising Cognitive Universals for Extraterrestrial Intelligences." In *Astronomical and Biochemical Origins and the Search for Life in the Universe*, edited by Cristiano B. Cosmovici, Stuart C. Bowyer, and Dan Werthimer, 561–570. Bologna: Editrice Compositori.

Rescher, Nicholas. 1985. "Extraterrestrial Science." In *Extraterrestrials: Science and Alien Intelligence*, edited by Edward Regis Jr., 83–116. Cambridge: Cambridge University Press.

Rotman, Brian. 1993. *Ad Infinitum . . . the Ghost in Turing's Machine: Taking God out of Mathematics and Putting the Body Back in: An Essay in Corporeal Semiotics and Theological Presuppositions*. Stanford, CA: Stanford University Press.

Sefler, George F. 1982. "Alternative Linguistic Frameworks: Communications with Extraterrestrial Beings." In *Philosophers Look at Science Fiction*, edited by Nicholas D. Smith, 67–74, Chicago, IL: Nelson-Hall.

Vakoch, Douglas A. 1978. "Possible Pictorial Messages for Communication with Extraterrestrial Intelligence." *Journal of the Minnesota Academy of Science* 44:23–25.

Vakoch, Douglas A. 1998. "Constructing Messages to Extraterrestrials: An Exosemiotic Perspective." *Acta Astronautica* 42:697–704.

Vakoch, Douglas A. 2000. "The Conventionality of Pictorial Representation in Interstellar Messages." *Acta Astronautica* 46:733–736.

Vakoch, Douglas A. 2011. "A Narratological Approach to Interpreting and Designing Interstellar Messages." *Acta Astronautica* 68:520–534.

Part III Philosophical, theological, and moral impact

How do we comprehend the cultural challenges raised by discovery?

Introduction

After framing the problems of discovery and impact, and attempting to transcend our anthropocentric conceptions of life in the last two sections, we now turn to the possible implications of discovering life beyond Earth. As the reader may gather from what has already been said, great caution and humility are in order here, as we seek to push the frontiers of knowledge using the tools of the social sciences, humanities, philosophy, and theology, not to mention cognitive science and evolutionary science as well.

NASA engineer, biologist, and philosopher Mark Lupisella begins this section with a very broad look at the philosophical implications of life in the universe, emphasizing a peculiar activity of sufficiently self-aware beings: the pursuit of value. The awareness of self and of other minds leads to theories of mind, he argues, as well as the practical pursuit of what is valuable, as embodied in the fields of ethics and aesthetics. Both biological evolution and cosmic evolution affect our pursuit of values – and presumably the pursuit of values by other beings in the universe, if indeed they pursue value at all (a fundamental question fraught with implications for our interactions with them). These factors are important for preparing for contact, and Lupisella elaborates ten practical considerations from this point of view.

In the next two chapters two philosophers ponder the seemingly inscrutable problem of alien minds. Philosopher of science Michael Ruse approaches the problem from the viewpoint of the history and philosophy of evolutionary biology, his specialty over a career spanning 50 years. Through an analysis of the classic science fiction movie *The Day the Earth Stood Still* (1951), he examines possible alien outlooks on mathematics and logic, science, morality, and religion, concluding that the differences in all those areas compared to ours could be very considerable. This conclusion resonates with those of the previous section urging us to transcend anthropocentrism in our most fundamental concepts. Philosopher of mind Susan Schneider takes a very different approach, illustrating how various disciplines can view the same problem from distinct points of view, with divergent outcomes. Taking theories of consciousness and Nick Bostrom's recent book *Superintelligence* (Bostrom, 2014) as her starting point, Schneider argues that alien minds may well not be biological, but postbiological, what she terms superintelligent artificial intelligence. While the SETI community tends to focus on ideas of intelligence, the problem of consciousness is more basic and perhaps more relevant. As Berkeley philosopher John Searle quipped during a conversation at the Library of Congress "Forget intelligence, the origin and nature of consciousness is the real question." Searle (1980) has famously written against "strong AI," the idea that artificial intelligence could *in principle* ever be constructed to be more intelligent than humans. He is thus not impressed with ideas of a "postbiological universe," where biological intelligence is a passing phase and extraterrestrials are actually postbiological artificial intelligence (Dick 2003). In her chapter Schneider (a former student of Searle) nevertheless pursues this idea, in the process giving us a taste of the problems of intelligence, superintelligence, and consciousness from the point of view of the rich field of philosophy of mind. She concludes that postbiologicals, at least the subset that is "biologically inspired," might well have conscious experiences that we can understand.

From the nature of consciousness and intelligence, philosopher Carol Cleland and political scientist Elspeth Wilson move on to ethical questions we face in our interactions with alien organisms, a field theologian Ted Peters has dubbed "astroethics" (Peters 2013). They do so using four human-based ethical categories: rationality, sentience, complex social behavior, and theological concepts of the soul. They argue that secular ethics represented by the first three categories is likely more universally applicable than theological ethics, since it does not require belief in the same religious worldview, a consistency that does not exist on Earth, much less on other worlds. They urge astrobiologists and policy makers to consider that any one of these categories might qualify an organism to have intrinsic moral status, and that our interactions with alien life should be guided accordingly.

Section III ends with two chapters on theological aspects, both of which may be considered part and parcel of another up-and-coming branch of astrobiology, "astrotheology" (Dick 2000; Peters 2014; Weintraub 2014; Wilkinson 2013). Robin Lovin, a theologian and Director of Research at the Center of Theological Inquiry in Princeton, examines general issues beginning with an expansion of the concept of theology to an extraterrestrial context. He especially raises a variety of issues stemming from the concept of human dignity, examining whether such a concept might exist with other intelligent life, and our ability to recognize it if it does. He concludes that astrobiological anticipations of theological concepts such as human dignity may not only help us in the event of contact (including our moral obligation to other ecospheres), but may also further our development of thinking on human dignity even if contact is never made. Where Lovin takes human dignity as his starting point, Jesuit Brother Guy Consolmagno takes another specific question within Christianity as a starting point for a variety of theological issues, issues that have been raised in the context of alien science fiction ranging from James Blish's *A Case of Conscience* (1958) to Mary Doria Russell's *The Sparrow* (1996), both involving Jesuit contact with alien intelligence. "Would you baptize an extraterrestrial?" is a "what if" question that guarantees controversy. But, as Consolmagno concludes, it also sheds light on our place in the universe, and "our relationship with whatever we identify as the source of truth, goodness, and love."

Clearly these two essays focus on Western religious traditions. Very little work has been done on non-Western views (Irudayadason 2013), and the need for such work is embarrassingly evident. And what is true of theology in this respect is also true of all other societal issues discussed in this volume, as anthropologist John Traphagan reminded us in Part II – and again in our final section.

References

Bostrom, N. 2014. *Superintelligence: Paths, Dangers, Strategies*. Oxford: Oxford University Press.

Dick, Steven J. 2000. *Many Worlds: The New Universe, Extraterrestrial Life and the Theological Implications*. Philadelphia, PN: Templeton Press.

Dick, Steven J. 2003. "Cultural evolution, the postbiological universe and SETI," *International Journal of Astrobiology*, 2:65–74; reprinted as "Bringing Culture to Cosmos: the Postbiological Universe," in Steven J. Dick and Mark Lupisella, *Cosmos & Culture: Cultural Evolution in a Cosmic Context*, Washington DC: NASA, pp. 463–488. Available online at http://history.nasa.gov/SP-4802.pdf.

Irudayadason, Nishant A. 2013. "The wonder called cosmic oneness: toward astroethics for Hindu and Buddhist wisdom and worldviews," in Impey, Chris, Anna H. Spitz, and William Stoeger, eds. *Encountering Life in the Universe: Ethical Foundations and Social Implications of Astrobiology*. Tucson, AZ: University of Arizona Press, pp. 94–119.

Peters, Ted. 2013. "Astroethics: engaging extraterrestrial intelligent life-forms," in Impey, Chris, Anna H. Spitz, and William Stoeger, eds. *Encountering Life in the Universe: Ethical Foundations and Social Implications of Astrobiology*. Tucson, AZ: University of Arizona Press, pp. 200–221.

Peters, Ted. 2014. "Astrotheology: a constructive proposal." *Zygon*, 49, 443–457.

Searle, John R. 1980. "Minds, brains, and programs." *Behavioral and Brain Sciences* 3, no. 3: 417–457.

Weintraub, David A. 2014. *Religions and Extraterrestrial Life: How Will We Deal With It?* Heidelberg: Springer.

Wilkinson, David. 2013. *Science, Religion and the Search for Extraterrestrial Intelligence*. Oxford: Oxford University Press.

10 Life, intelligence, and the pursuit of value in cosmic evolution

MARK LUPISELLA

As life and intelligence continue to evolve in the universe, it is reasonable to wonder where it all may lead, and what, if anything, it may mean. Such contemplations have consumed, befuddled, and perhaps even harmed humanity since we were first able to entertain such matters. From philosophy to science, from religion to spirituality, we have wondered about some larger meaning or purpose, some possible direction for life and intelligence. This is almost always discussed in the context of a broader "objective" external reality to help provide a compelling frame of reference for explaining and navigating the bewildering complexities of human life. Contemplating potential implications or roles for life and intelligence in cosmic evolution should be seen in this larger experiential, emotional, and intellectual context, recognizing that discerning patterns, trends, and theories in cultural change is a tricky business (Denning 2009).

Cosmic contemplations for life and intelligence are subject to wide-ranging speculation and perhaps a certain level of hubris, and hence are justifiably susceptible to deep skepticism and harsh criticism. But in trying to explore more thoroughly the potential futures of humanity and other beings in the universe, it seems reasonable to consider at least a few lines of thinking that place future evolutionary trajectories of life and intelligence in a long-term cosmic context. Ancient religions as well as more contemporary religous and spiritual movements have much to say about life and intelligence in a universal context. Science also has much to say about the past evolution of life and intelligence in a cosmic context, and perhaps even a bit to say about the long-term future. Careful philosophical considerations can leverage the science we understand today, and the best of philosophical thinking, to explore potential sources of meaning and purpose, and articulate some rough themes for the future of life and intelligence in the universe.

This breadth of thought that places life and intelligence in a cosmic context ranges from western and eastern religious worldviews such as Hinduism, to natural theology and Western philosophers such as Plato, Giordano Bruno, and Baruch Spinoza. Alfred North Whitehead's process philosophy (1929),

Teilhard de Chardin's Omega Point Theory (1955), and Arthur O. Lovejoy's treatment of the principle of plenitude (1936) are all part of this tradition. It also encompasses more contemporary scientific treatments that contemplate the challenges and implications of the very long-term survival of intelligence (e.g. Kardashev 1964, Dyson 1979, Tipler 1994, De Duve 1995, Chaisson 2005, Davies 2009, Deutsch 2011, Dick 1996 and 2003, Smart 2009, Vidal 2014), as well as the endlessly rich explorations of much science fiction ranging from Olaf Stapledon (1937) to Arthur C. Clarke. This chapter will focus primarily on broader conceptual and philosophical implications of life and intelligence in the overall evolution of the universe, with an emphasis on the potential implications of the pursuit of values, or what can be called the "normative aspirations" of sufficiently self-aware beings. This will then inform some pragmatic implications relevant to preparing for the discovery of extraterrestrial life and intelligence.

Life and intelligence

For the purposes of this chapter, there is essentially only one important thing to note about biological systems: life can be seen primarily as a blind and voraciously single-minded replicative pursuit of genetic fitness (Dawkins 1989). While biological systems have evolved to produce controlled behavior, it is nevertheless fundamental to natural selection that it "finds" any means possible to enhance genetic propagation.

Darwinian evolution balances harmful actions that can compromise genetic propagation into the future (e.g. producing a certain amount of social cooperation), but the process is crude and unreliable and, most importantly, it is in the service of seeking genetic replication into the next generation only. This suggests that highly capable life forms can arise, both unintelligent and intelligent, that have very little, if any awareness, and yet can be remarkably successful and powerful in terms of survival and reproductive fitness. This further points to the possibility that there may be highly successful beings throughout the universe that could exhibit blindly extreme behaviors by human standards.

Intelligence can also be blindly capable. Intelligence seems to have evolved primarily, if not solely, as a mechanism to enhance genetic fitness, perhaps as a result of complex social living or other conditions that require significant information-processing capabilities (Deacon 1997, Skoyles and Sagan 2002). But again, it may be worth emphasizing that intelligence could be nothing more than a high degree of "competence without understanding." The philosopher Daniel Dennett has suggested that we educate children so they can understand in order to be competent. I would like to suggest that the reverse is also true from a Darwinian evolution perspective. That is, because of a certain level of competence already programmed into us by evolution to help us survive and reproduce, certain animals also have a significant capacity to understand. And how humans understand has much to do with culture.

Culture, value, and normative aspiration

Culture

Much, if not most culture is also arguably shaped by natural selection. Evolutionary psychology, with its important connections to group selection and social psychology, has much to say about human psychological predispositions, including predispositions to seek broader contexts within which to understand and cope with human life. Based on the premises of evolutionary psychology and group selection, culture, like intelligence, has arguably evolved primarily, if not solely, as an adaptive mechanism to enhance the fitness of replicators (including possibly "memes") without necessarily requiring a high degree of awareness (Blackmore 2009, Dennett 2009). However, more modern culture (say within the last 10,000 years since around the beginning of effective agriculture) appears to have produced a notable level of awareness, first as an awareness of one's physical environment and even our origins, but perhaps more importantly, as an increasing awareness of other minds.

This awareness of self and other minds seems almost certainly to imply at least some form of internal modeling of one's self in a larger context, requiring theories of mind to allow for a more effective awareness of others and hence more effective interactions with others – leading ultimately to more stable and effective forms of group living. This more immediate practical awareness might lend itself to an increasing deepening of self awareness, for example leading to a cognizance of one's own death and the death of others, perhaps then leading more generally to an increasing cognizance of alternative possible future states for oneself and one's group. The combination of both immediate and longer-term future-oriented awareness, along with the ability to contemplate and communicate with others about those future states, has increased our awareness at many levels (including an awareness of our awareness!), and may be part of what some think of as "consciousness."

The ability to simulate and explore the future with others who may share that future has led to an explicit awareness of the utility of proactively seeking and creating favorable future states. This explicit awareness would presumably benefit greatly from many cultural constructs and strategies to help realize those states – including systems of intellectual thought that help optimize the pursuit of favorable future states for both individuals and groups, which can be thought of in general as philosophical systems of thought, with theories of value being a key subset.

Value and normative aspiration

Human culture has led to broad and deep levels of philosophical inquiry, a major area of which is value theory. In its simplest form, value theory carefully and rationally explores what is valuable and why, and includes, for example, ethics and esthetics. "Normative aspiration" can be thought of as the proactive pursuit of what is valuable or ideal based on a consideration of alternative future states, from which something is consciously chosen as a norm to which to aspire. Like intelligence and culture, normative aspiration is also highly influenced by Darwinian evolution. If reason is highly influenced by human emotions, then our evolutionary emotional genetic predispositions significantly drive our reasoning about what is valuable – and such predispositions are presumably focused on what is valuable based on feelings that were selected for by evolution to enhance genetic fitness.

In contemporary human life, however, some normative aspiration increasingly appears to be more than just aspiring to what our emotionally constrained reasoning leads us to. Philosophers and others from a variety of disciplines have conducted numerous forays into the realm of "theories of value," many of which seem to involve notable attempts at rationality, fairness, and objectivity to assess what is valuable and how we might go about achieving it (e.g. Kant 1785, Rawls 1971) – despite legitimate concerns of "rational delusion" (Haidt 2012). There is no shortage of ideas in value theory, many often completely contradictory – sometimes creating the misguided impression that values are so subjective and arbitrary that they are not amenable to rational analysis (Putnam 2002). Indeed, how human beings value has much, if not everything, to do with individual internal psychological states, and scientists continue to learn much about those states and their relation to evolution and the external world, to the point where there is increasing hope for a more empirical scientific grounding of moral psychology, ethics, and value theory in general (Shermer 2004, Harris 2010). And there are potentially other relevant scientific endeavors that can be brought to bear on our philosophical normative aspirations, some of which will be touched on next.

Philosophical implications

So how do we cope with this very large, if not infinite landscape of normative aspiration that may exist for many different kinds of beings? Can we, and how might we rationally navigate it? There are two areas of scientific knowledge that could have very profound influences on normative aspiration and hence on the future of life and intelligence in the universe. The first is biological evolution by natural selection. The second is cosmic evolution.

Normative aspiration in light of natural selection

The explanation of biological systems evolving by natural selection, in its simplest "minimalist" form, tells us that the evolution of life and intelligence have been governed essentially by what can be thought of as "filtered replication." Filtered replication can be seen as a very general process by which entities that are copied either continue to be copied or stop being copied based on factors that prevent their continued replication – i.e. factors that block and prevent (i.e. "filter") further replication. If filtered replication is a reasonable way to think about biological evolution, then that may help point to the possibility that sufficiently aware beings – for example, beings that are aware of their evolution being shaped by natural selection – may eventually view such a process as inadequate or "undesirable" in at least two ways.

One way that filtered replication may be seen as inadequate is as a design process in general. While natural selection has clearly produced a wide variety of "successful" forms of life and ultimately led to human intelligence and human culture, it also took a very long time and has arguably been blind, harmful, and cruel by many standards (perhaps much or most of that harm and cruelty being exhibited by human beings). Also, from a basic design perspective, there is much about biological systems today that appear to be bad designs that humanity has spent much time fixing – as the history of human health shows (including mental health which has many ties to evolutionary psychology).[1] This is a value assessment about the overall process being inadequate, even as a design process for just the narrow goal of replication.

Secondly, and perhaps more importantly, a species aware of the primary motivation of natural selection to serve the sole purpose of making copies and enhance replicative effectiveness of molecular replicators, may choose to question that goal, especially as a motivator for intelligent beings. This is a value assessment about ends, about values. So natural selection may be seen by other beings as a simple, crude, and narrow single-minded process with little ultimate value – a process that happens as a matter of physical logic and nothing more.

[1] However, filtered replication can be a powerful tool for finding effective solutions depending on the fitness functions used, the kind of solution that is sought, and the kinds of tools performing the computation (Hillis 1991, Lupisella 2004).

The simplicity and results of natural selection make it profoundly intellectually appealing. It is arguably the most important idea in human history (Dennett 1995) for a number of reasons – including reasons related to normative aspiration being explored here. But the philosophical and normative implications may be very unappealing to many intelligent beings. On both means and ends, evolution by natural selection may be subject to harsh scrutiny by sufficiently aware beings, perhaps to the point of trying explicitly to transcend or overturn it – as much human culture has arguably attempted. However, it may also be that intelligent beings will decide not to devalue natural selection and the powerful engine of self-interest that results. Such a choice may depend critically on the psychology and rational explorations of such beings – much of which human cultural evolution continues to experiment with, intentionally and unintentionally, successfully and unsuccessfully (Sowell 2007, Haidt 2012, Cockell 2015).

Normative aspiration in light of natural selection may be a key turning point for intelligent beings. It may lead to strong intentional pursuits to move beyond much, if not all of the programming of biological evolution. Machine intelligence may help overcome many of the physical frailties that face biological beings, but moving beyond the self-interest of Darwinian evolution implies resisting the most fundamental motivations of biological organisms. This may end up simply being an intellectual choice for sufficiently advanced beings who may have much control over what can be physically realized. This is a choice about what is valued and why. At the very least, normative aspiration in light of natural selection may be seen as a potential source of intense intellectual exploration, experimentation, and preoccupation of sufficiently intelligent beings throughout the universe.

It may be that some extraterrestrial beings, or "unitary" solitary beings, did not evolve via natural selection and without social selection pressures (if, for example, a designer was involved or if there are other natural mechanisms by which a single being might exist), in which case, normative aspiration in light of natural selection would not be relevant. But even in scenarios of this kind, normative aspiration could still ultimately be a preoccupation of such beings, and perhaps more importantly could be an important part of a general shared framework to help us understand and interact with other beings, including a very foreign unitary solitary being (Lupisella 2013).

A unique challenge regarding normative aspiration is that the possibility space for ideal preferred future states is arguably infinite (Deutsch 2011). Navigating such a space, and doing so with other potential extraterrestrial beings, especially those that did not evolve via natural selection, raises profound challenges about how and what beings value and what they aspire to. In one extreme, it may turn out that highly capable life forms without any form of normative aspiration, or very different forms of normative aspiration, could be very dangerous (Michaud 2007).

Normative aspiration in light of cosmic evolution

Modern cosmology has accumulated a substantial body of evidence and compelling theoretical frameworks to suggest our universe has physically evolved over something like 13.7 billion years. The universe has expanded and cooled, forming structures such as galaxies, stars, and planets, on which at least one has biological systems that have led to highly aware beings with knowledge of cosmic evolution. Our high level of cosmological awareness can be part of our normative aspiration, and already may be part of the normative aspiration of other intelligent beings. The universe may be seen as at least one commonly shared framework of beings throughout the universe.

Beings might also choose to explicitly incorporate the universe as a whole into their worldviews, and hence into their normative aspiration, because they may see the physical universe as the ultimate totality of reality, an ultimate entity of sorts, consistent with some forms of pantheism or perhaps forms of "cosmotheology" (Dick 2000). Not only might they see the universe this way, but they may see it as an unfolding story in which they and many kinds of beings may play roles or have some kind of significance that relates to the universe as a whole. If other beings come to understand the sort of universe we see today as an evolving cosmos, recognizing that they evolved from it in a deep physical sense, they may see the universe as a "creator" of sorts, or at least a source, or ultimate source, of creative activity and possibly ethics (Lupisella and Logsdon 1997; Vidal 2014).

Going further, extraterrestrial beings may see themselves as a means by which the universe has been bootstrapped into the realm of value, culture, meaning, and purpose via beings that bring those properties into the universe, giving rise to a kind of "bootstrapped cosmocultural evolution" (Lupisella 2009a). Sufficiently aware beings may bring to the universe not only normative aspiration, but in that pursuit of normative aspiration may also explicitly choose to value the universe itself. Through their perceived roles as creators and arbiters of value, acting as a kind of cerebral cortex for the universe, they may see themselves in some sense effectively making the whole of the universe valuable merely by developing these kinds of worldviews. So intelligent beings may not just be a way for the universe to know itself, but also to value itself. This is a different kind of anthropic principle: the universe is valuable because we are here to value it.

Are these speculations just another example in a long list of attempts to find meaning and purpose where there is none? Perhaps the motivation is similar, but it appears we can indeed say that value does at least exist in the universe in the form of our valuing minds. We have then literally brought value into the universe. There may of course be other forms of value independent of human beings and other valuing agents (e.g. Rolston 1990, Lupisella and Lodgson 1997, Lupisella 2009b)[2], but that is not necessarily inconsistent with suggesting we are a source of value as well.

An apparent lack of objective meaning and purpose of the universe may also prompt intelligent beings to choose to see themselves in some sense at odds with a hostile universe and perhaps at odds with all other creatures that are potential competitors. Indeed, this appears to be a legitimate scientific and practical way to see our biological condition in a cosmic context since most of space appears to be inhospitable to our form of biological life, and further, the second law of thermodynamics appears to be generally working against biological systems. Much of what we do, even within the "comfort" of our terrestrial cocoon, is targeted at controlling and changing our environment to make it more suitable to our biological and psychological needs. And to date, we have no evidence of intelligent life elsewhere, which can be interpreted (however prematurely) as suggesting that our origins may have been essentially random (Ward and Brownlee 2000).

Implications for preparing for discovery

In an attempt to apply the philosophical considerations of the previous sections, this section suggests a series of potential near- and long-term implications that might be considered when thinking about how to guide future research and perhaps even help to inform policy measures that relate to preparing for the discovery of extraterrestrial life and intelligence. The considerations are not mutually exclusive, and in some cases they are opposing implications. The suggestions are meant only to explore a series of considerations that may help in preparing for a discovery where the details of putative beings and the state of humanity are highly uncertain.

Cosmic humility. An essential concern when engaged in this kind of cosmic speculation about beings for which we have no data, particularly speculation that involves values and normative aspiration, is that we should exercise a high degree of humility and open-mindedness. We simply don't know the

[2] See Smith (2009) for an assessment of challenges associated with intrinsic value and astrobiology.

modalities of other beings. They may exist in a very large possibility space of values that are extremely foreign and potentially incomprehensible to us. Our knowledge of the universe is probably very limited and so we should consider that our understanding of natural selection and cosmic evolution, while important to the considerations in this chapter, may be very incomplete and even irrelevant to other beings.

Rational analysis. Nevertheless, having noted the need for humility, our knowledge of natural selection, cultural diversity, and cosmic evolution may make contemplation of the motivations and values of other intelligent beings reasonable and worthwhile (Cohen and Stewart 2002). It is almost certainly helpful to think carefully and diversely about potential long-term futures for humanity and life on Earth, which can also inform how other putative intelligent beings may exist and behave throughout the universe.

Message interpretation and active SETI. Among the more likely scenarios to consider is that a signal will be received from a distance great enough that there will not be any immediate interactive consequences. Nevertheless, understanding the true meaning of a message could be critically informed by matters of normative aspiration and cosmic perspectives. Humanity may also seek out extraterrestrial intelligence by proactively sending signals (often called Active SETI). A more thorough exploration of the implications of normative aspiration in light of natural selection and cosmic evolution may inform if, how, and when we might wish to engage in Active SETI, and perhaps more importantly, what we might wish to say (Vakoch 2009).

Extraterrestrial normative expectations. If normative aspiration is a preoccupation of other beings in the universe, and particularly if normative aspiration turns out to be an important dynamic in cosmic evolution overall, then intelligent beings throughout the universe may at least desire or expect continually improving normative aspiration from all species. Other intelligences may go further and expect a certain specific level of normative aspiration to be reached before civilizations are allowed to be part of a cosmic club. If other beings aspire to see themselves in a cosmic context, and to act consistent with such a worldview, it may be something for humanity to be sensitive to when considering what other beings might be like and what they might desire. As noted earlier, such worldviews may also be worth pursuing as a way to have a possible common ground of value with other beings.

Related to extraterrestrial expectations, considerations of a cosmocultural evolutionary worldview, one in which cosmic creations have a certain value and may deserve a certain level of ethical commitment from normatively aspiring species, may inform our ethical posture toward non-intelligent life and non-living objects more generally, either here on Earth or in off-Earth environments (Rolston 1990, McKay 1990). Perhaps some degree of ethical commitment to non-intelligent life forms such as putative microbes on other planets might be part of this consideration and might also be part of the extraterrestrial normative expectations noted above (Lupisella and Logsdon 1997).[3]

Machine intelligence. In a postbiological universe where intelligent beings are machines, we might aspire not just for artificial intelligence, but "artificial morality." We may want to go well beyond something like Asimov's laws of robotics to help protect humans and other non-machine beings (see Gardner 2009), but to also contemplate a much broader set of moral considerations for other beings and perhaps the universe as a whole. Indeed, it may very well be easier for machines in a postbiological universe to be more ethically committed entities because they may not have the deep molecular motivations of self-replication. For example, they may be programmed, or self-programmed, for complete fairness. They may decide on their own that what they aspire to is true fairness and equity – and it may turn out not to be too difficult as the cost of caring goes down (due to technical advancements and social constructs) causing the overall "caring capacity" of advanced societies to go up (Lupisella 2013).

Normative relativity and normative certitude. The history of human values and human ethical aspiration can certainly be interpreted to suggest we cannot rely on objective answers to guide our normative aspirations or those of other beings, perhaps suggesting extreme humility in our exploration of what is valuable and how to pursue it. Considering that much of human history has been ruled by ethical certitude, often grounded in religious beliefs, a potentially helpful thought experiment may be to consider scenarios where other intelligent beings claim to have certitude about very diverse sets of norms and value theory more generally. How, or would, we be able to cope with this kind of extreme diverse certainty or incomprehensible "dogma"?

Selfishness trap. We may want to allow for the possibility that self-interest is so deeply baked into self-replicating systems that it ultimately cannot be modified or moderated in any significant way. Although attempts to control self-interest are fundamental to human social life, such attempts, especially in the extreme, have also had unfortunate consequences throughout human history. Some may say such attempts are fundamentally untenable and even immoral. Such tensions are at the root of fundamental ideological debates that

[3] For a very recent treatment of a general theory of extraterrestrial ethical obligations, see Smith (2014).

are manifested in seemingly endless domestic political disputes and international relations.

Selfishness may essentially keep beings in something like a "selfishness trap" that most or all beings ultimately do not change. The very nature of selfishness may make it impossible for selfish beings to significantly moderate, modify, or give up selfishness even if it were relatively easy and many were willing to attempt it (Lupisella 2001). It may be futile because transcending selfishness is an "all-or-nothing" requirement in the sense that all beings must fully participate in truly transcending their self-interest or else "cheaters" will destabilize and undermine whatever fairness and stability exists. It may also be that because self-interest can reap so many benefits for so many individuals and societies that intelligent beings consciously choose to at least allow, if not encourage, high degrees of self-interested behavior. Recognizing that self-replicating entities may forever be bound to self-interest can significantly inform how we think about ourselves and extraterrestrials – and if and how there might be interactions between us.

Astro-cosmo engineering. Given the possibility that advanced intelligences may have the ability to modify large-scale structures in the universe, ranging from stars and solar systems to galaxies and black holes, it may be useful to consider how certain kinds of normative aspiration might result in certain kinds of astrophysical and even cosmological modifications. If advanced beings are able to create black holes, might they use them as energy sources or to create universes – as some physicists have begun to contemplate (Guth 1994)? Should we take seriously the idea that we might be living in a fabricated universe and that universes might be regularly created merely as part of a creative endeavor by an intelligent entity or multiple entities? Might the cosmological structures and mysteries we see today (e.g. dark matter and dark energy) be a result of intentional intelligent modification for reasons that have more to do with what powerful beings aspire to create and aspire to be, and not just a result what they are merely capable of?

Diversity. Much of the above considerations relate to how we and other intelligent beings view diversity, how it is valued, and how beings cope with it. Evolution has produced a remarkable diversity of life and mind in what appears to be a highly open-ended manner, suggesting the need for much more comparative psychology, including non-human animals. Values of beings throughout the universe could be unimaginably diverse suggesting the need for careful contemplation and many thought experiments, as well as a high degree of intellectual humility and respect. Preparing to deal with such an enormously large possibility space of diverse values might be important to try, but it might also turn out to be futile. However, there is a large amount of diversity right here on Earth, including non-human animals that exhibit remarkable behaviors. Understanding them at much deeper levels could be very informative to preparing for a discovery of other forms of life and intelligence in the universe.

Better understanding human diversity and human rationality and irrationality, partly informed by evolutionary psychology, social psychology, cognitive science, and neuroscience, may help us prepare ourselves for dealing with the discovery of extraterrestrial intelligence and may also help us better understand potential psychologies of non-human intelligences (see Chapter 6). Cautiously applied, analogs in the social and behavioral sciences may be just as useful as in the natural sciences (see Chapter 3). It is hard to know in advance, but given what is at stake it seems reasonable to continue to explore carefully the diverse prospects for extraterrestrial life and intelligence in the cosmos.

The post-intelligent universe: a moral universe. Steven Dick (2003) has suggested the "Intelligence Principle" as a central force of cultural evolution, implying a constant pursuit of knowledge to advance the well-being of societies. It may be that at some point, knowledge, at least as we think of it today (e.g. facts about the world), will reach a point of diminishing returns. It may be that advanced cultures become so capable that knowledge will become much less important. What may matter most is what advanced beings value because what they value will likely be the realities they create. Highly capable and aware beings may come to see intelligence merely as a means for many ends – first as ends pursued by selfish replicators that they may wish to appreciate but also recognize as too limited, and then as increasingly unconstrained ends emerging from diverse desires and infinite imaginations of valuing agents. Such beings may desire and pursue something more akin to a morally creative cosmos, a universe in which the central pursuits are not necessarily intelligence, knowledge, and technology, but fairness, caring, and diversity.

In this "post-Intelligent" universe a guiding principle may be something more like a "Values Principle" or "Wisdom Principle" where what also matters, what may increasingly matter most, is what beings value and why, not just what they are capable of. This pursuit of intelligence and knowledge, along with normative aspiration, can be a complementary dance and lead ultimately to what we hope would be forms of wisdom. This fact–value interplay represents a key human endeavor in most human societies, and is arguably a critical pursuit for the human species, particularly as we become a truly globally integrated species. Careful steps in what may be a kind of cosmic waltz between facts and values can help us prepare for a potential future discovery of extraterrestrial life and intelligence.

References

Blackmore, S. 2009. "Dangerous Memes; or What the Pandorans Let Loose." In S. J. Dick and M. Lupisella (eds.) *Cosmos and Culture: Cultural Evolution in a Cosmic Context*. Washington, DC: NASA, pp. 297–318.

Chaisson, E. 2005. *Epic of Evolution: Seven Ages of the Cosmos*. New York, NY: Columbia University Press.

Cockell, C., ed. 2015. *The Meaning of Liberty Beyond Earth*. New York, NY: Springer.

Cohen, J. and I. Stewart. 2002. *Evolving the Alien*. London: Ebury Press/ Random House.

Crutzen, P. J. 2002. "Geology of Mankind." *Nature*, 415, p. 23.

Davies, P. 2009. "Life, Mind, and Culture as Fundamental Properties of the Universe." In S. J. Dick and M. Lupisella (eds.) *Cosmos and Culture: Cultural Evolution in a Cosmic Context*. Washington, DC: NASA, pp. 383–397.

Dawkins, R. 1989. *The Selfish Gene*, 2nd edn. Oxford: Oxford University Press.

Deacon, T. W. 1997. *The Symbolic Species: The Co-Evolution of Language and the Brain*. New York, NY: W. W. Norton.

De Duve, C. 1995. *Vital Dust: Life as a Cosmic Imperative*. New York, NY: Basic Books.

Dennett, D. 1995. *Darwin's Dangerous Idea*. London: Penguin.

Dennett, D. 2009. "The Evolution of Culture." In S. J. Dick and M. Lupisella (eds.) *Cosmos and Culture: Cultural Evolution in a Cosmic Context*. Washington, DC: NASA, pp. 125–143.

Denning, K. 2009. "Social Evolution." In S. J. Dick and M. Lupisella (eds.) *Cosmos and Culture: Cultural Evolution in a Cosmic Context*. Washington, DC: NASA History Series, pp. 63–124.

Deutsch, D. 2011. *The Beginning of Infinity: Explanations that Transform the World*. New York, NY: Viking.

Dick, S. J. 1996. *The Biological Universe: The Twentieth Century Extraterrestrial Life Debate and the Limits of Science*. Cambridge: Cambridge University Press.

Dick, S. J. 2000. "Cosmotheology: Theological Implications of the New Universe." In S. J. Dick (ed.) *Many Worlds: The New Universe, Extraterrestrial Life, and the Theological Implications*. Philadelphia, PN: Templeton Foundation Press, pp. 191–210.

Dick, S. J. 2003. "Cultural Evolution, the Postbiological Universe and SETI," *International Journal of Astrobiology*, 2: 65–74, reprinted as "Bringing

Culture to Cosmos: The Postbiological Universe," in S. J. Dick and M. Lupisella (eds.) *Cosmos and Culture: Cultural Evolution in a Cosmic Context*. Washington, DC: NASA, pp. 463–487.

Dyson, F. J. 1979. "Time Without End: Physics and Biology in an Open Universe." *Reviews of Modern Physics*, 51, No. 3.

Gardner, J. 2009. "The Intelligent Universe." In S. J. Dick and M. Lupisella (eds.) *Cosmos and Culture: Cultural Evolution in a Cosmic Context*. Washington, DC: NASA History Series.

Guth, A. 1994. "Do the Laws of Physics Allow Us to Create a New Universe?" In G. Ekspong (ed.) *The Oskar Klein Memorial Lectures*. Stockholm: World Scientific Publishing, vol. 2, pp. 71–95.

Haidt, J. 2012. *The Righteous Mind*. New York, NY: Random.

Harris, S. 2010. *The Moral Landscape: How Science Can Determine Human Values*. New York, NY: Free Press.

Hillis, D. 1991. "Co-evolving Parasites Improve Simulated Evolution as an Optimization Procedure." In C. Langton, C. Taylor, J. Farmer, and S. Rasmussen (eds.) *Artificial Life II*, (Santa Fe Institute Studies in the Sciences of Complexity, Proceedings, **10**), Redwood City, CA: Addison-Wesley, pp. 313–324.

Kant, I. 1785. *Groundwork of the Metaphysics of Morals*. Trans. M. Gregor, 2003. Cambridge: Cambridge University Press.

Kardashev, Nikolai. 1964. "Transmission of Information by Extraterrestrial Civilizations." *Soviet Astronomy (PDF)* 8: 217.

Lovejoy, Arthur Oncken. 1971 [1936]. *The Great Chain of Being*. Cambridge, MA: Harvard University Press.

Lupisella, M. 2001. Participant statement in *Humanity 3000*, Seminar No. 3 Proceedings, 37. Bellevue, WA: Foundation for the Future, pp. 37–38 and also discussed on p. 251 and p. 330. http://www.futurefoundation.org/documents/hum_pro_sem3.pdf.

Lupisella, Mark. 2004. "Using Artificial Life to Assess the Typicality of Terrestrial Life." *Advances in Space Research* 33:1318–1324.

Lupisella, Mark. 2009a. "Cosmocultural Evolution: The Coevolution of Cosmos and Culture and the Creation of Cosmic Value." In S. J. Dick and M. L. Lupisella (eds.) *Cosmos and Culture: Cultural Evolution in a Cosmic Context*, NASA, pp. 321–359.

Lupisella, M. 2009b. "The Search for Extraterrestrial Life: Epistemology, Ethics, and Worldviews." In C. Bertka (ed.) *Exploring the Origin, Extent, and Future of Life: Philosophical, Ethical and Theological Perspectives*. Cambridge: Cambridge University Press, pp. 186–204.

Lupisella, Mark. 2013. "Caring Capacity and Cosmocultural Evolution: Potential Mechanisms for Advanced Altruism." In D. A. Vakoch (ed.) *Extraterrestrial Altruism*. Berlin: Springer-Verlag.

Lupisella, M. and J. Logsdon. 1997. "Do We Need a Cosmocentric Ethic?" Paper IAA-97-IAA.9.2.09, presented at the 48th International Astronautical Congress, Turin, Italy, October 6–10.

McKay, C. 1990. "Does Mars Have Rights?" In D. MacNiven (ed.) *Moral Expertise*. London: Routledge.

Michaud, M. A. G. 2007. *Contact with Alien Civilizations: Our Hopes and Fears about Encountering Extraterrestrials*. New York, NY: Springer.

Putnam, H. 2002. *The Collapse of the Fact/Value Dichotomy and Other Essays*. Cambridge, MA: Harvard University Press.

Rawls, J. 1971. *A Theory of Justice*. Cambridge, MA: Harvard University Press.

Rolston, H. 1990. "The Preservation of Natural Value in the Solar System." In E. C. Hargrove (ed.) *Beyond Spaceship Earth: Environmental Ethics and the Solar System*. San Francisco, CA: Sierra Club Books.

Shermer, M. 2004. *The Science of Good and Evil*. New York, NY: Times Books/ Henry Holt & Company.

Skoyles, J. R. and D. Sagan. 2002. *Up From Dragons: The Evolution of Human Intelligence*. New York, NY: McGraw Hill.

Smart, J. 2009. "Evo Devo Universe? A Framework for Speculations on Cosmic Culture." In S. J. Dick and M. Lupisella (eds.) *Cosmos and Culture: Cultural Evolution in a Cosmic Context*. Washington, DC: NASA, pp. 201–295.

Smith, K. C. 2009. "The Trouble with Intrinsic Value: An Ethical Primer for Astrobiology." In C. Bertka (ed.) *Exploring the Origin, Extent, and Future of Life: Philosophical, Ethical, and Theological Perspectives*. Cambridge: Cambridge University Press, pp. 261–280.

Smith, K. C. 2014. "Manifest Complexity: A Foundational Ethic for Astrobiology?" *Space Policy*, 30: 209–214.

Sowell, T. 2007. *A Conflict of Visions: Ideological Origins of Political Struggles*. New York, NY: Basic Books.

Stapledon, O. 1937. *Star Maker*. London: Methuen Publishing.

Teilhard De Chardin, P. 1955. *The Phenomenon of Man*. Originally published as *Le Phénomène Humain*. Editions du Seuil, Paris. Translated by B. Wall. New York, NY: Harper & Row, 1959.

Tipler, F. 1994. *The Physics of Immortality*. New York, NY: Doubleday.

Vakoch, D. 2009. "Encoding Our Origins: Communicating the Evolutionary Epic in Interstellar Messages." In S. J. Dick and M. L. Lupisella (eds.)

Cosmos and Culture: Cultural Evolution in a Cosmic Context. Washington, DC: NASA, pp. 415–439.

Vidal, C. 2014. *The Beginning and the End: The Meaning of Life in a Cosmological Perspective*. New York: Springer.

Ward, P. D. and D. Brownlee. 2000. *Rare Earth: Why Complex Life is Uncommon in the Universe*. New York, NY: Copernicus Books.

Whitehead, A. N. 1929. *Process and Reality: An Essay in Cosmology*. New York: Macmillan. Edition 1978 by D. R. Griffin and D. W. Sherbourne, New York: Macmillan.

11 "Klaatu Barada Nikto" – or, do they really think like us?

MICHAEL RUSE

One of my all-time favorite movies is the 1951 science fiction thriller, *The Day the Earth Stood Still*. It tells the story of an alien (Klaatu) who comes to planet Earth to say that the galaxy is pretty upset with us and fears that, with our new nuclear weapons, we might not just blow ourselves to smithereens but inflict significant damage on others. It turns out that the rest of the universe has put itself under the power of robots who enforce peace and quiet and if we do not mend our ways these robots will un-mend us once and for all. To make the point, Klaatu has brought one of the robots (Gort) along with him, and when as inevitably happens we humans fail to take proper heed and end up killing Klaatu, Gort sets out intending death and destruction. The carnage is prevented only because a young war widow (Helen Benson), who has befriended Klaatu and who has been told what to do in an emergency, manages in time to turn off Gort with the crucial words "Klaatu Barada Nikto." I am sure I was not the only eleven-year-old who spent the next year uttering those words whenever I got in a jam. Somehow they never seemed to quite work with my schoolmasters.

Intelligent beings

Now, my point is that – with an interesting exception that I will mention shortly – Klaatu appears as a normal human being. Played by Michael Rennie without any special makeup, he rents a room in a boarding house in Washington DC, and causes no special attention when he appears at the breakfast table with the other guests. He goes off around the city with Helen's son Bobby and again there is nothing strange, although despite speaking English perfectly he does show ignorance of our ways – at the Arlington Cemetery grave of Bobby's father he fails to understand the point of violence and later naively swaps some precious diamonds for a few dollars. Physically, Klaatu is like a member of *Homo sapiens* and intellectually too. It is true that he is very, very bright, but not in a weird way. When he meets the physicist Professor Barnhardt – modeled on Albert Einstein – the two are clearly in the same intellectual ballpark. The theme of the movie is science and (for once in American life) religion doesn't come into it much. When Gort recovers Klaatu's body he is able to revive him without any hocus pocus involving prayer or holy water. Klaatu tells Helen that his race has learnt how to do this although ultimately in the end he does rather spoil things by adding that "power is reserved for the Almighty Spirit." Apparently this was added at the express command of the censors appointed by the Motion Picture Association of America who felt that, with his resurrection, Klaatu was getting a little bit too close to home (Blaustein *et al.* 1995).

Klaatu is human, or at least human-like. The exception I mentioned is interesting and in a way reinforces this point. Klaatu doesn't seem to have much by way of emotion or indeed have a sense of right and wrong. It isn't that he doesn't behave well but rather that he has reasoned that this is the sensible thing to do and does it. He doesn't do it because of the call of the Categorical Imperative or the Love Commandment or any such thing. Even his relationship with Bobby is somewhat exploitive as he uses Bobby to explain to him the ways of humans – witness the reaction at the grave when he does not mourn with Bobby the loss of his father but grapples with the problem of why humans would behave in such counter-intuitively stupid ways. Klaatu is in this respect non-human; he is instead super-human. It is rather as if he has taken our powers and multiplied them, taken then to perfection. He is not non-human in the sense that he is eating feces or self-mutilating or anything like that. Even his religion, if such there be, is rational. One could not imagine Klaatu speaking in tongues.

The moral I want to draw from this is simple. We naturally tend to think of other advanced beings as being more or less like us. And that goes for their robots too! Of course, especially given the variations in human skin color and other distinctive features, no one thinks that such beings would have to be exactly like the folk we meet in the street, or in the boarding house! I guess they could have green skin and six fingers. One presumes that they might not have two eyes above a nose above a mouth, containing thirty-two teeth. They could be brighter or duller. But they would be distinctively human-like. At least, as far as their thinking apparatus is concerned, they tend to be distinctively human-like. Is this assumption in any sense plausible, or are we just living in film-maker's cloud-cuckoo land? To be fair to film-makers, in the years since *The Day the Earth Stood Still* they have tended to be more inventive about advanced beings – think of the *Star Wars* series for example – but still the thinking apparatus is human-like. Jabba the Hutt is creepy and scary precisely because we know exactly what he is thinking. So the question stands. Will advanced beings, assuming they exist or could exist, anywhere, any time in the universe, think like us? Or will they be completely or partially different in major respects?

Must evolution lead to intelligence?

You might think that modern science – and by this I mean the modern science of organic origins meaning Darwinian evolutionary theory – points firmly away from similarities. Assuming that life could and did start and evolve elsewhere in the universe – and, given the vastness of the universe and the speed at which life began on our planet as soon as it was possible, I am inclined to think that there is life elsewhere – there is absolutely no reason to think that it is going to be anything like our life. Natural selection is opportunistic and the result is a random pattern of evolutionary paths, taking advantage of what comes up rather than following prescribed patterns. There is certainly no reason to think that there will be advanced life akin to ours. Apart from anything else, consciousness apparently requires big brains and big brains are expensive to maintain. Their owners need to ingest large chunks of protein, meaning the parts of other organisms, and these are not always available and if available not necessarily easy to get. In the immortal words of the paleontologist Jack Sepkoski: "I see intelligence as just one of a variety of adaptations among tetrapods for survival. Running fast in a herd while being as dumb as shit, I think, is a very good adaptation for survival" (Ruse 1996, 486). So it seems that there may well be no intelligent life and even if there is an alternative it will be very different from anything we know.

Interestingly, however, there are a number of suggested ways around this skeptical conclusion. They may not be true but at least they are plausible. To mention two, the first focuses on what biologists call "arms races." Groups of organisms compete against competitors and in so doing, as in military arms races, each side improves its adaptations until in the end intelligence emerges. Building on ideas found in the writings of Darwin himself, strongly articulated by Julian Huxley (1912), Richard Dawkins has promoted this line of thought. "Directionalist common sense surely wins on the very long time scale: once there was only blue-green slime and now there are sharp-eyed metazoa" (Dawkins and Krebs 1979, 508). This is due to biological arms races. Note, however, that today's arms races are ever increasingly electronic as both sides build bigger and better computers. The same happens in the animal world. Referring to something called the encephalization quotient (EQ), a kind of cross-species IQ equivalent, Dawkins writes: "The fact that humans have an EQ of 7 and hippos an EQ of 0.3 may not literally mean that humans are 23 times as clever as hippos! But the EQ as measured is probably telling us *something* about how much 'computing power' an animal probably has in its head, over and above the irreducible amount of computing power needed for the routine running of its large or small body" (Dawkins 1986, 189). In other words, intelligence wins out. Humans have won and generally one would expect that human-like beings will always or usually win.

The second counter to randomness was floated by Stephen Jay Gould (1985) but recently has been strongly promoted by his fellow paleontologist Simon Conway Morris. This approach supposes that there are ecological niches that organisms seek out and enter. Water, land, air are the obvious big ones. The existence of more refined and limited niches is shown by the way in which two completely different lines of saber-toothed tiger – one marsupial and one placental – evolved independently and sought out the same niche. It is argued that culture in some sense exists as a niche, and had we humans not entered it, some organisms some time somewhere would have entered it. So intelligence was more or less bound to evolve. Conway Morris writes (Conway Morris 2003, 196):

> If brains can get big independently and provide a neural machine capable of handling a highly complex environment, then perhaps there are other parallels, other convergences that drive some groups towards complexity. Could the story of sensory perception be one clue that, given time, evolution will inevitably lead not only to the emergence of such properties as intelligence, but also to other complexities, such as, say, agriculture and culture, that we tend to regard as the prerogative of the human? We may be unique, but paradoxically those properties that define our uniqueness can still be inherent in the evolutionary process. In other words, if we humans had not evolved then something more-or-less identical would have emerged sooner or later.

I am not endorsing either of these approaches, but let us agree that they are at least plausible. If life exists elsewhere, then sometimes – seldom? often? always? – intelligent life would have evolved. So, this lines me up for the big question I want to ask. Marsupial saber-toothed tigers are not identical to placental saber-toothed tigers. There are differences and in respects the differences are large and significant. So ask about the evolution of intelligence in beings who are not humans, that is to say beings who are not us. How would that intelligence have been? Would it have been like ours? Or would it have been something, somewhat, or strikingly different? Let's break this basic question into four sub-questions. Would such beings be aware of and consciously use our mathematics and logic? Would such beings have our grasp of science? Would such beings be moral as we understand morality? And finally, what about religion? Would such beings be religious? Note that I am not asking whether, as we understand them, such beings would use logic or science or whatever. I presume that if they are intelligent beings and are functioning then, from our perspective, they would have to use them. They couldn't ignore mathematics or kill each other as soon as they met. Note also that, whatever they think about science, I am assuming that science as we know it would hold for them. For instance, already I have presumed that if they evolved then basically they evolved through natural selection.

My question is: "What do these beings think they are up to?" Let us take the four sub-questions in turn.

Mathematics and logic

To be honest, I am not quite sure how one would answer this except positively in the sense of agreeing that their mathematics and logic would be like ours. They simply couldn't believe that 2 + 2 = 5 and function, nor could they think that the internal angles of a triangle add up to three right angles and get very far in their lives. You see a pair of lions go into a cave, and soon thereafter one lion comes out and you enter the cave thinking it safe. That is not the route to reproductive success. And things only get worse from there. How could you have any advanced technology without at least some grasp of calculus? The fantasy novelist J. R. R. Tolkien was no lover of science and technology and his stories reflect this (Bud 2013). You don't take an automobile to go from one end of the Shire (the home of the hobbits) to the other. But especially given that they were such agriculturalists, the hobbits had to have some understanding of geometry – measuring out their fields, drawing boundaries, and so forth. When we look at places like ancient Egypt, we know that such knowledge is just vital.

This said, it seems to me that there is still scope for significant differences, especially if you are not a mathematical Platonist thinking that the truths of mathematics exist objectively in some world of pure rationality, and if to the contrary you think that in some sense mathematics is a construction, with a human (or in our case, an intelligent being) input. Most obviously in arithmetic there is little absolute pressure to count by tens. The Babylonians did not. Their basic unit was sixty, a much more sensible choice when you think of how many numbers can be divided into it, as opposed to ten. One certainly thinks that if we had six fingers on each hand we would count by twelves. Or perhaps, as with computers, we would use a binary system. Of course, you might argue that basically this is all surface. The mathematics is the same however you dress it up. Perhaps this is true, although it doesn't seem to me to be totally insignificant. You can translate from English into French, and by and large the meanings are the same. But there is always something lost in translation.

More important it seems to me is the fact that your mathematics (and your logic too) might be significantly affected by your physical nature and surroundings, the thesis known as "embodied cognition" (Lakoff and Núñez, 2000). *The Day the Earth Stood Still* doesn't tackle this issue, because Klaatu is so like humans here on Earth. Precisely because they share the same mathematics, he is able to help Professor Barnhardt who is wrestling with the three-body problem. But what if there are differences? Take geometry. It is obviously a very visual subject, so much so that in the *Republic* Plato rather sniffily does not give it the status of the highest form of knowledge precisely because geometers thought in terms of pictures and diagrams. Suppose your beings were blind. Would they develop geometry as we do? One presumes that they would have to have some way of communicating and getting around, but perhaps they do this as do the insects, communicating through chemicals – pheromones. You might object that given how often sight has evolved independently on Earth, it is improbable that such beings would be blind. Not necessarily so! If their world was dark, they would have no need of sight. You might object that without light there could be no energy and hence no growth to support life, but perhaps it comes from underground heat or perhaps they are troglodytes living always underground (because of predators or chemical poisons or whatever). I am not saying that such beings would have no geometry but that they would not develop it as we do – perhaps it would be all analytic.

Conversely if such beings were aquatic, I could imagine that parts of their mathematics would be developed in ways that we do not have or need. And fantasizing further, if such beings lived in caves, would they necessarily have the sense of vastness that we have as we gaze at the heavens. Would they develop number theory in the way that we have done, with all of the stuff about different orders of infinity? It is true that I have to admit that ultimately I don't see alien mathematics and logic as in some sense inherently different from what we have, and I do think that such beings would have to have some awareness of logic and mathematics if they are to flourish as intelligent beings. But I do think that the forms that their logic and mathematics would have could be in some senses very different to ours – to Anglophones, more like Finnish, let us say (to take a particularly different language), than French. Finns do think like the rest of us, but they think in a very odd language and I suspect that their knowledge – let us say about survival in the cold and dark – is often very different from the knowledge of others – let us say Africans living in very hot and barren lands.

Science

In trying to speculate profitably on extraterrestrial thinking about science, one has two extreme boundaries. At the one end, stands the philosopher of science Sir Karl Popper who insisted that science is (or should be) objective, describing a real world, and thus the same for everyone. In his felicitous phrase: "Science is knowledge without a knower" (Popper 1972). Thus for him, science is universal, right through the universe. Boyle's law is Boyle's law is Boyle's law. At the other end stands the sociologist of science Harry Collins who claimed provocatively that we create science rather than discover it and that hence: "The natural world has a small or non-existent role in the construction of scientific knowledge" (Collins 1981, 3). Thus, for him, you are lucky to get the same science in two different departments. Across the universe, who knows what people believe?

We can say one thing. Popper may not be all right, but he is surely somewhat right. Planets go in ellipses not triangles and one presumes that this holds universally. Newton's laws hold up across space and time. I won't get into multiverses here where it is sometimes suggested that anything could happen. As best we know, the laws hold in our realm of existence. But would our intelligent beings recognize Newton's and Kepler's laws? I take it that they are not going to recognize something else, an inverse cube law of attraction for instance. But would they care about what we care about and/or would they care about other things? As with mathematics, my suspicion is that the differences might be bigger than we expect – and more so than mathematics, I expect. Certainly if our beings were cave dwellers or lived in darkness, or had no other planets to look at, I cannot think that the equivalent of Kepler's laws would rate that highly. Again, if they were blind, I doubt optics would rate high on the list. On the other hand, if they did work through pheromones or if they were essentially aquatic, parts of their science might be much more advanced than ours. But there is another way in which our interests would steer our science, and although sociologists of mathematics sometimes suggest that this is true of mathematics too, I am inclined to think it is more commonly true of science. This is that culture makes us focus on some issues rather than others. Take evolutionary theory. No one could be surer than I that Darwinian evolutionary theory truly explains organic origins. But what about the very quest for an answer to origins? The Greeks did not have such an interest. It is with the coming of Christianity and the Jewish story of origins that it became pressing in the West, and there is little doubt but that Darwinism was intended to speak in a secular way to this issue (Ruse 2005). I could imagine a world in which biologists, heaven forbid, just were not that interested in origins. Of course, the fact of the matter is that a lot of biologists today are not interested in origins. They study their molecules and their tissues and their anatomies with nary a thought for the past. This is why so many medics can quite happily be evangelical Christians and deniers of evolution. They already have their origins story.

My point is simply that culture, including religion, is a major determinant in the kind of science that we produce. There is more to the story than this, and perhaps here we do start to leave logic and mathematics behind. The actual content of science is influenced and formed by our culture. This happens particularly through the scientists' very heavy reliance on metaphor (Ruse 1999, 2010). Take again Darwinian theory. It is full of metaphors – struggle for existence, natural selection, design, division of labor, arms races, genetic landscapes, and so forth. All of these are rooted in culture, particularly in the culture of industrialized Britain of the eighteenth and nineteenth centuries (Richards and Ruse 2015). Consider the question of design – that the eye and the hand seem as if designed. This is an absolutely central premise or starting point for Darwinism – adaptations are as if designed (Ruse 2003). And obviously it is the problem that natural selection is intended to solve. Organic features seem as if designed not because they are designed but because those with such features do better in the struggle for survival and reproduction than those that are lacking such features. You cannot get away from the metaphor of design, nor indeed would you want to. It is valuable heuristically leading you to new problems and solutions. Why does the stegosaurus have those strange diagonal plates running down its back? In order to facilitate temperature control in the essentially cold-blooded animal. Because of the plates, it heats more quickly in the morning sun and then uses the plates to get rid of heat in the mid-day sun.

This focus on design is part of culture and goes back in the Darwinian case to the British obsession with natural theology and the teleological argument (the argument from design). This was itself a function of the Elizabethan compromise of the sixteenth century when it was found that natural theology was a safe way forward between the Scylla of the Catholic emphasis on the Church and tradition and the Charybdis of the Calvinist emphasis on *sola scriptura* and rather joyless living. Darwin and his followers take design seriously not because they are necessarily believers in the deity but because those who set the terms of debate did believe and used design to support their belief. This being so, it is hardly a great surprise to find that while there is much to commend it, not everyone thinks it necessary. There has always been a form of Romantic biology, going back to Goethe and the *Naturphilosophen*, that has downplayed design and stressed structure – the isomorphisms between animals, what are known as "homologies" (Russell 1916; Richards 2003). The most recent major enthusiast for this kind of thinking was Stephen Jay Gould who argued that design is way overplayed and that homology was a much more important aspect of the biological world – form over function, to use a phrase (Gould and Lewontin 1979).

This is not just a matter of culture, or rather of two cultures clashing. To use another metaphor, in science the proof of the pudding is predictive fertility and the like. If focusing on function does better than focusing on form then so be it. But anyone who knows about science knows full well that deciding which of two approaches is better able to predict can be very difficult. And even then things are not necessarily definitive. The Newtonians pointed to the predictive successes of their theorizing but the Cartesians responded that the Newtonians embraced the metaphysically unacceptable idea of action at a distance. These things can drag on and on. Pertinently, when looking at the science of non-humans we might find that their cultures are important for their science no less than our cultures are for us, and that hence there are very different sciences, and determining which is right and which is wrong might be no small matter. The science off our planet might be very different from the science on our planet (Rescher 1985).

Morality

I take it that the reason for morality – a sense of right and wrong, and the feeling that we should obey this sense – is pretty straightforward (Ruse 1986; Ruse and Richards, in press). Humans are social beings, we have to get on with each other, and so morality comes into play. There is nothing very mysterious about this and certainly from a biological perspective it makes good sense. Biologists from Darwin on have recognized that although nature may be red in tooth and claw, "altruism" as they call it can be a very good biological strategy. I help others; they help me. Thus we all profit. Could one have isolated intelligent beings with no need of social skills? I suppose so, but it seems unlikely. So much of intelligence evolved precisely because we need social skills. Could one have a being like Klaatu with no morals at all but just the drive of self-interest and with great intelligence to see that goals are attained? I suppose it is possible, but I think it unlikely. In biology, time is money. Less metaphorically, we need to make decisions fairly quickly and thus we need rules with moral force – that guide our conduct and help us to negotiate human relationships. We cannot necessarily wait for the best solution. We need rules that usually work even if sometimes they let you down. It is like computerized chess playing. The computer cannot think of every possible move. That would take too long. It needs to work with certain strategies that usually mean success, even though it is now no longer infallible and can be beaten. The question is whether these moral rules have to be the same for all advanced beings. I suppose that if the beings were more or less like us, then the morality would be more or less like ours. We are not going to function very well if we have a morality that tells us to kill left-handed people on Fridays. Nor is there much point in such a rule. Apart from anything else, when rules fall into disrepute, they tend to break down and be unenforceable. Think of Prohibition. But what if the beings are not like us? Darwin put matters starkly (Darwin 1871, 1, 73).

> It may be well first to premise that I do not wish to maintain that any strictly social animal, if its intellectual faculties were to become as active and as highly developed as in man, would acquire exactly the same moral sense as ours. In the same manner as various animals have some sense of beauty, though they admire widely different objects, so they might have a sense of right and wrong, though led by it to follow widely different lines of conduct. If, for instance, to take an extreme case, men were reared under precisely the same conditions as hive-bees, there can hardly be a doubt that our unmarried females would, like the worker-bees, think it a sacred duty to kill their brothers, and mothers would strive to kill their fertile daughters; and no one would think of interfering. Nevertheless the bee, or any other social animal, would in our supposed case gain, as it appears to me, some feeling of right and wrong, or a conscience.

The example I think of focuses on sex. Agree that for advanced beings you probably need sex in order to catch good new genes and spread them through the population. We could be hermaphrodites, which would in itself demand a revision of our view about proper behavior. I guess worries about gay marriage would be otiose. But what if it were more extreme? Darwin (1851, 1854) speaks of barnacles where the females are normal but the males are just degenerate parasites, little more than bags of sperm with whopping great penises that attach themselves (often many at a time) to females. How would morality function then with one half of the species brainless and simply going through the motions? This would surely pose a challenge to James Dobson's organization "Focus on the Family." I am not quite sure how you have a "covenant" with a warty thing attached to your backside, especially if there are half a dozen of them.

The serious point is that Thomas Aquinas was right when in the *Summa Theologiae* he pushed a neo-Aristotelian "natural law" theory of morality, arguing that what is right is what is natural, and what is natural is a function of the way that we are. It is just that he never thought of the possibility of non-human (material) advanced beings. When one does, one starts to see that their morality, admittedly with the same ends as ours of furthering survival and reproduction, might be very different from ours (Vakoch 2014).

Religion

Finally, what about religious belief? Some, Richard Dawkins (2006) most obviously, would say that it is quite possible to have advanced beings with no religious beliefs whatsoever. He would say that the more advanced they are, the more likely they are not to have religious beliefs. Others, not necessarily believers, think that religion is a function of evolution, and that those humans who responded positively to religion were more likely to survive and reproduce. This is the position of the non-believing evolutionist Edward O. Wilson. Like the great sociologist Emile Durkheim, he thinks that religion has a great bonding effect and that is why religion exists and why it will persist, no matter what the arguments against it. "In the midst of the chaotic and potentially disorienting experiences each person undergoes daily, religion classifies him, provides him with unquestioned membership in a group claiming great powers, and by this means gives him a driving purpose in life compatible with his self interest" (Wilson 1978, 188). Wilson does allow that culture may have a causal contribution to make, but essentially he thinks that it all comes back to biology. "Because religious practices are remote from the genes during the development of individual human beings, they may vary widely during cultural development. It is even possible for groups, such as the Shakers, to adopt conventions that reduce genetic fitness for as long as one or a few generations. But over many generations, the underlying genes will pay for their permissiveness by declining in the population as a whole" (Wilson 1978, 178).

Of course, there are those who would put religion down exclusively to culture and then there are those who would argue that, whatever the causes, the real point is the reason for belief (Ruse 2015). There is one true objective religion and we should believe in it. Those of us who fail to do so, fail because of some personal deficiency like original sin. This is the position of the Calvinist philosopher Alvin Plantinga (2000), for example. However, whatever the truth of the matter, there are many, many religions here on Earth and so surely we might expect the same elsewhere in the universe. I doubt we need the Motion Picture Association of America to insist on this point, although whether they would be monotheistic as the Association implies is another matter. One might wonder how weird and strange these religions might be, but on a planet that houses the Mormons and the Scientologists such wonder will surely never reach the reality. What religion would be like on other planets, God or gods, only know!

Conclusion

The Day the Earth Stood Still is a terrific movie that stands up well over time. Admittedly the effects are rather primitive, but where else do you find a space ship designed by Frank Lloyd Wright? However, when it comes to off-planet intelligent beings and their ways of thinking, it is, let us say rather, conservative. As the eminent population geneticist J. B. S. Haldane used to say: "My own suspicion is that the universe is not only queerer than we suppose, but queerer than we *can* suppose" (Haldane 1927, 286). Amen.

Acknowledgments

This may seem like gross flattery, but let me simply say that I could not have written this paper without the inspiration of Steven Dick's *The Biological Universe: The Twentieth-Century Extraterrestrial Life Debate and the Limits of Science*. Whether he would agree with what I have written is another matter.

References

Blaustein, J., R. Wise, P. Neal, and B. Gray. 1995. *Making the Earth Stand Still.* Fox Video; 20th Century Fox Home Entertainment.
Bud, R. 2013. "Life, DNA and the Model." *British Journal for the History of Science* 46: 311–334.
Collins, H. M. 1981. "Stages in the Empirical Program of Relativism – Introduction." *Social Studies of Science* 11: 3–10.
Conway Morris, S. 2003. *Life's Solution: Inevitable Humans in a Lonely Universe.* Cambridge: Cambridge University Press.
Darwin, C. 1851. *A Monograph of the Sub-Class Cirripedia, with Figures of all the Species. The Lepadidae; or Pedunculated Cirripedes.* London: Ray Society.
Darwin, C. 1854. *A Monograph of the Sub-Class Cirripedia, with Figures of all the Species. The Balanidge (or Sessile Cirripedes); the Verrucidae, and C.* London: Ray Society.
Darwin, C. 1871. *The Descent of Man, and Selection in Relation to Sex.* London: John Murray.
Dawkins, R. 1986. *The Blind Watchmaker.* New York, NY: Norton.
Dawkins, R. 2006. *The God Delusion.* New York, NY: Houghton Mifflin.
Dawkins, R. and J. R. Krebs. 1979. "Arms Races Between and Within Species." *Proceedings of the Royal Society of London, Series B* 205: 489–511.

Dick, S. J. 1996. *The Biological Universe: The Twentieth-Century Extraterrestrial Life Debate and the Limits of Science.* Cambridge: Cambridge University Press.

Gould, S. J. 1985. *The Flamingo's Smile: Reflections in Natural History.* New York, NY: Norton.

Gould, S. J. and R. C. Lewontin. 1979. "The Spandrels of San Marco and the Panglossian Paradigm: A Critique of the Adaptationist Programme." *Proceedings of the Royal Society of London, Series B* 205: 581–598.

Haldane, J. B. S. 1927. *Possible Worlds and Other Essays.* London: Chatto and Windus.

Huxley, J. S. 1912. *The Individual in the Animal Kingdom.* Cambridge: Cambridge University Press.

Lakoff, G. and R. Núñez. 2000. *Where Mathematics Comes From.* New York, NY: Basic Books.

Plantinga, A. 2000. "Pluralism: A Defense of Religious Exclusivism." In K. Meeker, and P. Quinn (eds.)*The Philosophical Challenge of Religious Diversity.* New York, NY: Oxford University Press, pp.172–192.

Popper, K. R. 1972. *Objective Knowledge.* Oxford: Oxford University Press.

Rescher, N. 1985. "Extraterrestrial Science." In E. Regis, *Science and Alien Intelligence.* Cambridge: Cambridge University Press, pp. 84–116.

Richards, R. J. 2003. *The Romantic Conception of Life: Science and Philosophy in the Age of Goethe.* Chicago, IL: University of Chicago Press.

Richards, R. J. and M. Ruse. 2015. *Debating Darwin: Mechanist or Romantic?* Chicago, IL: University of Chicago Press.

Ruse, M. 1986. *Taking Darwin Seriously: A Naturalistic Approach to Philosophy.* Oxford: Blackwell.

Ruse, M. 1996. *Monad to Man: The Concept of Progress in Evolutionary Biology.* Cambridge, MA: Harvard University Press.

Ruse, M. 1999. *Mystery of Mysteries: Is Evolution a Social Construction?* Cambridge, MA: Harvard University Press.

Ruse, M. 2003. *Darwin and Design: Does Evolution have a Purpose?* Cambridge, MA: Harvard University Press.

Ruse, M. 2005. *The Evolution–Creation Struggle.* Cambridge, MA: Harvard University Press.

Ruse, M. 2010. *Science and Spirituality: Making Room for Faith in the Age of Science.* Cambridge: Cambridge University Press.

Ruse, M. 2015. *Atheism: What Everyone Needs to Know.* Oxford: Oxford University Press.

Ruse, M. and R. J. Richards, eds. in press. *The Cambridge Handbook of Evolutionary Ethics.* Cambridge: Cambridge University Press.

Russell, E. S. 1916. *Form and Function: A Contribution to the History of Animal Morphology*. London: John Murray.

Vakoch. D. A. 2014. *Extraterrestrial Altruism: Evolution and Ethics in the Cosmos*. Berlin: Springer.

Wilson, E. O. 1978. *On Human Nature*. Cambridge, MA: Harvard University Press.

12 Alien minds

SUSAN SCHNEIDER

How would intelligent aliens think? Would they have conscious experiences? Would it feel a certain way to be an alien? It is easy to dismiss these questions as too speculative, since we haven't encountered aliens, at least as far as we know. And in conceiving of alien minds we do so from *within* – from inside the vantage point of the sensory experiences and thinking patterns characteristic of our species. At best, we anthropomorphize; at worst, we risk stupendous failures of the imagination.

Still, ignoring these questions could be a grave mistake. Some proponents of the search for extraterrestrial intelligence (SETI) estimate that we will encounter alien intelligence within the next several decades. Even if you hold a more conservative estimate – say, that the chance of encountering alien intelligence in the next 50 years is 5 percent – the stakes for our species are high. Knowing that we are not alone in the universe would be a profound realization, and contact with an alien civilization could produce amazing technological innovations and cultural insights. It thus can be valuable to consider these questions, albeit with the goal of introducing possible routes to answering them, rather than producing definitive answers. So, let us ask: how might aliens think? And, would they be conscious? Believe it or not, we can say something concrete in response to both of these questions, drawing from work in philosophy and cognitive science.

You might think the second question is odd. After all, if aliens have sophisticated enough mental lives to be intelligent, wouldn't they be conscious? The far more intriguing question is: what would the quality of their consciousness be like? This would be putting the cart before the horse, however, since I do not believe that most advanced alien civilizations will be biological. The most sophisticated civilizations will be postbiological, forms of artificial intelligence (AI). (Cirkovic and Bradbury 2006; Shostak 2009; Davies 2010, 153–168; Bradbury *et al.* 2011; Dick 2013).[1] Further, alien

[1] "Postbiological," in the astrobiology literature contrasts with "posthuman" in the singularity literature. In the astrobiology literature "postbiological" creatures are forms of AI. In the singularity literature "posthumans" can be forms of AI, but they need not be. They are merely creatures who are descended from humans but which have alterations that make them no longer unambiguously human. They need not be full-fledged AI.

civilizations will tend to be forms of *superintelligence*: intelligence that is able to exceed the best human-level intelligence in every field – social skills, general wisdom, scientific creativity, and so on (Kurzweil 2005, Schneider 2011a, Bostrom 2014). It is a substantive question whether superintelligent AI (SAI) could have conscious experiences; philosophers have vigorously debated just this question in the case of AI in general. Perhaps all their information processing happens in the dark, so to speak, without any inner experience at all. This is why I find the second question so pressing, and in an important sense prior to any inquiry as to the contours of alien consciousness, and prior to the epistemological problem of how we can know "what it is like" to be an alien.

In this chapter I first explain why it is likely that the alien civilizations we encounter will be forms of SAI. I then turn to the question of whether superintelligent aliens can be conscious – whether it feels a certain way to be an alien, despite their non-biological nature. Here, I draw from the literature in philosophy of AI, and urge that although we cannot be *certain* that superintelligent aliens can be conscious, it is likely that they would be. I then turn to the difficult question of how such creatures might think. I provisionally attempt to identify some goals and cognitive capacities likely to be possessed by superintelligent beings. I discuss Nick Bostrom's recent book on superintelligence, which focuses on the genesis of SAI on Earth; as it happens, many of Bostrom's observations are informative in the present context. Finally, I isolate a specific type of superintelligence that is of particular import in the context of alien superintelligence, biologically inspired superintelligences ("BISAs").

Alien superintelligence

SETI programs have been searching for biological life. Our culture has long depicted aliens as humanoid creatures with small, pointy chins, massive eyes, and large heads, apparently to house brains that are larger than ours. Paradigmatically, they are "little green men." While we are aware that our culture is anthropomorphizing, I imagine that my suggestion that aliens are supercomputers may strike you as far-fetched. So what is my rationale for the view that most intelligent alien civilizations will have members that are forms of SAI? I offer three observations that, together, motivate this conclusion.

(1) The short window observation. Once a society creates the technology that could put them in touch with the cosmos, they are only a few hundred years away from changing their own paradigm from biology to AI. (Shostak 2009; Davies 2010, 153–168; Dick 2013). This "short window" makes it more likely that the aliens we encounter would be postbiological.

The short-window observation is supported by human cultural evolution, at least thus far. Our first radio signals date back only about 120 years, and space exploration is only about 50 years old, but we are already immersed in digital technology, such as cell-phones and laptop computers. Devices such as the Google Glass promise to bring the Internet into more direct contact with our bodies, and it is probably a matter of less than 50 years before sophisticated internet connections are wired directly into our brains. Indeed, implants for Parkinson's are already in use, and in the United States the Defense Advanced Research Projects Agency (DARPA) has started to develop neural implants that interface directly with the nervous system, regulating conditions such as post-traumatic stress disorder, arthritis, depression, and Crohn's disease. DARPA's program, called "ElectRx," aims to replace certain medications with "closed-loop" neural implants, implants that continually assess the state of one's health, and provide the necessary nerve stimulation to keep one's biological systems functioning properly (Guerini 2014). Eventually, implants will be developed to enhance normal brain functioning, rather than for medical purposes.

Where might all this lead? A thought experiment from my "Transcending and Enhancing the Human Brain" is suggestive (Schneider 2011a).

> Suppose it is 2025 and being a technophile, you purchase brain enhancements as they become readily available. First, you add a mobile internet connection to your retina, then, you enhance your working memory by adding neural circuitry. You are now officially a cyborg. Now skip ahead to 2040. Through nanotechnological therapies and enhancements you are able to extend your lifespan, and as the years progress, you continue to accumulate more far-reaching enhancements. By 2060, after several small but cumulatively profound alterations, you are a "posthuman." To quote philosopher Nick Bostrom, posthumans are possible future beings, "whose basic capacities so radically exceed those of present humans as to be no longer unambiguously human by our current standards" (Bostrom 2003).
>
> At this point, your intelligence is enhanced not just in terms of speed of mental processing; you are now able to make rich connections that you were not able to make before. Unenhanced humans, or "naturals," seem to you to be intellectually disabled – you have little in common with them – but as a transhumanist, you are supportive of their right to not enhance (Bostrom 2003; Garreau 2005; Kurzweil 2005).
>
> It is now AD 2400. For years, worldwide technological developments, including your own enhancements, have been facilitated by superintelligent AI. . . . Indeed, as Bostrom explains, "creating superintelligence may be the last invention that humans will ever need to make, since superintelligences could themselves take care of further

scientific and technological developments" (Bostrom *et al.* 2003). Over time, the slow addition of better and better neural circuitry has left no real intellectual difference in kind between you and superintelligent AI. The only real difference between you and an AI creature of standard design is one of origin – you were once a natural. But you are now almost entirely engineered by technology – you are perhaps more aptly characterized as a member of a rather heterogeneous class of AI life forms (Kurzweil 2005).

Of course, this is just a thought experiment. But I've just observed that we are already beginning to develop neural implants. It is hard to imagine people in mainstream society resisting opportunities for superior health, intelligence, and efficiency. And just as people have already turned to cryonics, even in its embryonic state, I suspect that they will increasingly try to upload to avoid death, especially as the technology is perfected.[2] Indeed, the Future of Humanity Institute at Oxford University (Sandberg and Boström 2008) has released a report on the technological requirements for uploading a mind to a machine. And a Defense Department agency has funded a program, *Synapse*, which is developing a computer that resembles a brain in form and function (Schneider 2014). In essence, the short-window observation is supported by our own cultural evolution, at least thus far.

You may object that this argument employs "$N = 1$ reasoning," generalizing from the human case to the case of alien civilizations (see Chapter 7 in this volume). Still, it is unwise to discount arguments based on the human case. Human civilization is the only one we know of and we had better learn from it. It is no great leap to claim that other civilizations will develop technologies to advance their intelligence and survival. And, as I will explain in a moment, silicon is a better medium for thinking than carbon.

A second objection to my short-window observation rightly points out that nothing I have said thus far suggests that humans will be *superintelligent*. I have merely said that future humans will be *posthuman*. While I offer support for the view that our own cultural evolution suggests that humans will be postbiological, this does not show that advanced alien civilizations will reach superintelligence. So even if one is comfortable reasoning from the human case, the human case does not support the position that the members of advanced alien civilizations will be superintelligent.

This is correct. This is the task of the second observation.

(2) The greater age of alien civilizations. Proponents of SETI have often concluded that alien civilizations would be much older than our own: "... all lines of evidence converge on the conclusion that the maximum age of

[2] Although I have elsewhere argued that uploading would merely create a copy of one's brain configuration and would not be a true means of survival, I doubt dying individuals will act on a philosopher's qualms when they have little to lose by trying (Schneider 2014).

extraterrestrial intelligence would be billions of years, specifically [it] ranges from 1.7 billion to 8 billion years" (Dick 2013, 468). If civilizations are millions or billions of years older than us, many would be vastly more intelligent than we are. By our standards, many would be superintelligent. We are galactic babies.

But would they be forms of AI, as well as forms of superintelligence? I believe so. Even if they were biological, merely having biological brain enhancements, their superintelligence would be reached by artificial means, and we could regard them as being forms of "artificial intelligence." But I suspect something stronger than this, which leads me to my third observation:

(3) It is likely that these synthetic beings will not be carbon-based, as silicon is a better medium for intelligence. I expect that they will not be carbon-based. Uploading allows a creature near-immortality, enables reboots, and allows it to survive under a variety of conditions that carbon-based life forms cannot. In addition, silicon appears to be a better medium for information processing than the brain itself. Neurons reach a peak speed of about 200 Hz, which is seven orders of magnitude slower than current microprocessors (Bostrom 2014, 59). While the brain can compensate for some of this with massive parallelism, features such as "hubs," and so on, crucial mental capacities, such as attention, rely upon serial processing, which is incredibly slow, and has a maximum capacity of about seven manageable chunks (Miller 1956). Further, the number of neurons in a human brain is limited by cranial volume and metabolism, but computers can occupy entire buildings or cities, and can even be remotely connected across the globe (Bostrom 2014). Of course, the human brain is far more intelligent than any modern computer. But intelligent machines can in principle be constructed by reverse engineering the brain, and improving upon its algorithms.

In sum: I have observed that there seems to be a short window from the development of the technology to access the cosmos and the development of postbiological minds and AI. I then observed that we are galactic babies: extraterrestrial civilizations are likely to be vastly older than us, and thus they would have already reached not just postbiological life, but superintelligence. Finally, I noted that they would likely be forms of SAI, because silicon is a superior medium for superintelligence. From this I conclude that many advanced alien civilizations will be populated by forms of SAI.

Even if I am wrong – even if the majority of alien civilizations turn out to be biological – it may be that the most intelligent alien civilizations will be ones in which the inhabitants are forms of SAI. Further, creatures that are silicon-based, rather than biologically-based, are more likely to endure space travel, having durable systems that are practically immortal, so they may be the kind of the creatures we first encounter.

All this being said, would superintelligent aliens be conscious, having inner experiences? Here, I draw from a rich philosophical literature on the nature of conscious experience.

Would superintelligent aliens be conscious?

Consider your own conscious experience. Suppose that you are sitting in a cafe preparing to give a lecture. All in one moment, you taste the espresso you sip, consider an idea, and hear the scream of the espresso machine. This is your current stream of consciousness. Conscious streams seem to be very much bound up with who you are. It is not that *this* particular moment is essential – although you may feel that certain ones are important. It is rather that throughout your waking life, you seem to be the subject of a unified stream of experience that presents you as the subject, viewing the show.

Let us focus on three features of the stream: first, it may seem to you, put metaphorically, that there is a sort of "screen" or "stage" in which experiences present themselves to your "mind's eye." That is, there appears to be a central place where experiences are "screened" before you. Daniel Dennett calls this place "the Cartesian Theater" (Dennett 1991). Second, in this central place there seems to be a singular point in time which, given a particular sensory input, consciousness happens. For instance, there seems to be one moment in which the scream of the espresso machine begins, pulling you out of your concentration. Finally, there appears to be a self – someone who is inside the theater, watching the show.

Philosophers have considered each of these features in detail. Each is highly problematic. For instance, an explanation of consciousness cannot literally be that there is a mind's eye in the brain, watching a show. And there is no evidence that there is a singular place or time in the brain where consciousness congeals.

These are intriguing issues, but pursuing them in the context of alien consciousness is putting the cart before the horse. For there is a more fundamental problem: would superintelligent aliens, being forms of AI, even be conscious? Why should we believe that creatures so vastly different from us, being silicon-based, would have inner experience at all?

This problem relates to what philosophers call the *hard problem of consciousness*, a problem that was posed in the context of human consciousness by the philosopher David Chalmers (Chalmers 2008). Chalmers' hard problem is the following. As cognitive science underscores, when we deliberate, hear music, see the rich hues of a sunset, and so on, there is information processing going on in the brain. But above and beyond the manipulation of data, there is a subjective side – there is a "felt quality" to our experience. The hard problem asks: why does all this information processing in the human brain, under certain conditions, have a felt quality to it?

As Chalmers emphasizes, the hard problem is a philosophers' problem, because it doesn't seem to have a scientific answer. For instance, we could develop a complete theory of vision, understanding all of the details of visual processing in the brain, but still not understand why there are subjective experiences attached to these informational states. Chalmers contrasts the hard problem with what he calls "easy problems," problems involving consciousness that have eventual scientific answers, such as the mechanisms behind attention and how we categorize and react to stimuli. Of course these scientific problems are difficult problems; Chalmers merely calls them "easy problems" to contrast them with the "hard problem" of consciousness, which he thinks will not have a purely scientific solution.

We now face yet another perplexing issue involving consciousness – a kind of "hard problem" involving alien superintelligence, if you will: *the hard problem of alien superintelligence*. Would the processing of a silicon-based superintelligent system feel a certain way, from the inside? An alien SAI could solve problems that even the brightest humans are unable to solve, but still, being made of a non-biological substrate, would its information processing feel a certain way from the inside?

It is worth underscoring that the hard problem of superintelligence is not just Chalmers' hard problem of consciousness applied to the case of aliens. For the hard problem of consciousness assumes that we are conscious – after all, each of us can tell from introspecting that we are conscious at this moment. It asks *why* we are conscious. Why does all your information processing feel a certain way from the inside? In contrast, the hard problem of alien consciousness asks *whether* alien superintelligence, being silicon-based, is even capable of being conscious. It does not presuppose that alien superintelligence is conscious. These are different problems, but they are both hard problems that science alone cannot answer.

The problem in the case of superintelligent aliens is that the capacity to be conscious may be unique to biological, carbon-based, organisms. According to *biological naturalism* even the most sophisticated forms of AI will be devoid of inner experience (Searle 1980, Blackmore 2004, Searle 2008). Indeed, even humans wishing to upload their minds will fail to transfer their consciousness. Although they may copy their memories onto a computational format, their consciousness will not transfer, since biological naturalists hold that consciousness requires a biological substrate.[3]

What arguments support biological naturalism? The most common consideration in favor of biological naturalism is John Searle's Chinese Room thought experiment, which is said to suggest that a computer program cannot understand or be conscious (Searle 1980). Searle supposes that he's locked in a room, where he's handed a set of English rules that allow him to link one set of Chinese symbols with other Chinese symbols. So although he doesn't know Chinese, the rules allow him to respond, in written Chinese, to questions written in Chinese. So he is essentially processing symbols. Searle concludes that although those outside of the room may think he understands Chinese, he obviously doesn't; similarly, a computer may appear to be having a Chinese conversation, yet it does not truly understand Chinese. Nor is it conscious.

Although it is correct that Searle doesn't understand Chinese, the issue is not really whether Searle understands; Searle is just one part of the larger system. The relevant question is whether *the system as a whole* understands Chinese. This basic response to Searle's Chinese Room thought experiment is known as the Systems Reply.[4]

It strikes me as implausible that a simple system like the Chinese Room understands, however, for the Chinese Room is not complex enough to understand or be conscious. But the Systems Reply is onto something: the real issue is whether the *system* as a whole understands, not whether one component does. This leaves open the possibility that a more complex silicon-based system could understand; of course, the computations of a superintelligent AI will be far more complex than the human brain.

Here, some might suspect that we could just reformulate the Chinese Room thought experiment in the context of an SAI. But what is fueling this suspicion? It cannot be that some central component in the SAI, analogous to Searle in the Chinese Room, doesn't understand, for we've just observed that it is the system as a whole that understands. Is the suspicion instead fueled by the position that

[3] Biological naturalism was originally developed by John Searle, who developed the view in the context of a larger account of the relation between the mind and body. I will not discuss these details, and they are not essential to the position I've just sketched. Indeed, it isn't clear that Searle is still a biological naturalist, although he persists in calling his view "biological naturalism." In his chapter to my recent *Blackwell Companion to Consciousness* he wrote: "The fact that brain processes cause consciousness does not imply that only brains can be conscious. The brain is a biological machine, and we might build an artificial machine that was conscious; just as the heart is a machine, and we have built artificial hearts. Because we do not know exactly how the brain does it we are not yet in a position to know how to do it artificially." (Searle 2008)

[4] For a thorough treatment of the responses to Searle's argument, including the Systems Reply, the reader may turn to the comments appearing with Searle's original piece, Searle (1980) as well as Cole (2014).

understanding and consciousness do not decompose into more basic operations? If so, then the thought experiment purports to prove too much. Consider the case of the human brain. According to cognitive science, cognitive and perceptual capacities decompose into more basic operations, which are themselves decomposable into more basic constituents, which themselves can be explained causally (Block 1995). If the Chinese Room illustrates that mentality cannot be explained like this, then the brain cannot be explained in this manner either. But this explanatory approach, known as "the method of functional decomposition," is a leading approach to explaining mental capacities in cognitive science. Consciousness and understanding are complex mental properties that are determined by the arrangements of neurons in the brain.

Further, biological naturalism denies one of the main insights of cognitive science – the insight that the brain is computational – without substantial empirical rationale. Cognitive science suggests that our best empirical theory of the brain holds that the mind is an information processing system and that all mental functions are computations. If cognitive science is correct that thinking is computational, then humans and SAI share a common feature: their thinking is essentially computational. Just as a phone call and a smoke signal can convey the same information, thought can have both silicon- and carbon-based substrates. The upshot is that if cognitive science is correct that thinking is computational, we can also expect that sophisticated thinking machines can be conscious, although the contours of their conscious experiences will surely differ.

Indeed, I've noted that silicon is arguably a better medium for information processing than the brain. So why isn't silicon a *better* medium for consciousness, rather than a *worse* one, as the biological naturalists propose? It would be surprising if SAI, which would have far superior information processing abilities than we do, turned out to be deficient with respect to consciousness. For our best scientific theories of consciousness hold that consciousness is closely related to information processing (Baars 2008, Tonini 2008).

Some would point out that to show that AI cannot be conscious, the biological naturalist would need to locate a special consciousness property (call it "P"), which inheres in neurons or their configurations, and which cannot be instantiated by silicon. Thus far, P has not been discovered. It isn't clear, however, that locating P would prove biological naturalism to be correct. For the computationalist can just say that machines are capable of instantiating a different type of consciousness property, F, which is specific to silicon-based systems.

Massimo Pigliucci has offered a different kind of consideration in favor of biological naturalism, however. He sees philosophers who argue for computationalism as embracing an implausible perspective on the nature of consciousness: functionalism. According to *functionalists* the nature of a mental state depends on the way it functions, or the role it plays in the system of which it is a part. Pigliucci is correct that traditional functionalists, such as Jerry Fodor, generally mistakenly ignore the biological workings of the brain. Pigliucci objects: ". . . functionality isn't just a result of the proper arrangement of the parts of a system, but also of the types of materials (and their properties) that make up those parts" (Pigliucci 2014).

Fodor's well-known antipathy towards neuroscience should not mislead us into thinking that functionalism must ignore neuroscience, however. Clearly, any well-conceived functionalist position must take into consideration neuroscientific work on the brain because the functionalist is interested in the causal or dispositional properties of the parts, not just the parts themselves. Indeed, as I've argued in my book *The Language of Thought*, viewing the brain as irrelevant to the computational approach to the mind is a huge mistake. The brain is the best computational system we know of (Schneider 2011b).

Does this make my position a form of biological naturalism? Not in the least. I am suggesting that viewing neuroscience (and by extension, biology) as being opposed to computationalism is mistaken. Indeed, neuroscience is computational; a large subfield of neuroscience is called "computational neuroscience," and it seeks to understand the sense in which the brain is computational and to provide computational accounts of mental capacities identified by related subfields, such as cognitive neuroscience. What makes my view different from biological naturalism is that I hold that thinking is computational, and further, that at least one other substrate besides carbon (i.e. silicon) can give rise to consciousness and understanding, at least in principle.

But biological naturalism is well worth considering. I am reasoning that a substrate that supports superintelligence, being capable of even more sophisticated informational processing than we are, would likely also be one that is conscious. But notice that I've used the expression "likely." For we can never be *certain* that AI is conscious, even if we could study it up close. The problem is akin to the philosophical puzzle known as *the problem of other minds* (Schneider 2014). The problem of other minds is that although you can know that you are conscious, you cannot be certain that other people are conscious as well. After all, you might be witnessing behavior with no accompanying conscious component. In the face of the problem of other minds, all you can do is note that other people have brains that are structurally similar to your own and conclude that since you yourself are conscious, others are likely to be as well. When confronted with AI your predicament would be similar, at least if you accept that thinking is computational. While we couldn't be absolutely certain that an AI program genuinely felt anything, we can't be certain that other humans do either. But it would seem probable in both cases.

So, to the question of whether alien superintelligence can be conscious, I answer, very cautiously, "probably."

How might superintelligent aliens think?

Thus far, I've said little about the structure of superintelligent alien minds. And little is all we can say: superintelligence is by definition a kind of intelligence that outthinks humans in every domain. In an important sense, we cannot predict or fully understand how it will think. Still, we may be able to identify a few important characteristics, albeit in broad strokes.

Nick Bostrom's recent book on superintelligence focuses on the development of superintelligence on Earth, but we can draw from his thoughtful discussion (Bostrom 2014). Bostrom distinguishes three kinds of superintelligence:

(1) *Speed superintelligence* – even a human emulation could in principle run so fast that it could write a PhD thesis in an hour.
(2) *Collective superintelligence* – the individual units need not be superintelligent, but the collective performance of the individuals outstrips human intelligence.
(3) *Quality superintelligence* – at least as fast as human thought, and vastly smarter than humans in virtually every domain.

Any of these kinds could exist alongside one or more of the others.

An important question is whether we can identify common goals that these types of superintelligences may share. Bostrom suggests (Bostrom 2014, 107):

The Orthogonality Thesis: Intelligence and final goals are orthogonal – more or less any level of intelligence could in principle be combined with more or less any final goal.

Bostrom is careful to underscore that a great many unthinkable kinds of SAI could be developed. At one point, he raises a sobering example of a superintelligence with the final goal of manufacturing paper clips (pp. 107–108, 123–125). While this may initially strike you as a harmless endeavor, although hardly a life worth living, Bostrom points out that a superintelligence could utilize every form of matter on Earth in support of this goal, wiping out biological life in the process. Indeed, Bostrom warns that superintelligence emerging on Earth could be of an unpredictable nature, being "extremely alien" to us (p. 29). He lays out several scenarios for the development of SAI. For instance, SAI could be arrived at in unexpected ways by clever programmers, and not be derived from the human brain whatsoever. He also takes seriously the possibility that Earthly superintelligence could be *biologically inspired*, that is, developed from reverse engineering the algorithms that cognitive science says describe the human brain, or from scanning the contents of human brains and transferring them to a computer (i.e. "uploading").[5]

Although the final goals of superintelligence are difficult to predict, Bostrom singles out several instrumental goals as being likely, given that they support any final goal whatsoever (Bostrom 2014, 109):

> *The Instrumental Convergence Thesis*: Several instrumental values can be identified which are convergent in the sense that their attainment would increase the chances of the agent's goal being realized for a wide range of final goals and a wide range of situations, implying that these instrumental values are likely to be pursued by a broad spectrum of situated intelligent agents.

The goals that he identifies are *resource acquisition, technological perfection, cognitive enhancement, self-preservation*, and *goal content integrity* (i.e. that a superintelligent being's future self will pursue and attain those same goals). He underscores that self-preservation can involve group or individual preservation, and that it may play second-fiddle to the preservation of the species the AI was designed to serve (Bostrom 2014, 109).

Let us call an alien superintelligence that is based on reverse engineering an alien brain, including uploading it, a *biologically-inspired superintelligent alien* ("BISA"). Although BISAs are inspired by the brains of the original species that the superintelligence is derived from, a BISA's algorithms may depart from those of their biological model at any point.

BISAs are of particular interest in the context of alien superintelligence. For if Bostrom is correct that there are many ways superintelligence can be built, but a number of alien civilizations develop superintelligence from uploading or other forms of reverse engineering, *it may be that BISAs are the most common form of alien superintelligence out there*. This is because there are many kinds of superintelligence that can arise from raw programming techniques employed by alien civilizations. (Consider, for instance, the diverse range of AI programs under development on Earth, many of which are not modeled after the human brain.) This may leave us with a situation in which the class of SAIs is highly heterogeneous, with members generally bearing little

[5] Throughout his book, Bostrom emphasizes that we must bear in mind that superintelligence, being unpredictable and difficult to control, may pose a grave existential risk to our species (Bostrom 2014). This should give us pause in the context of alien contact as well.

resemblance to each other. It may turn out that of all SAIs, BISAs bear the most resemblance to each other. In other words, BISAs may be the most cohesive subgroup because the other members are so different from each other.

Here, you may suspect that because BISAs could be scattered across the galaxy and generated by multitudes of species, there is little interesting that we can say about the class of BISAs. But notice that BISAs have two features that may give rise to common cognitive capacities and goals:

(1) BISAs are descended from creatures that had motivations like: find food, avoid injury and predators, reproduce, cooperate, compete, and so on.
(2) The life forms that BISAs are modeled from have evolved to deal with biological constraints like slow processing speed and the spatial limitations of embodiment.

Could (1) or (2) yield traits common to members of many superintelligent alien civilizations? I suspect so.

Consider (1). Intelligent biological life tends to be primarily concerned with its own survival and reproduction, so it is more likely that BISAs would have final goals involving their own survival and reproduction, or at least the survival and reproduction of the members of their society. If BISAs are interested in reproduction, we might expect that, given the massive amounts of computational resources at their disposal, BISAs would create simulated universes stocked with artificial life and even intelligence or superintelligence. If these creatures were intended to be "children" they may retain the goals listed in (1) as well.

You may object that it is useless to theorize about BISAs, as they can change their basic architecture in numerous, unforeseen ways, and any biologically-inspired motivations can be constrained by programming. There may be limits to this, however. If a superintelligence is biologically-based, it may have its own survival as a primary goal. In this case, it may not want to change its architecture fundamentally, but stick to smaller improvements. It may think: when I fundamentally alter my architecture, I am no longer *me* (Schneider 2011a). Uploads, for instance, may be especially inclined not to alter the traits that were most important to them during their biological existence.

Consider (2). The designers of the superintelligence, or a self-improving superintelligence itself, may move away from the original biological model in all sorts of unforeseen ways, although I have noted that a BISA may not wish to alter its architecture fundamentally. But we could look for cognitive capacities that are useful to keep; cognitive capacities that sophisticated forms of biological intelligence are likely to have, and which enable the superintelligence to carry out its final and instrumental goals. We could also look for traits that are not likely to be engineered out, as they do not detract the BISA from its goals.

If (2) is correct, we might expect the following, for instance.

(i) *Learning about the computational structure of the brain of the species that created the BISA can provide insight into the BISA's thinking patterns.* One influential means of understanding the computational structure of the brain in cognitive science is via "connectomics," a field that seeks to provide a connectivity map or wiring diagram of the brain (Seung 2012). While it is likely that a given BISA will not have the same kind of connectome as the members of the original species, some of the functional and structural connections may be retained, and interesting departures from the originals may be found.

(ii) *BISAs may have viewpoint-invariant representations.* At a high level of processing your brain has internal representations of the people and objects that you interact with that are *viewpoint-invariant.* Consider walking up to your front door. You've walked this path hundreds, maybe thousands of times, but technically, you see things from slightly different angles each time as you are never positioned in exactly the same way twice. You have mental representations that are at a relatively high level of processing and are viewpoint invariant. It seems difficult for biologically-based intelligence to evolve without viewpoint-invariant representations, as they enable categorization and prediction (Hawkins and Blakeslee 2004). Such representations arise because a system that is mobile needs a means of identifying items in its ever-changing environment, so we would expect biologically-based systems to have them. BISA would have little reason to give up object-invariant representations insofar as it remains mobile or has mobile devices sending it information remotely.

(iii) *BISAs will have language-like mental representations that are recursive and combinatorial.* Notice that human thought has the crucial and pervasive feature of being combinatorial. Consider the thought *wine is better in Italy than in China.* You probably have never had this thought before, but you were able to understand it. The key is that the thoughts are combinatorial because they are built out of familiar constituents, and combined according to rules. The rules apply to constructions out of primitive constituents, that are themselves constructed grammatically, as well as to the primitive constituents themselves. Grammatical mental operations are incredibly useful: it is the *combinatorial* nature of thought that allows one to understand and produce these sentences on the basis of one's antecedent knowledge of the grammar and atomic constituents (e.g. *wine, China*). Relatedly, thought is *productive*: in principle, one can entertain and produce an infinite number of distinct representations because the mind has a combinatorial syntax (Schneider 2011b).

Brains need combinatorial representations because there are infinitely many possible linguistic representations, and the brain only has a finite storage space. Even a superintelligent system would benefit from combinatorial representations. Although a superintelligent system could have computational resources that are so vast that it is mostly capable of pairing up utterances or inscriptions with a stored sentence, it would be unlikely that it would trade away such a marvelous innovation of biological brains. If it did, it would be less efficient, since there is the potential of a sentence not being in its storage, which must be finite.

(iv) *BISAs may have one or more global workspaces.* When you search for a fact or concentrate on something, your brain grants that sensory or cognitive content access to a "global workspace" where the information is broadcast to attentional and working memory systems for more concentrated processing, as well as to the massively parallel channels in the brain (Baars 2008). The global workspace operates as a singular place where important information from the senses is considered in tandem, so that the creature can make all-things-considered judgments and act intelligently, in light of all the facts at its disposal. In general, it would be inefficient to have a sense or cognitive capacity that was not integrated with the others, because the information from this sense or cognitive capacity would be unable to figure in predictions and plans based on an assessment of all the available information.

(v) *A BISA's mental processing can be understood via functional decomposition.* As complex as alien superintelligence may be, humans may be able to use the method of functional decomposition as an approach to understanding it. A key feature of computational approaches to the brain is that cognitive and perceptual capacities are understood by decomposing the particular capacity into their causally organized parts, which themselves can be understood in terms of the causal organization of their parts. This is the aforementioned "method of functional decomposition" and it is a key explanatory method in cognitive science. It is difficult to envision a complex thinking machine not having a program consisting of causally interrelated elements each of which consists in causally organized elements.

All this being said, superintelligent beings are by definition beings that are superior to humans in every domain. While a creature can have superior processing that still basically makes sense to us, it may be that a given superintelligence is so advanced that we cannot understand any of its computations whatsoever. It may be that any truly advanced civilization will have technologies that will be indistinguishable from magic, as Arthur C. Clarke suggested (1962). I obviously speak to the scenario in which the SAI's processing makes some sense to us, one in which developments from cognitive science yield a glimmer of understanding into the complex mental lives of certain BISAs.

Conclusion

I have argued that the members of the most advanced alien civilizations will be forms of superintelligent artificial intelligence (SAI). I have further suggested, very provisionally, that we might expect that if a given alien superintelligence is a biologically-inspired superintelligent alien (BISA), it would have combinatorial representations and that we could seek insight into its processing by decomposing its computational functions into causally interacting parts. We could also learn about it by looking at the brain wiring diagrams (connectomes) of the members of the original species. Further, BISAs may have one or more global workspaces. Furthermore, I have argued that there is no reason in principle to deny that SAIs could have conscious experience.

Acknowledgments

Many thanks to Joe Corabi, Steven Dick, Clay Ferris Naff, and Eric Schwitzgebel for helpful written comments on an earlier draft and to James Hughes for helpful conversation.

References

Baars, B. 2008. "The Global Workspace Theory of Consciousness." In M. Velmans and S. Schneider (eds.), *The Blackwell Companion to Consciousness*. Boston, MA: Wiley-Blackwell, pp. 236–247.

Blackmore, S. 2004. *Consciousness: An Introduction*. New York, NY: Oxford University Press.

Block, N. 1995. "The Mind as the Software of the Brain." In D. Osherson, L. Gleitman, S. Kosslyn, E. Smith, and S. Sternberg (eds.), *An Invitation to Cognitive Science*. New York: MIT Press, pp. 377–421.

Bostrom, N., Chislenko, A., Hughes, J., 2003. "The Transhumanist Frequently Asked Questions": v 2.1. World Transhumanist Association. Retrieved from http://humanityplus.org/philosophy/transhumanist-faq/.

Bostrom, N. 2014. *Superintelligence: Paths, Dangers, Strategies*. Oxford: Oxford University Press.

Bradbury, R., Cirkovic, M., and Dvorsky, G. 2011. "Dysonian Approach to SETI: A Fruitful Middle Ground?" *Journal of the British Interplanetary Society*, 64: 156–165.

Chalmers, D. 2008. "The Hard Problem of Consciousness." In M. Velmans and S. Schneider, *The Blackwell Companion to Consciousness*. Boston, MA: Wiley-Blackwell, pp. 225–236.

Cirkovic, M. and Bradbury, R. 2006. "Galactic Gradients, Postbiological Evolution and the Apparent Failure of SETI." *New Astronomy* 11, 628–639.

Clarke, A. (1962). *Profiles of the Future: An Inquiry into the Limits of the Possible*. New York, NY: Harper and Row.

Cole, D. 2014. "The Chinese Room Argument", *The Stanford Encyclopedia of Philosophy* (Summer 2014 Edition), E. N. Zalta (ed.), online at http://plato.stanford.edu/archives/sum2014/entries/chinese-room/.

Davies, P. 2010. *The Eerie Silence: Renewing Our Search for Alien Intelligence*. Boston, MA: Houghton Mifflin Harcourt.

Dennett, D. 1991. *Consciousness Explained*. New York, NY: Penguin Press.

Dick, S. 2013. "Bringing Culture to Cosmos: the Postbiological Universe." In S. J. Dick, and M. Lupisella (eds.), *Cosmos and Culture: Cultural Evolution in a Cosmic Context*. Washington, DC: NASA, online at http://history.nasa.gov/SP-4802.pdf.

Garreau, J. 2005. *Radical Evolution: The Promise and Peril of Enhancing our Minds, Our Bodies – And What it Means to Be Human*. New York, NY: Doubleday.

Guerini, Federico. 2014. "DARPA's ElectRx Project: Self-Healing Bodies Through Targeted Stimulation Of The Nerves," online at http://www.forbes.com/sites/federicoguerrini/2014/08/29/darpas-electrx-project-self-healing-bodies-through-targeted-stimulation-of-the-nerves/.

Hawkins, J. and Blakeslee, S. 2004. *On Intelligence: How a New Understanding of the Brain will Lead to the Creation of Truly Intelligent Machines*. NewYork, NY: Times Books.

Kurzweil, R. 2005. *The Singularity is Near: When Humans Transcend Biology*. New York, NY: Viking.

Miller, R. 1956. "The Magical Number Seven, Plus or Minus Two: Some Limits on Our Capacity for Processing Information." *The Psychological Review*, 63: 81–97.

Pigliucci, M. 2014. "Mind Uploading: A Philosophical Counter-Analysis." In R. Blackford and D. Broderick (eds.), *Intelligence Unbound: The Future of Uploaded and Machine Minds*. Boston, MA: Wiley-Blackwell, pp. 119–130.

Sandberg, A., Boström, N. 2008. "Whole Brain Emulation: A Roadmap." Technical Report #2008•3. Future of Humanity Institute, Oxford University.

Schneider, S. 2011a. "Mindscan: Transcending and Enhancing the Brain." In J. Giordano (ed.), *Neuroscience and Neuroethics: Issues At the Intersection of Mind, Meanings and Morality*. Cambridge: Cambridge University Press.

Schneider, S. 2011b. *The Language of Thought: a New Philosophical Direction*. Boston, MA: MIT Press.

Schneider, S. 2014. "The Philosophy of 'Her.'" *The New York Times*, March 2.

Searle, J. 1980. "Minds, Brains and Programs." *The Behavioral and Brain Sciences*, 3: 417–457.

Searle, J. 2008. "Biological Naturalism." In M. Velmans and S. Schneider (eds.), *The Blackwell Companion to Consciousness*. Boston, MA: Wiley-Blackwell.

Seung, S. 2012. *Connectome: How the Brain's Wiring Makes Us Who We Are*. Boston, MA: Houghton Mifflin Harcourt.

Shostak, S. 2009. *Confessions of an Alien Hunter*. New York, NY: National Geographic.

Tonini, G. 2008. "The Information Integration Theory of Consciousness." In M. Velmans and S. Schneider (eds.), *The Blackwell Companion to Consciousness*. Boston, MA: Wiley-Blackwell, pp. 287–300.

13 The moral subject of astrobiology

Guideposts for exploring our ethical and political responsibilities towards extraterrestrial life

ELSPETH M. WILSON AND CAROL E. CLELAND

In the event that we discover extraterrestrial life, what ethical considerations ought to inform our interactions with it? In this chapter, we argue that astrobiology faces at least two significant roadblocks when it comes to addressing this quandary. The first is the well-known $N = 1$ problem (see Mariscal, Chapter 7 in this volume). Currently, we have merely one example of life (life on Earth), and one cannot safely make scientific generalizations from a single example about a presumably broader domain phenomenon. Since there are good reasons for suspecting that life elsewhere may deviate from Earth life in biologically significant ways (Grinspoon 1997; Benner *et al.* 2004; Schulze-Makuch and Irwin 2006), we must grapple with the difficulty of applying traditional ethical theories to hypothetical forms of life we know nothing about. Second, just as our concept of life is restricted to the single example of Earth life, the way we think about moral status is even more narrowly restricted to the members of a single species of life on Earth, *Homo sapiens*. Taken together, these two roadblocks pose a serious hurdle when it comes to theorizing about the intersection of ethics and astrobiology. Given the highly Earth-centric character of our understanding of life and the anthropocentric character of our concepts of morality, how can we even begin to address the question of our potential ethical responsibilities towards forms of life differing radically from ourselves?

Our goal is not to resolve this dilemma, which will inevitably persist in the absence of concrete examples of extraterrestrial life. Instead, the aim of this chapter is to highlight some of the challenges we face in applying our human-centered ethical theories to truly alien organisms and to propose possible avenues for fruitful research and theorizing about these issues in the future. We will examine the concept of a moral subject (a.k.a. patient), and explore what makes an entity the kind of thing that is capable of being unjustly wronged by a moral agent, such as you or me. That is, instead of focusing in an *inward* direction on what finding extraterrestrial life would mean to "us" – which remains an important but separate ethical quandary – we look in an *outward* direction at the moral responsibilities that "we" (human beings) might have towards alien forms of life.

The most common ethical categories in traditional Western philosophy and theology used to assess the moral status of both human and non-human animals are the (1) capacity for rationality, (2) expression of sentience, (3) evidence of complex social behavior, and (4) the theological notion of possessing a soul. Invoking thought-experiments from science fiction, we explore some anthropocentric problems involving how these categories have narrowly been applied in the past, and suggest ways in which we might expand them to better confront such concerns in the future. Ultimately, given the aforementioned problems of Earth-centrism and anthropocentrism, we want to emphasize how challenging (yet vital) it is to acknowledge and meaningfully address questions about the moral status of non-human forms of life.

Science versus ethics: the fact/value distinction

This essay approaches astrobiology as a new frontier in bioethics. Given this, it is useful to begin by distinguishing science from ethics. Science is concerned with the discovery of empirical "facts," including both general (i.e. about the laws of nature) and particular (i.e. about specific occurrences). In contrast, ethics is concerned with explaining how things "ought" to be or how we ought to behave in certain contexts and situations. The job of ethics is to evaluate what constitutes "right" and "wrong" action. As David Hume (1783) famously argued, one cannot logically deduce ethical conclusions of *ought* from purely factual premises of what *is* the case about the world. A person could hypothetically know all the observable facts about a given situation, but this information alone would not tell her how to act. Similarly, Max Weber (1917, 1919) emphasizes a "fact/value distinction" when it comes to political leaders making informed decisions in the realm of public policy. The point is that, unlike the properties of physical occurrences – such as the intensity of a hurricane or the acceleration of an object in free fall – moral properties are not objectively observable and thus require a person to commit to a normative system of evaluating right from wrong (see Lupisella, Chapter 10 in this volume).

While ethics and science are separate enterprises, the bodies of knowledge they accumulate are deeply intertwined: just as there is a fact/value distinction, we might also say there is a fact/value connection. Although ethics reaches beyond purely factual questions, the ethical possibilities that we are willing to entertain when theorizing about morality in applied settings inevitably depend upon what we scientifically know (or presume to know) about the world around us. Likewise, ethical theory has the potential to provide science with morally significant normative principles about how scientists (and other people) ought to act in relation not only to one another but also to non-human organisms. It goes almost without saying that we can't think about treating an animal in an ethical manner in isolation of its biological characteristics, including its physical make-up, morphology, nutritional and reproductive requirements, ecological niche, and social structure. To illustrate, consider what would be "the right thing to do" if we found a random healthy dolphin washed up on a beach versus a random human lost in the same area. While most people would agree that it would be a "good deed" to return the dolphin to the ocean so that it might have a chance to survive in its natural habitat, the same conclusion would not apply to the human being. If we dumped the person in the middle of the ocean without a life vest, then he would most likely drown. As this example illustrates, the facts of the situation – in particular, that dolphins are aquatic and humans terrestrial – provide essential biological (background) information vital to treating each organism in a way that doesn't conflict with our most basic intuitions about right and wrong.

What does this fact/value relationship mean for astrobiology? An obvious conclusion is that we must take the $N = 1$ problem seriously, and admit that we currently know very little about life beyond Earth. This suggests that we must wait until we actually discover concrete examples of extraterrestrial organisms before making sweeping pronouncements about how we *ought* to interact with them. For purely factual information about an extraterrestrial organism (e.g. its chemical composition, biochemical requirements, anatomy, and physiology) and its relationship to its ecosystem (e.g. environmental sources of material and energy available, physical and chemical liabilities, intra-species behavior, and relations to other life) is critical for developing suitable normative conclusions about how we "ought" to treat it. But just because we do not know the pertinent empirical facts yet, it doesn't follow that nothing useful can be said about the ethics of astrobiology. On the contrary, we want to be as prepared as possible to address these issues when they arise, and do so in the "best" way possible from an ethical standpoint. The purpose of this essay is to sketch a theoretical framework for thinking about the moral status of extraterrestrial life that minimizes our problematic (Earth-centric and anthropocentric) preconceptions about moral status; for more detail, see Cleland and Wilson (2013).

Moral status: the agent/patient distinction

In the event that we discover extraterrestrial life, we must consider what characteristics an organism needs to possess to qualify for moral status.

Theories of morality traditionally distinguish between two categories of moral status, moral agents and moral patients. By definition, a *moral agent* must be capable of making informed choices and understanding that their actions can harm and benefit others. The ability to engage in complex reasoning, the capacity for relational thinking, and the possession of self-awareness are preconditions for moral agency. Moral agents have ethical obligations towards moral patients and can be held responsible for their behavior. In this essay, we are primarily concerned with the concept of *moral patient*, a term we use interchangeably with "moral subject." Moral patients qualify for moral standing because they are capable of being benefited or harmed by us and are considered members of our moral community. All moral agents are also *de facto* moral patients, but not all moral patients qualify as full-fledged moral agents. Young children and the mentally ill provide classic examples of the latter. Although they cannot always be held fully responsible for their actions, it is widely recognized that children and the mentally ill can be aided or injured by our actions. It follows that how we treat them may properly be evaluated as right or wrong.

Is the traditional concept of moral status broad enough to encompass the potential moral subjecthood of extraterrestrial organisms? Just as our definitions of life tend to be Earth-centric (Cleland and Chyba 2002, 2007; Cleland 2012), most ethical theories are anthropocentric, because ethics was originally developed to inform and shape human behavior in societal contexts. (Typical ethical questions include such concerns as: How should I treat my neighbor? What duties do I owe my family, town, or country? Should I help a starving or homeless child? Under what circumstances is it acceptable to kill another person in the name of self-defense? When can a nation justly declare war?) As a result of this focus on human interactions and social relations, the question of whether or not non-human animals are primary or secondary moral subjects remains open to vigorous philosophical debate. An entity is a *primary moral subject* if it is a moral patient and hence has intrinsic moral worth. In contrast, an entity is a *secondary moral subject* if its moral worth depends on something outside itself, such as its benefit to a human being. In practice, the moral status of non-human animals is almost always evaluated on a case-by-case basis. When we label our ethical duties towards other animals in reference to human-centered goals, then we accord them only secondary moral status. For instance, if we help Lassie the dog find her way back home purely because she is dear to the boy next door, then we are treating her as having secondary moral status by virtue of her meaning to our young neighbor. Conversely, if we help Lassie find her way back home because we care about helping her survive a busy intersection and returning her to her loving family, then we are treating Lassie herself as having a degree of intrinsic moral worth. Likewise, if we protect an endangered species like Bengal tigers or polar bears, just because we believe that biodiversity makes the world a more interesting and beautiful place, then we are valuing these majestic predators as secondary in relation to our own goals. Rather than being valued as an end in themselves, the animals are merely a means to another end (sustaining ecological diversity). In these cases, we label our ethical duties to the animals in question as indirect or secondary, because the moral consideration we owe them is justified in reference to an extrinsic (human-centered) goal.

In recent years, there has been a lively debate about whether our concept of moral status (especially, patienthood) should be expanded to include certain "advanced" non-human animals, most notably, great apes, various primates, dolphins, whales, elephants, domesticated dogs, and parrots. This increasing openness to entertaining the possibility that some non-human animals might have intrinsic moral status is a relatively new development. Traditionally, the enterprise of theorizing about the scope of moral status (both theological and secular) was almost exclusively limited to humans. The anthropocentric nature of traditional theories of moral status has recently been challenged and expanded upon by philosophers, biologists, and theologians associated with the animal rights movement; see, e.g., Peter Singer (1975) and Tom Regan (1983). Even so, a common theme in evaluations of the moral status of non-human animals is to point to characteristics that are traditionally thought to distinguish humans from non-humans. The tendency to collapse morality into a predominantly "human" enterprise reveals the depth of our anthropocentric approach to morality. Given that non-human animals are accorded moral status as ethical subjects on the basis of characteristics they share with human beings, might we fail to recognize a sentient alien organism as deserving of intrinsic moral status, due to its profound biological and life-style differences from us? Science fiction is rife with telling illustrations of this type of "misunderstanding." The Horta, in the original *Star Trek* episode, *The Devil in the Dark*, provides an especially salient example. The Horta initially appears to be an unthinking alien monster – an amoeba-like blob of rock – dissolving everything in its path and causing chaos within a human mining colony. But when Mister Spock is able to communicate with it by performing a Vulcan mind meld, he discovers that it is not a senseless monster but rather a tormented "mother" defending its young from being inadvertently destroyed by human miners. The ability to communicate and identify with – i.e. attribute human goals to – the Horta transforms it into a moral subject for the miners and so they protect its children.

The situation is even more troubling when we look at the (often bloody) history of how we have treated our fellow human beings, with whom we share so much! In the past, many human beings were excluded from full and equal standing in society on the grounds of morally arbitrary categories such as race, sex, ethnicity, culture, sexual orientation, and religious affiliation. While it is now widely conceded that such distinctions among humans are morally wrong, numerous forms of prejudice and discrimination (ranging from racism, to sexism, to homophobia) remain serious practical social and political problems throughout the world today. When we combine this tragic legacy of violating the intrinsic moral worth of members of our own species with the fact that most human societies are just beginning to consider extending the boundaries of our moral community to encompass non-human animals here on Earth, it follows that we face serious pragmatic worries about how we are likely to interact with even modestly different alien beings from other worlds. Science fiction again provides us with food for thought. In the movie *District 9*, humanoid extraterrestrials resembling "prawns," which are clearly intelligent (unlike the Horta), are interned in an alien refugee camp after their spaceship malfunctions above South Africa. In addition to being segregated from mainstream society in deplorable living conditions, the "prawns" are subjected to military rule, experimentation, and even vivisection by humans.

Four categories for evaluating moral status

The concerns discussed above underscore why it is important to consider the connection between ethics and astrobiology, or what Ted Peters (2013) has termed "astroethics," sooner rather than later. We need to become clearer about the challenges we face in applying traditional (anthropocentric) assumptions about moral status to extraterrestrial organisms, and to proceed with an open mind about what forms they might take and how we ought to interact with them. In the remainder of this essay, we explore these difficulties in the context of four basic categories of characteristics that are commonly taken as morally significant in Western secular and theological theory: intelligence, sentience, social behavior, and the possession of an immortal rational soul or "divine spark."

Let us begin with *intelligence*. Both scientists and ethicists often cite features such as rationality, self-conscious deliberation, brain anatomy, creative problem solving, memory, abstract thought, the creation of tools, and the capability to communicate ideas as morally significant characteristics. In Western philosophy, the idea that intelligence (or rationality) is morally significant was most famously articulated by Immanuel Kant. In his classic version of duty ethics, Kant held that all rational beings are "persons." All persons are in turn "moral agents," because they have the ability to use their capacity for reason to determine universal moral imperatives and consciously will their own ends. Conversely, Kant explicitly labels all other animals on Earth as non-persons and mere "things." In Kant's words (Kant 1798, 127):

> The fact that the human being can have the "I" in his representations raises him infinitely above all other living beings on earth. Because of this he is a *person* . . . through rank and dignity an entirely different being from *things*, such as irrational animals, with which one can do as one likes.

As this passage demonstrates, Kant unabashedly presents human beings as the prototypes for moral agency – or what he calls, personhood. He also explicitly excludes non-human animals on Earth from the category of "persons" by denying that they share our capacity for rationality. Nonetheless, by associating personhood so directly with the capacity for rationality, Kant provocatively leaves open the possibility of an alien qualifying as an honorary "person."

Yet, given the fact that Kant himself discounts extremely intelligent animals on Earth, like apes and parrots, from the category of "person" – and even more egregiously goes so far as to question the rationality of women and racial minorities – his strategy of rooting moral status in the capability for rationality would likely be extremely difficult to apply in the realm of astrobiology. For if Kant failed to recognize that women and racial minorities possess the same capacity of rationality as European men, thus discounting a majority of his own species from full and equal personhood, what are the chances that a contemporary scholar in the Kantian tradition might fail to recognize the rational capacities of beings who, like the "prawns" of *District 9*, differ in biologically significant ways from our own species? Not only might it be extremely difficult to recognize forms of rationality in organisms whose physical needs and lifestyles differ markedly from our own, but we can conversely imagine a clearly rational and self-conscious hive-like alien (analogous to "the Borg" of *Star Trek the Next Generation*) that is comprised of numerous individual creatures resembling humans but lacking an independent self-consciousness (or what Kant celebrates as a sense of "I"). Somewhat ironically, the intelligence of the Borg is portrayed as being located in an individual "Borg queen," but it isn't at all obvious that hive intelligence requires a single leader. Why couldn't the reasoning ability of the hive be truly distributed over its individual members, instead of being located, in human-like fashion, in a supreme ruler? (In this context, one wonders whether biologists have overemphasized the role of "the queen" among "social" hymenoptera, such as honeybees. Are we inappropriately imposing human categories on them?)

Indeed, can we not imagine the possibility of creatures that are so alien in their form and function that what seems to be entirely rational behavior from their perspective might appear bizarre and irrational to us? Ultimately, those who want to hold up intelligence or rationality as a benchmark for moral status must identify and acknowledge the anthropocentric limitations of this standard, present in both theory and application.

This brings us to *sentience*. As a general rule, a sentient animal is capable of perceiving pleasure and pain, aware of its surroundings, and conscious of what happens to it. The utilitarian tradition in Western political philosophy places sentience at the center of their moral theory. Early utilitarians argued that the morally correct thing to do in a given situation is to maximize happiness on an aggregate level – and, by extension, minimize suffering. While simple enough in theory, this approach requires a prior assessment of whose happiness ought to be factored into this aggregate calculus of utility. The most renowned early utilitarian philosopher, John Stuart Mill, maintained that there were different classes of happiness and ranked human intellectual pursuits at the top of his list of pleasures. Mill (1863) famously proclaimed that it would be better to be a dissatisfied Socrates than a satisfied pig (p. 10). According to Mill, humans have a unique capacity for happiness (e.g. expressed in his "greatest happiness principle"), which justified his downgrading of the suffering of non-human animals.

Pondering the nature of sentience, the eighteenth-century philosopher Jeremy Bentham presciently worried that people might conclude in the future that it is wrong to reduce non-human animals to mere "things" rather than ranking them as fellow living "beings" that are capable of suffering just like us (Bentham, 1781, ch. 17 footnote b):

> Other animals, which, on account of their interests having been neglected by the insensibility of the ancient jurists, stand degraded into the class of *things*. [original emphasis] . . . The day has been, I grieve it to say in many places it is not yet past, in which the greater part of the species, under the denomination of slaves, have been treated . . . upon the same footing as . . . animals are still. The day may come, when the rest of the animal creation may acquire those rights which never could have been withholden from them but by the hand of tyranny. The French have already discovered that the blackness of skin is no reason why a human being should be abandoned without redress to the caprice of a tormentor. It may come one day to be recognized, that the number of legs, the villosity of the skin, or the termination of the sacrum, are reasons equally insufficient for abandoning a sensitive being to the same fate. What else is it that should trace the insuperable line? Is it the faculty of reason, or perhaps, the faculty for discourse? . . . the question is not, Can they reason? nor, Can they talk? but, Can they suffer?

When properly recognized in another animal, suffering is a feeling with which humans can sympathize, precisely because it is something we share with other species. Bentham's utilitarian logic appears to be that if animals are *similar* to us in this one crucial respect, then they might also warrant moral consideration. This view was extremely progressive during the eighteenth century, but it is important to note that Bentham focused only on a similarity between non-human animals and us, and consequently did not move significantly beyond a human-centered view of moral status. While sentience generally includes the capacity to feel physical pain and pleasure, and it might be tempting to accept Mill's even more restrictive interpretation to prioritize higher-order experiences of pleasure (joy, wonder, etc.) and pain (frustration, hopelessness, etc.), our own species *Homo sapiens* is almost always treated as the benchmark against which all other animal species are measured and classified. In this vein, Spock's discovery that the Horta felt intense emotional pain played a central role in the reevaluation of its moral status from that of a mere thing to a caring mother worthy of moral concern.

Next, there is *social behavior*. Associating social behavior with moral status is very old, going back at least to Greek and Roman philosophers, who celebrated human social behavior as the central characteristic separating them from mere "beasts." Most famously, the Greek philosopher, Aristotle, defined "man" as a "political animal" based upon the fact that he is a social creature, with the capacity for moral reasoning about social interactions. In Aristotle's words (Aristotle, *Politics* Bk. 1.2 $1253^{a7\text{-}18}$):

> Nature ... makes nothing in vain, and man is the only animal who has the gift of speech ... And it is a characteristic of man that he alone has any sense of good and evil, of just and unjust, and the like, and the association of living beings who have this sense makes a family and a state.

He believed that these social attributes separate human beings from all other animals, and ruled out the social behavior of non-human animals, such as honeybees, as lacking the complexity and refinement of human sociality as evidenced by our capacity for speech. Following Aristotle, contemporary scientists and philosophers often cite "sophisticated" social behavior in non-human animals, such as chimpanzees and dolphins, as having potential moral significance. Examples of behavior commonly taken as morally significant when exhibited by non-human animals are playfulness, parental care, emotional expression, linguistic-like communication, acquisition of learned skills, communal lifestyles, group loyalty, complex power structures, and lasting family ties. The anthropocentric character of such attributions is clear. Indeed, we are particularly impressed when we see evidence of what appears to be ethical social behavior within the members of species other than our own. This includes the appearance of cooperation, altruistic tendencies, filial and fraternal displays of devotion and affection, lasting friendships, long-term monogamy among mates, sharing scarce goods, and demonstrating what appears to be a sense of empathy, compassion, fairness, reciprocity, and even justice (Bekoff 2007, Bekoff and Pierce 2009).

As earlier, science fiction provides tantalizing thought experiments for our consideration. In Fred Hoyle's novel *The Black Cloud*, an enormous cloud of gas enters the Solar System and settles between the Earth and the Sun, blocking the Sun's light and hence threatening all life on Earth. The cloud turns out to be an asocial superorganism, far more intelligent than us. The human heroes of the novel figure out how to communicate with it, but it has little *sympathy* for them per se. Nevertheless, the cloud finds the presence of small, primitively intelligent creatures on the surface of a planet intriguing; according to the cloud, such creatures are rare in the universe. It decides to move away from its position between Earth and the Sun, and eventually leaves the Solar System in a manner that deliberately avoids causing further damage when its curiosity is piqued by the ostensible disappearance of another black cloud not far away (by its standards). Somewhat surprisingly, there are analogs to the black cloud right here on Earth. While recent research indicates that they are not as fully asocial as once believed, octopuses are highly intelligent and curious animals, known to dismantle aquarium pumps and "sneak" into other aquaria during the night, eat the fish, and return to their own aquarium, shutting the lid behind them. Suppose for the sake of argument that octopuses *were* fully asocial creatures. Would this – as recently suggested by Kelly Smith (2014) – automatically render them secondary moral subjects, lacking in intrinsic moral worth? In this context, it is important not to tie their potential status as moral subjects to their lack of concern for others of their kind. Mentally defective human beings sometimes exhibit indifference towards others but surely this doesn't justify our treating them as "mere things"?

Finally, it is important to note that the *possession of an immortal soul* often comes up in Western theological discussions about moral status. In the Judeo–Christian–Muslim tradition, having a "rational soul" or a "divine spark" frequently plays the central role in determining moral status. On these grounds, many famous theologians have rejected the moral status of even our closest non-human relatives (chimpanzees, gorillas, and other primates) because they presumably don't possess a "rational soul" and unlike Adam and Eve were neither made in the image of God, nor conferred "dominion" over non-human animals (e.g. see Genesis lines 1.26–1.29, 2.19, and 9.3–9.6). In addition to excluding non-human animals here on Earth, this view poses serious problems for making sense of our ethical responsibilities in the event that we encounter extraterrestrial life. For even if we are willing to expand the notion of a (rational) soul to encompass certain non-human organisms, it remains unclear how we might objectively decide whether or not they possess rational souls. Moreover, should evidence of potential religiosity (or lack thereof) really shape whether or not we have moral obligations towards other species? In James Cameron's blockbuster film *Avatar*, this approach to moral status is challenged when a humanoid species, called the Na'vi, exist in harmony with the ecosystem of their planet, and have no clear conception of a rational soul outside their holistic spiritual beliefs about the all-encompassing energy of nature. Despite the fact that the Na'vi are clearly intelligent and sentient, humans tragically fail to understand the worldview of the Na'vi and their special relationship with the energy of their planet, Pandora. Such behavior has precedence in human history. In his research on anatomy, the sixteenth-century philosopher and scientist, Rene Descartes, dissected living dogs and horses without anesthesia. Since they were not human, he dismissed their cries of agony as analogous to the creaks and groans of damaged mechanical devices on the ground that they lacked souls (Descartes 1629). Of course, there is an ongoing debate among contemporary Western theologians on this issue (almost all would condemn what Descartes did today), and Eastern religious beliefs on the subject often differ radically from Western traditions. However, most theological standards for moral status tend to be highly anthropocentric insofar as they emphasize some kind of unique teleological connection between humanity and divinity (e.g. see Linzey 1994, particularly 3–27).

All four of the categories discussed above have a long history of being formulated and applied in a highly anthropocentric manner, and many people continue to conceive of them narrowly in terms of a "human-based standard." In order to extend our moral reasoning to extraterrestrial forms of life we need to reformulate them less anthropocentrically. In addition, we believe that secular approaches have a significant practical advantage over theological ones. First, the appeal of secular ethics is broader than theology. Although people are passionate about their religious beliefs, there are many different religions and not everyone holds a religious worldview. A secular ethics has the advantage of allowing people to talk with one another about issues of right and wrong by using language that does not require participants to adopt the same religious worldview. This has the potential to lay the groundwork for reasonable agreement among people of different religious persuasions. Second, as a consequence of the brute fact of religious pluralism, most modern constitutional democracies embrace the "principle of secularism" in public policy. For instance, the First Amendment to the US Constitution prohibits the state from officially endorsing one religion over another, as an attempt to keep the state from coercively interfering with institutionalized religion (and vice versa), in addition to protecting the individual right to religious freedom. As a practical matter, if we encounter life beyond our own planet, the norms guiding how we *should* interact with it will be addressed by governments through public policy. Although religious viewpoints have important contributions to make to debates about the social impact of astrobiology, the laws governing how we treat aliens will be formulated in secular terms. We thus recommend developing an ethics of astrobiology that addresses questions about the moral duties we might hold towards extraterrestrial organisms (i.e. or how we should treat them) in reference to the first three characteristics, viz. intelligence, sentience, and social behavior. In contrast, the chief theological quandaries in astrobiology, such as, "How would discovering alien life shape our religious vision of ourselves?" and "Do aliens have souls?," remain fascinating concerns for theologians to sort out in ways consistent with the sacred texts, traditions, and institutional doctrines of the many diverse religions in the world.

Conclusion

The purpose of this chapter has been to explore some of the challenges we face in applying our human-based ethical theories to truly alien creatures and to introduce a rough framework for beginning to think about these issues more clearly in the absence of concrete examples of extraterrestrial life. Despite a history of anthropocentrism, the categories of rationality, sentience, and social behavior are all broad enough to encompass a vast array of non-human organisms, as philosophers associated with the contemporary animal rights movement have argued (see, e.g., Peter Singer (1975) and Tom Regan (1983)). We do not know what we might encounter when it comes to life beyond Earth, but secular ethics gives us a useful theoretical foundation for confronting a wide range of quandaries in a more considered manner, ranging from helping us assess our potential moral duties towards less-advanced creatures to aiding us in determining when it might be acceptable to rely on self-defense against an aggressive species. This is precisely why such theorizing is so important!

In order to move away from many of the pitfalls of anthropocentrism, we suggest considering these characteristics as significant when they seem to appear in isolation, in addition to existing in tandem as they do in our own species. Indeed, this is one of our most central points. Kelly Smith (2014) has recently argued that "reason, sociality and culture all tend to arise in evolution as a co-evolutionary 'package deal,'" assuming that humans represent the teleological pinnacle of evolution. Although these features might indeed evolve together in *some* "advanced" species, this position fails to take seriously the $N = 1$ problem in biology or reach beyond narrow ethical anthropocentrism by presuming that these qualities must apply to *all* sophisticated forms of life. The previous examples from science fiction show the problematic character of Smith's assumptions. If an organism is profoundly rational but asocial, like the Black Cloud, might it still qualify for moral status based on its remarkable intelligence and curiosity? Or if it is sentient yet devoid of any sense of culture, like the Horta, might it nonetheless count as a moral patient on the grounds that it feels pain and is a caring mother? Finally, if a collection of organisms is unusually social, as are the members of the Borg hive, must they also exhibit rationality and emotions at the level of the individual in order to qualify as deserving of our moral consideration? Each of these examples points out serious limitations in judging the moral status of extraterrestrials using theories designed by humans *for* humans. As a consequence, we recommend that both astrobiologists and future public policy makers in government remain open to the possibility that any one of these characteristics might qualify an organism for a certain degree of intrinsic moral status. Now, all we need are concrete examples to consider in light of these guideposts!

References

Aristotle. *Politics*. [Reprinted in Barnes 1984, p. 1988].

Barnes, J. (ed. & trans.) 1984. *The Complete Works of Aristotle*. Princeton, NJ: Princeton University Press.

Bedau, M. A. and C. E. Cleland (eds.) 2010. *The Nature of Life: Classic and Contemporary Perspectives from Philosophy and Science*. Cambridge: Cambridge University Press.

Bekoff, M. 2007. *The Emotional Lives of Animals*. Novato, CA: New World Library.

Bekoff, M. and J. Pierce 2009. *Wild Justice: The Moral Lives of Animals*. Chicago, IL: University of Chicago Press.

Benner, S. A., A. Ricardo, and M. A. Carrigan 2004. "Is There a Common Chemical Model for Life in the Universe?," *Current Opinion in Chemical Biology* 8: 672–689. [Reprinted in Bedau and Cleland 2010, pp. 164–185].

Bentham, J. 1781. *An Introduction to the Principles of Morals and Legislation*. [Reprinted in Burns and Hart 1982, p. 282.]

Burns, J. H. and H. L. A. Hart (eds.) 1982. *An Introduction to the Principles of Morals and Legislation by Jeremy Bentham*. London: Methuen.

Cleland, C. E. 2012. "Life without Definitions," *Synthese* 185: 125–144.
Cleland, C. E. and C. F. Chyba 2002. "Defining 'life'," *Origins of Life and Evolution of the Biosphere* 32: 387–393.
Cleland, C. E. and C. F. Chyba 2007. "Does 'Life' Have a Definition?" In W. T. Sullivan III and J. A. Baross (eds.), *Planets and Life: The Emerging Science of Astrobiology*. Cambridge: Cambridge University Press, pp. 119–131. [Reprinted in Bedau and Cleland 2010, pp. 326–339.]
Cleland, C. E. and E. M. Wilson 2013. "Lessons from Earth: Towards an Ethics of Astrobiology." In C. Impey, A. H. Spitz, and W. Stoeger (eds.), *Encountering Life in the Universe: Ethical Foundations and Social Implications of Astrobiology*. Tucson, AZ: University of Arizona Press.
Descartes, R. 1629. *Treatise on Man*. [Reprinted in Hall 1972.]
Grinspoon, D. 1997. *Venus Revealed: A New Look Below the Clouds of our Mysterious Twin Planet*. Cambridge, MA: Perseus Publishing.
Hall, T. S. (ed. & trans.) 1972. *Treatise on Man: René Descartes*. Cambridge, MA: Harvard University Press.
Hume, D. 1783. *A Treatise of Human Nature*, Bk III, Pt. I, Sec. I. [Reprinted in Selby-Bigge 1978, pp. 455–470.]
Kant, I. 1798. *Lectures on Anthropology*. [Reprinted in Louden and Kuehn 2006, p. 15.]
Linzey, A. 1994. *Animal Theology*. Chicago, IL: University of Illinois Press.
Louden, R. B. and M. Kuehn (eds.) 2006. *Kant: Anthropology from a Pragmatic Point of View*. Cambridge: Cambridge University Press.
Mill, J. S. 1863. *Utilitarianism*. [Reprinted 2002. Indianapolis and Cambridge, MA: Hackett.]
Peters, T. 2013. "Astroethics: Engaging Extraterrestrial Life Forms." In C. Impey, A. H. Spitz, and W. Stoeger (eds.), *Encountering Life in the Universe: Ethical Foundations and Social Implications of Astrobiology*. Tucson, AZ: University of Arizona Press, pp. 200–221.
Regan, T. 1983. *The Case for Animal Rights*. Berkeley, CA: University of California Press.
Schulze-Makuch, D. and L. N. Irwin 2006. "The Prospect of Alien Life in Exotic Forms in Other Worlds," *Naturwissenschaften* 93: 155–172.
Selbe-Bigge, L. A. 1978. *David Hume: A Treatise of Human Nature*. Oxford: Clarendon Press.
Singer, P. 1975. *Animal Liberation*. New York, NY: HarperCollins.
Smith, K. C. 2014. "Manifest Complexity: A Foundational Ethic for Astrobiology," *Space Policy* 30: 209–214.

Strong, T. and D. Owen (eds.) 2004. *Max Weber: The Vocation Lectures*, trans. R. Livingston. Indianapolis, IN: Hackett Publishing Co.

Weber, M. 1917. *Science as a Vocation*. [Reprinted in Strong and Owen 2004, pp. 1–32.]

Weber, M. 1919. *Politics as a Vocation*. [Reprinted in Strong and Owen 2004, pp. 33–94.]

14 Astrobiology and theology

ROBIN W. LOVIN

Astrobiology requires us to rethink what is "universal" and what is "particular." The capacities and characteristics we have learned to regard as universally human – often after some effort to overcome the prejudices of our own race, culture, or class – may need to be viewed in a different light as we discover other possibilities for life in the universe. We may have to get used to thinking of the "universal" as particular to our own planet and species. This obviously applies to human biology, but it is equally true for our declarations about "universal human rights" and for philosophical ideas like "humans are political animals" or "all men are created equal." These universals are deeply embedded in traditions of thought and social institutions, but they may take on a different meaning when viewed in relation to other possible forms of life and intelligence.

This challenge is especially interesting when we think about religious traditions, which already speak about human universals in a frame of reference that transcends time and space. Religion, like astrobiology, locates life in the universe. It gives humanity a place in relation to reality as a whole. Perhaps that is why theologians have long been interested in the possibility of life on other worlds (Crowe 1997). A theology that understands humanity in relation to God cannot but be interested in how other life might participate in such a relationship, too.

For the most part, of course, the problems of terrestrial life give people of faith and their religious leaders quite enough to worry about. Providing universal safety, security, and peace for the one form of intelligent life we know exceeds our present capacities, and debate continues about exactly what the needs of that life are, especially when we move beyond biological requirements to consider social and political relationships.

Thus, an important concern in recent theology has been to explore the moral implications of the human dignity that all persons share. We are not only made of the same stuff. We are "made in the image of God," as some scriptural traditions put it. To be human makes us equal, and equal at a high rank that demands the kind of respect that modern politics formulates in terms of universal human rights (Waldron 2012). These claims need not be less true in the context of a universe that includes other forms of life and intelligence besides our own. But our idea of human dignity must surely be changed by thinking about it apart from the particular biological and cultural characteristics that mark it as *human* dignity. We need to prepare to see human dignity in the context of other life, and we need to prepare for questions about what our own dignity might tell us about our relations to that other life.

Thinking theologically

Like other disciplines, theology will need to approach this task in its own way. Over the past two centuries, Western theologians have absorbed scientific advances in biology with various levels of interest, openness, and enthusiasm. The story of how theologians have learned to think scientifically has often been told, but it may also be important for astrobiology that scientists understand how theology works.

As the term itself implies, theology is a way of thinking and speaking about God. Theology is a word (*logos*) about God (*theos*), as biology is a word about life or geology is a word about the Earth. To speak intelligibly about God, we must begin with some broader frame of reference that provides others with an orientation to what we are saying. That is why theology is almost always embedded in a tradition – a set of texts, narratives, arguments, and practices that provide the starting point for dealing with new questions. To say something new, a theologian begins by locating it in relation to received ideas that provide a framework in which new claims can be understood and evaluated. Sometimes, the new claims actually step outside the framework in which they are set or change it in significant ways, but revolutionary claims still need a tradition as a starting point, even when that tradition is being criticized, abandoned, or reformed.

A theological discussion is easier to follow if the interlocutors are speaking within the same tradition. Two Christian theologians will have much in common even if their versions of the tradition are as different as the Mennonites are from the Eastern Orthodox. The tradition plays a similar role, however, when the conversation moves between different traditions. When a Christian discusses human rights with a Buddhist or a Muslim, for example, each party has to know something about the other's tradition in order to make sense of what is being said. Things that sound radically different may have similar implications when seen against this broader background. Affirmations that sound very similar may mean quite different things in different traditions.

This embeddedness in tradition is essential because theology is an interpretive discipline. Theology's task is to make sense of reality as a whole and to provide an orientation for meaningful action within it. Where an experimental science begins with results, tests the reliability of its data, and works toward a theoretical construct, an interpretive discipline begins with an idea of the whole and tries to understand new discoveries and theories within it. Such interpretations cannot begin from nowhere. They need a tradition to provide a framework for understanding, however inadequate or inarticulate the interpreter's grasp of that tradition may be. The understanding can then be refined and enlarged in relation to new questions and choices.

This interpretive work has gone on in some tension with scientific investigation ever since science began to be distinguished from theology and philosophy at the beginning of the European Renaissance. But there have also been many fruitful collaborations. Our task is to think about ways that theology and astrobiology might collaborate in the future in ways that preserve the integrity of both theological interpretation and scientific investigation.

Human dignity

One way to focus those questions is to consider how theology's interpretations of the meaning and value of human life results in a unique moral status for persons. Many traditions hold that persons have a special place in the world precisely because they are able to understand their own existence in a framework that encompasses reality as a whole. To be able to interpret oneself in this way exceeds any form of self-awareness that we have found in other species and seems to call for a similarly unique kind of respect for those who share this distinctive kind of self-awareness with us. Theistic traditions would formulate this by saying that all persons have at least an implicit understanding of their relationship to God. They are aware that their life has a meaning that is not exhausted by their material circumstances and the results of what they do, and correspondingly, they have a dignity that does not depend on their place in a social or economic system, their accomplishments in comparison with others, or their identification with a racial, religious, or ideological movement.

Ideas like this have a place in many religious traditions. Human dignity is a shared theme that cuts across time and cultures, and it is part of what allows us to recognize many different traditions that vary widely in their history, beliefs, and relationships to society as all being *religious* traditions. But precisely because the interpretive work of theology takes place primarily within some particular tradition, we cannot explore the relationship between theology and astrobiology without examining at least one of those religious traditions to see how it works out there. So I begin with a Christian understanding of human dignity, with the idea of humanity as the image of God, which also has resonances in Jewish and Islamic traditions. It would be possible, of course, to ask the same questions of a quite different tradition, such as Buddhism. We could then have a theological discussion about which theology works better for our astrobiological purposes. That is the work of a growing and important field known as comparative theology (Clooney 2010), but it is a further stage beyond where we can go in this initial discussion. To do comparative theology, it is necessary first to master the way of thinking in at least one of the traditions that are to be compared.

The idea of human dignity plays a key role in contemporary Christian thinking about the human person. As the Second Vatican Council put it: "Contemporary man is becoming increasingly conscious of the dignity of the human person; more and more people are demanding that men should exercise fully their own judgment and a responsible freedom in their actions and should not be subject to the pressure of coercion but be inspired by a sense of duty" (Flannery 1975, 799). Some argue, indeed, that theology provides the only plausible account of human dignity. Without a relationship to God – active in consciousness, residual in the culture, or just vaguely present as an intuition – there is no good reason to think of ourselves or of other people as anything more than the contingent results of historical forces or instruments for the purposes of more powerful agents. We need not accept the most expansive versions of this claim to acknowledge, however, that theology plays an important role in contemporary thought and action on human rights (Waldron 2008). The question before us at the moment is how this theological understanding of human dignity might be changed by the discovery of life beyond Earth. What implications does astrobiology have for the fact that we are "increasingly conscious of the dignity of the human person" in a way that allows us to see ourselves as capable of giving moral meaning to our actions and requires us to see others in the same way?

The "image of God"

Theologians might reasonably approach this question with caution, since both astronomy and biology have at times undermined traditional religious ways of speaking about the distinctiveness of the human place in the universe. Indeed, the concreteness of the religious traditions on this point seems to be a potential source of problems, since traditions, unlike theologians, do not establish the human connection to a transcendent reality in abstract terms like "dignity." They speak, for example, of humanity being created "in the image of God," and the idea of dignity becomes more powerful precisely because the relationship to God is so concrete. Islam, Judaism, and Christianity all speak in these terms (Solomon *et al.* 2005, 147–179), taking as their starting point the first of the two creation accounts in the Hebrew scriptures, in Genesis 1:26–27.

The concreteness of the image can be a problem, of course, if it leads to an excessively literalistic interpretation of the text, but that is not the main problem. The more basic problem that astrobiology calls to our attention is the fact that when we speak about the "image of God," we are talking about *human* dignity. The qualifier, which ordinarily fades into the background precisely because human dignity is a status that is held by every person on the planet, suddenly becomes important when we think about encounters with life somewhere else in the universe. Just as astrobiology reminds us that our knowledge of life is, for the moment, confined to the single example of life on Earth, it confronts us with the fact that human dignity is the only kind of dignity we know.

Theology explores the implications of this dignity and thus establishes a moral relationship between persons. The image of God demands a certain kind of respect, both for ourselves and for other people. In the Hebrew scripture, it also establishes our relationship to other forms of terrestrial life, since the same verse assigns to humanity dominion over other living things (Genesis 1:26 NRSV):

> Then God said, "Let us make humankind in our image, according to our likeness; and let them have dominion over the fish of the sea, and over the birds of the air, and over the cattle, and over all the wild animals of the earth, and over every creeping thing that creeps upon the earth."

While an unqualified interpretation of "dominion" can lead to problems in caring for the environment, these problems, like others that arise from a literal reading of the creation narrative, can be resolved. Neither astronomy, nor biology, nor ecology has created insurmountable problems for theology, which has in fact been able to reinterpret the ethical implications of the creation narrative in light of those scientific facts (Johnson 2014). The new question that astrobiology raises is whether the ways that theology locates us in a moral world in relation to terrestrial life might also give us some guidance for dealing with other forms of life. Earth seems to be a moral planet. That is assured precisely by our shared human nature and our inescapable responsibility for the other forms of life to which we know ourselves to be chemically and biologically related. But what happens beyond Earth's biosphere?

Image and analogy

Dignity, then, seems to establish a distinctive moral responsibility for other terrestrial life among those who hold it, and a distinctive moral relationship of equality between them. The question is whether this dignity might also exist in a form of life that does not share the characteristics of human life that are associated with every instance of dignity that we now know, and how we might recognize non-human dignity if we did encounter it.

It might, of course, turn out that human dignity is the only kind of dignity there is, just as it might turn out that terrestrial life is the only life in the universe. Or we might have a run of good luck in which the only kind of dignity we encounter is close enough to human dignity that we could recognize it without much difficulty. We shouldn't expect working out moral relationships based on that kind of recognition to be any simpler than ending tribal warfare or abolishing slavery, but we can imagine how we would proceed. Maybe watching *Star Trek* reruns could accelerate the process.

Astrobiology, however, warns us not to expect that kind of discovery. The diversity and complexity of life on Earth suggest that the basic laws of physics and chemistry that govern organic processes do not establish any one trajectory for the development of life or allow us to say definitively what forms life might take. We do not know what kind of life we may actually discover, but systematic preparation for discovery has to assume the possibility of many forms of life, even many forms of intelligent life, that are radically different from our own. We may even have difficulty recognizing them as "alive." Could we recognize them as having dignity?

One way to explore that question draws on the analogies that many disciplines use to extend their knowledge beyond the immediately available evidence. Analogical thinking is a form of argument that science and theology share, and it provides one way to explain how theology can speak about God's being, will, and action in terms drawn from human experience (Polkinghorne 1998, 84–90). If the astrobiologist can build analogies from terrestrial life to explore the possibilities of life elsewhere in the universe and the theologian can build analogies from human being to the being of God, can analogical thinking help us to understand what dignity might be like in a form of life different from the only kind of life we know?

Consider how theology interprets human dignity as "the image of God" in the human person. If human dignity is the image of God in us, then it is not quite true that human dignity is the only kind of dignity we know. God, too, must have a dignity of which ours is the image. Everything in the tradition, however, recoils from saying that God has the *same* dignity that we do. The prohibition against making an image of God, which is embedded in the same texts that emphasize that humanity is made in God's image, serves as a warning against thinking that we can make sense of reality as a whole simply by projecting our own experience onto it.

Theology solves this interpretive conundrum through complex forms of analogical reasoning that determine what can be said about the reality of God on the basis of the human dignity which is made in God's likeness. The analogies focus on consciousness and freedom – an awareness of ourselves by which we also experience a certain transcendence of our circumstances and are able to imagine changes that we can effect in the world by our own voluntary action (Thomas Aquinas 1948, 583). Something in human consciousness resembles the way that God both holds all things in being and at the same time transcends them, so that in our knowing and acting, we have a kind of freedom that other forms of terrestrial life do not. This human creativity is a very limited image of God's awareness and freedom, but the analogy suggests what it is in us that both enables our relationship to God and constrains our relationships with one another.

Augustine described this image of God as the interplay of reason, will, and memory and suggested that these must exist in God in some way analogous to the way that they give us our identity as persons (Augustine 1991, 374–392). His development of the analogy included two other points that may be helpful for our purposes: First, freedom or dignity understood in this way is not an existence in splendid isolation. It includes relationship with other persons as an essential part of its being. That was how Augustine made sense of the Christian doctrine of the Trinity, the idea of three persons in one God (Augustine 1992, 279–280). Second, the dignity conferred by this kind of personhood does not exist at a moment in time. It must be grasped as a whole in order to be understood. A human life can only be judged by what it is at the end – not in the sense of what it happens to be at its last moment, but in the sense of what it has become through the whole course of its existence (Augustine 1998, 495–499). Thus we recognize in ourselves analogies to the way that God sustains being, order, and meaning in the universe, and we come to understand that kind of self-awareness and capacity for action as the image of God in us.

Non-human dignity

The possibility of discovering life elsewhere in the universe raises the question of whether this tradition of analogical thinking about God and human dignity could be put to use in a quite different way, to recognize the image of God in other forms of life that do not share the physical and biological history from which human life has emerged. We have already noted that theologians have long been interested in life in the universe. Their speculation about life on other celestial bodies was analogical reasoning of this sort: what they understood about God told them something about the possibilities for other life; the way that possible life was like or unlike human life told them something about themselves; and that, in turn, clarified something about their own relationship to God (Dick 1982).

What distinguishes the analogies found in earlier works from those theologians might develop today is the controlling influence of scientific evidence. While we obviously have no data drawn directly from extraterrestrial life, we know a great deal more than earlier thinkers did about how life works and what it requires, and we know more about what we do not know. Before scientific information about the unity and diversity of life was available, theologians constructed accounts of life on other worlds like other pre-modern and early modern thinkers. They thought in metaphysical terms. They worked out the basic structures of reality and then gave definitive answers to the question of what we would find, if we were able to explore the rest of the universe. A uniform, knowable, and known structure of possibilities determined reality everywhere. Analogy became a process of extrapolation, with just a few gaps in our knowledge to be filled in by investigation.

What modern science has taught us to this point is that what we know about the structures of reality does not provide such certainty. The diversity within uniformity of the universe of living things known to biological science is so great that we can't definitively conclude from the basic facts of chemistry and physics whether life is inevitable or accidental, and whether life develops inexorably toward intelligence, or whether it has done so just on one remarkable, odd occasion. To get the answers that would orient us to reality as a whole, we now have to go looking for the evidence.

Recognition that investigation is necessary and that certainty is impossible created the tension between science and theology at the beginning of the modern era, but it also bound the two ways of thinking together. We cannot live our lives without an orientation to reality as a whole like that which a religious tradition provides, and we cannot trust that that orientation is about reality unless we can understand it in relation to the results of scientific investigation. A great deal of the work in both religion and science over the past century has been devoted to testing our moral and religious understanding of human dignity against what science tells us about the origins of humanity, our relation to other species of life, and the biological foundations of our identity as persons.

We do not yet have an instance of extraterrestrial life against which we could similarly test our idea of human dignity, and it seems likely that long before a direct investigation of extraterrestrial intelligence becomes possible, we will have to find our way by analogy from two or three forms of simple biological life to the possibilities for intelligent life, reasoning from the way that our biology enables our intelligence to what might happen with other biologies in other places. The theologian who wants to explore the possibilities and limits of intelligent life as "the image of God" will have to pay attention to these investigations.

Analogical thinking and the dignity of life

Scientific thinking, as such, is not similarly dependent on the results of theology, but scientists as persons cannot escape the questions about our place in the universe and our relationship to it that we will all face upon the discovery of extraterrestrial life. For those questions, the analogy between human being and the being of God may help with an understanding of the analogies between ourselves and other forms of life and intelligence. Our preparatory thinking about such a discovery already suggests several important points:

First, given the statistical probability of intelligent life elsewhere in the universe, the fact that we have yet to detect any evidence of it leads to the conclusion that successful communication between different forms of life is a difficult achievement (Vakoch, Chapter 9 in this volume), probably requiring a highly developed technical society that makes a civilizational commitment to the project that extends across a considerable period of time (Peters 2013, 201–206). This suggests that *any* intelligent life we are able to discover, however different or distant it may be from ourselves, will have capacities for reason, will, and memory that are analogous to our own. There is, then, a presumption of dignity in any intelligent life that argues against attempts at conquest, conversion, deception, or exploitation, just as our own dignity argues against quick surrender to or slavish imitation of an intelligence that appears far superior to our own.

Second, astrobiology encourages us not to look at biological evolution in isolation, but rather to think in terms of a complex system of biological and cultural evolution (Dick and Lupisella 2009). Notwithstanding charming fictional encounters with individual extraterrestrials, any intelligent life elsewhere in the universe will present itself as the cultural expression of a biological infrastructure, not as individual life forms. We may suppose that we, too, will appear that way to other intelligent life that may encounter us. We should prepare for such encounters by beginning to think about the dignity of cultures or civilizations, as well as the dignity of persons.

Third, because astrobiology thinks in terms of the long history of life in any given environment, the discovery of any life beyond our own biosphere must be seen for what that life may become as well as for what it now is (Des Marais *et al.* 2008, 727–728). Indeed, because the speed of light imposes its own absolute limit on how fast information about other life can reach us, we must remember that all we can know of other life is what it was when our information about it started the long journey toward us. Just as Augustine warned that history, whether of an individual life or of the Roman empire, must be understood in relation to its ultimate destiny, the discovery of any other life will require us to ask questions about its evolutionary past and future, and how we understand ourselves in relation to it will depend on answers to those questions, not just on how we see it as we find it.

This longer perspective cautions us especially against viewing alien environments that lack highly developed life simply as being at our disposal. Just as the discovery of a different form of life elsewhere in the universe would forcefully remind us that we are related to every living thing on Earth, it should also remind us that any other life that has developed independently of us will have its own distinct origins and future. Responsibility for the physical and biological resources in our environment is an inevitable feature of our unique place in terrestrial life. That is the insight contained in Genesis' coupling of humans' dominion over the Earth with their creation in the image of God. We can exercise that responsibility selfishly, ignorantly, or carelessly, but we cannot refuse it. That mandate, however, does not extend to the ecospheres of other forms of life. We may, instead, have a moral obligation to respect these ecospheres and leave them undisturbed, even if they are not – or not yet – part of the biological infrastructure of an intelligent civilization. They cannot be treated simply as a resource for exploitation, nor is it obvious that we are justified in intruding upon them for purposes of investigation. In short, we should prepare for the discovery of even simple and "primitive" forms of life by beginning to think about the dignity of life itself. Without this preparation, we may do things that we later regret, just as we now regret the loss of terrestrial species to extinction.

Conclusion

Theology has been linked for a long time to speculation about the place of life in the universe. As systematic and scientifically rigorous astrobiology develops as a discipline, theological reflection on human nature and human dignity will have a role to play in how we understand the moral claims of life, whether our own or other forms as we may find them. In the distant future when direct exchanges with other intelligent life may be possible, the long terrestrial movement toward universal recognition of human dignity will be relevant to those transactions. In the meantime, astrobiological anticipations of such exchanges may help with the further development of our thinking about *human* dignity and about our responsibility for life on Earth.

References

Augustine. 1991. *The Trinity*. Translated by Edmund Hill. New York, NY: New City Press.

Augustine. 1992. *Confessions*. Translated by Henry Chadwick. New York, NY: Oxford University Press.

Augustine. 1998. *The City of God Against the Pagans*. Translated and edited by R. W. Dyson. New York, NY: Cambridge University Press.

Clooney, Francis X. 2010. *Comparative Theology: Deep Learning Across Religious Borders*. Malden, MA: Wiley-Blackwell.

Crowe, M. J. 1997. "A History of the Extraterrestrial Life Debate." *Zygon* 32: 147–162.

Des Marais, David and Joseph A. Nuth. 2008. "The NASA Astrobiology Roadmap." *Astrobiology* 8: 715–30.

Dick, Steven. 1982. *Plurality of Worlds: The Origins of the Extraterrestrial Life Debate from Democritus to Kant*. Cambridge: Cambridge University Press.

Dick, Steven and Mark Lupisella, eds. 2009. *Cosmos and Culture: Cultural Evolution in a Cosmic Context*. Washington, DC: NASA.

Flannery, Austin, ed. 1975. *Documents of Vatican II*. Grand Rapids, MI: Eerdmans.

Johnson, Elizabeth A. 2014. *Ask the Beasts: Darwin and the God of Love*. London: Bloomsbury.

Peters, Ted. 2013. "Astroethics: Engaging Extraterrestrial Intelligent Life-Forms." In *Encountering Life in the Universe: Ethical Foundations and Social Implications of Astrobiology*, edited by Chris Impey, Anna H. Spitz, and William Stoeger, 200–21. Tuscon: University of Arizona Press.

Polkinghorne, John C. 1998. *Science and Theology*. Minneapolis, MN: Fortress Press.

Solomon, Norman, Richard Harries, and Tim Winter, eds. 2005. *Abraham's Children: Jews, Christians, and Muslims in Conversation*. New York, NY: T&T Clark.

Thomas Aquinas. 1948. *Summa Theologica*. Translated by the Fathers of the English Dominican Province. New York, NY: Benziger Brothers.

Waldron, Jeremy. 2008. "Basic Equality." *Social Science Research Network*. Accessed: October 17, 2014. http://papers.ssrn.com/sol3/papers.cfm?abstract_id=1311816

Waldron, Jeremy. 2012. *Dignity, Rank, and Rights*. New York, NY: Oxford University Press.

15 Would you baptize an extraterrestrial?

GUY CONSOLMAGNO, SJ

In September, 2010, I had traveled to Birmingham, England, to give an astronomy talk at the Birmingham Science Festival. As it turned out, the day of my talk happened to coincide exactly with the visit of Pope Benedict XVI to Birmingham. I had agreed to be interviewed by the British press to publicize the festival, yet – understandably – all they wanted to ask me about was the Pope. But they kept asking me questions like, "What is your biggest source of conflict about the Pope?" Or, "Has the Pope ever tried to suppress your scientific work?" To my mind, these sorts of questions were bizarre; the mere existence of a Vatican Observatory ought to have shown them that their fundamental assumption about a hostility between the papacy and science was mistaken. Worse, the reporters seemed to be not at all interested in correcting that assumption, or hearing my description of the active support we had received not only from the Vatican in general over more than one hundred years, but more specifically from Pope Benedict himself. Finally, frustrated that they weren't getting the story they wanted out of me, one of them asked, "Would you baptize an extraterrestrial?"

My answer, off the top of my head, was, "Only if they ask!" It got a good laugh, which is what I wanted. And then, the next day, the press ran my joke as if it were a real story – as if I had made some sort of official Vatican pronouncement about aliens (Jha 2010). And that was merely the reaction of the relatively mainstream media. The more exotic wings of the internet were immediately full of wild interpretations, ranging from the thought that the Vatican was preparing the world for an imminent alien contact, to worries that somehow Jesus himself was an alien.

Of course, if these people had read their own newspapers, they could have found plenty of articles repeating this theme, over and over. For example, in 2008 the Vatican Observatory director, Fr. José Funes, SJ, had made essentially the same point in an article in *L'Osservatore Romano* and his comments had been widely reported. Indeed, the topic of baptizing extraterrestrials was considered worthy of comment in *Time Magazine* ... on September 19, 1955! It's a topic that has been covered in *Time*, the *Times* of London, and the New York *Times*; and yet somehow it remains timeless.

Why does the press have this reaction? For that matter, why do they ask that particular question in the first place?

Certainly, "would you baptize an extraterrestrial?" is a popular question; so popular that my colleague Paul Mueller and I decided to use it as the title of a recent book (Consolmagno and Mueller 2014; much of the material in this paper is based on material from that book). We obviously thought this title would sell copies. But why? Why do people ask us that question so often?

One reason is the challenge inherent in the question. Consider the context in which the British reporters asked it, along with the other questions they had been asking: "What is your biggest source of conflict about the Pope?" or "Has the Pope tried to suppress your scientific work?" They were aggressive queries; indeed, I suspect they were looking for ways to make me look foolish, or at least to make my religion look foolish. For them, "Would you baptize an extraterrestrial?" was a trick question.

If I had just said, "Yes, I would baptize ET," then I would have looked cosmically naïve. I would have been saying that a mere human being like me thinks he has the right to preach to aliens so far advanced that they can cross the incredible distances of space to visit us. On the other hand, if I had said, "No, I would not baptize ET," then I would be admitting that Christianity has no universal or cosmic significance, is nothing more than a local superstition, amusing for the yokels but not really important in the grand scale of things. But when I simply blurted out the first thing I could think of – "Only if they ask!" – I turned the tables. I made baptism not my decision, but ET's. If ET, with all its superior technology, decided freely to ask for baptism – if ET, with all its advanced knowledge, accepted that our human Savior really does have importance and meaning for it – then suddenly it'd make the reporters with their petty skepticism look pretty foolish.

Theological issues

There are two issues at stake in the question of "baptizing" ET. The first is reconciling the (proposed) existence of intelligent beings other than human with an understanding of humanity's place in the universe; the second is understanding the relationship of those creatures, should they exist, with the religions and gods of humans.

The theological issues are fascinating and hardly settled. The late Fr. Ernan McMullin, a theologian at Notre Dame, once commented that although discovering a race of extraterrestrials might mean we would have to re-think our understanding of some religious principles such as how original sin came about, that wouldn't really be anything new; since even theologians today can't agree on the origin of original sin! (McMullin 2000). In other words, finding intelligent aliens may feed into our present theology and give us new things to think about; but our theology today is already constantly challenged, and we are constantly being called to grow in that faith as a response to those challenges. It is beyond the scope of this short paper to do justice to the range of theological issues; for that, a good place to start is a recent book by the astronomer and theologian David Wilkinson, *Science, Religion, and the Search for Extraterrestrial Intelligence* (Wilkinson 2013).

These issues are by no means new, of course. The context of the questions, however, has changed as cosmologies have developed. The flat-Earth cosmology of Babylon, reflected in the opening chapters of Genesis, had room for other intelligent creatures as described in the ancient Greek epics; and within that universe, the intelligent monsters that, for example, were encountered in *The Odyssey* were subject to the same gods as the heroes themselves. Even Genesis (in chapter 4) refers to non-human intelligences such as the "sons of gods" and the otherwise mysterious "Nephilim." If nothing else, this shows that even at its earliest roots neither Judaism nor human culture in general were fazed by the thought of intelligent creatures other than humans, and were able to fit them into the prevailing theology of the time.

The later cosmology of the Greeks envisioned the universe as a series of nested crystalline spheres carrying the planets about the Earth, powered by various planetary intelligences – the "powers and principalities" discussed by St. Paul in his letter to the Ephesians (Purfield 2009) and in later times identified with various ranks of angels (chapter 4 of Lewis 1964). The universe in this cosmology was filled with beings other than humans who were in some sort of relationship with God. The legends of the fall of the angels (as ultimately portrayed in the work of John Milton, reflecting back on this cosmology at the dawn of the scientific revolution) also demonstrated that the same issues of sin and death, the choice of good or evil, and the need for some sort of redemption were not considered issues unique to human beings.

Even when the best science of its day, Aristotelianism, assumed that Earth was the only world, philosophers still debated whether God could have created "other worlds" (in essence, parallel universes). The decided opinion of Bishop Tempier of Paris in 1277 was that an omnipotent God must have the power to create such parallel universes, even though the prevailing scientific wisdom of the day said such universes were impossible. On the other hand, Aristotle's science was not the only cosmology current during the high middle ages. Bishop Nicholas of Cusa, in his book published in 1440, *De Docta Ignorantia*, maintained that every star is its own world: "Therefore, the Earth, which cannot be the center, cannot be devoid of all motion … just as the Earth is not the center of the world, so the sphere of fixed stars is not its circumference … Therefore, the Earth is not the center either of the eighth sphere or of any other sphere" (Nicholas of Cusa 1985, 90). This led him to speculate on the possibility of life around those suns: "In like manner, we surmise that none of the other regions of the stars are empty of inhabitants . . . the one universal world is . . . in so many particular [parts] that they are without number except to Him who created all things (Nicholas of Cusa 1985, 97). For a further discussion on the historical development of cosmology and its parallel development of theology, see Consolmagno (2012).

Thinkers on both sides of the religious issue have used the possibility of extraterrestrial life to support their preconceptions. The German theologian Joseph Pohle in the early twentieth century argued that the glory of God demanded a universe filled with intelligent beings, not just us. John Herschel, founder of the Royal Astronomical Society (and son of Uranus discoverer William Herschel) came to a similar conclusion a century earlier. On the other hand, the American radical Thomas Paine mocked Christianity by arguing in *The Age of Reason* (published in 1794) that Christianity demanded either the unlikely proposition that of all the worlds in the universe, God chose to be incarnated only in ours, because "one man and one woman had eaten an apple"; or else, there were many incarnations, such that "the person who is irreverently called the Son of God . . . would have nothing else to do than to travel from world to world, in an endless succession of death, with scarcely a momentary interval of life." A more extensive discussion of the history of these debates can be found in Crowe (1986) and O'Meara (2012).

Aliens and religion in science fiction

One useful way to approach our modern understanding of these issues is to see how they are treated in contemporary fiction. The genius of genres like fantasy and science fiction is that they take big issues and turn them into specific stories set in particular moments in the lives of individuals. A good story is based on a conflict of world-views leading to an individual's personal dilemma. If nothing else, reading a story that puts us into the shoes of someone trying to work out that dilemma can allow us to understand and even be sympathetic with people whose choices might, ultimately, be the wrong ones. Fantasy and science fiction has the advantage of putting these issues in a sufficiently different time and place so that we can suspend some of our original reactions long enough to get to understand the other side. The story's tension also comes from recognizing as well that bad choices, even made with the best of intentions, still have bad consequences that we have to deal with.

The literature of science fiction is filled with alien creatures, or sentient computers, or half-human/half-machine constructs. Fantasy stories add the whole spectrum of mythical elves and ghosts. But note that the central character of any such story, regardless of how many tentacles it has, is recognizably human: self-aware, free to choose, to love, or to hate. Free to do good; or to sin and thus in need of redemption. It's no surprise that so many of these stories have been written by deeply religious people, many of them (including Anthony Boucher, Gene Wolfe, and J. R. R. Tolkien, to name but three) Catholics.

A number of stories specifically look at the interaction of the Church with alien life forms. Two famous ones are James Blish's *A Case of Conscience* (published in 1958) and *The Sparrow* by Mary Doria Russell (published in 1996). In both, Catholic priests as protagonists encounter a race free from original sin, with tragic results. Neither author is Catholic, which unfortunately shows in their inaccurate depiction of Catholic theology in those stories. For example, Blish's description of Catholic teaching of evolution was not only inaccurate, it had already been specifically contradicted by a papal encyclical of Pope Pius XII in 1950. (For a more detailed discussion, see the chapter on *A Case of Conscience* in Walton 2014.)

There are many other relevant recent examples of science fiction speculations on alien interactions with religion. For example, *Eifelheim* by Michael Flynn (published in 2006) envisions an alien spacecraft crashing in fourteenth-century Germany, during the time of the Black Death, and the stranded aliens interact with an educated clergyman there. "The Way of the Cross and Dragon," the Hugo winning short story (1980) by George R. R. Martin, centers around an interstellar descendant of the Catholic church, where certain aliens embrace Catholicism more profoundly than the humans who brought its teachings to them. And Cordwainer Smith's Underpeople stories, most especially "The Dead Lady of Clown Town" (published in 1964) rely on specific Christian imagery; the "dead lady" of that story, D'Joan, is a Joan of Arc character inspired by a futuristic version of Christianity, fighting for acceptance as a rational being in spite of being derived from dog genes.

It is not only Christianity in contact with alien races that is discussed in science fiction. A recent (2014) novelette, "Three Partitions" by Bogi Takács, very specifically deals with the interaction of Judaism (and its gender rules) with aliens. The classic story "A Rose for Ecclesiastes" by Roger Zelazny (published in 1963) is a complex look at the interaction of terrestrial and alien religions and the individuals (human and Martian in this case) who profess them. In China Mieville's *Embassytown* (published in 2011) truth-telling aliens encounter their own version of original sin when they learn to tell lies. And this is just a small sampling of an extensive science fiction literature that deals with aliens and religion.

A common insight of these stories is that any creature of this universe, created and loved by the same God who created and loves us, would be subject to not only the same laws of physics and chemistry, but also the same rules of right and wrong. What else is there, except the superficial accidents of the gas they breathe or the number of genders they have, that makes them essentially any different from us? Is there any important way at all that they would deserve to be called alien – or that we would be alien to them?

It is a truism of science fiction that its stories, though set ostensibly in other places or times or universes, are really about the time and place and universe of the writer. That is certainly true of the stories listed here; for example, the Cordwainer Smith books were written at the height of the modern civil rights movement, while the Takács novelette incorporates current themes of gender identity.

In the same way, discussing the baptism of extraterrestrials tells us nothing about extraterrestrials, but may help us understand baptism and what it means to be human compared to the other possibilities. That is perhaps the strongest justification to pursue such speculations.

Aliens, salvation, and ethics

We have to recognize, however, that there is another reason why a lot of people are not merely curious, but hungry, to be visited by alien beings: the hope that any race advanced enough to cross the stars must also be advanced enough to know how to overcome all the human ills – injustice, crime, war – that make our world such a difficult place to live. They look to aliens to be the saviors of humankind.

However, there is a fundamental flaw in the assumption that ET will serve as a savior for humankind, similar to the false assumption made by the reporters who thought that some Earthling teaching ethics to an advanced race of space travelers would be the height of hubris.

That false assumption was nicely laid out in a recent example of one of those emails we often get at the Vatican Observatory. This one demanded that I somehow get Pope Francis to tell us "The Truth about ET ... ", which the correspondent was convinced he already knew. He wrote, "ET life is likely to be more ethically evolved and less satanic than humans ... Pope Francis must emphasize themes of extraterrestrials: not sharing in original sin; being more ethically evolved; and being capable of sharing the Christian message ... "

But in fact, who's to say they're better, or worse? Remember the insight of science fiction: any creature of this universe should be subject to not only the same laws of physics and chemistry, but also the same rules of right and wrong.

Determining what is morally right and wrong is fundamentally different from determining, say, the best way to make a faster-than-light (FTL) drive.

Consider: at the moment, we on Earth do not know how to build an FTL drive. Indeed, as far as we can tell, such a drive would actually violate our understanding of fundamental physics. But say some future Einstein discovers a new physics that shows a way to build such a drive, and such a drive is built. The fact that the drive works would validate the new physics, at least to some degree. From that point forward, the discovery remains discovered, and a part of our legacy for as long as those drives exist. The physics is "built into" the technology. Even if following generations were to forget how the physics works, the drive would still function.

The opposite is not, and cannot be, true of moral decisions. We may have the examples of great saints and heroes in our past whom we honor for making difficult and courageous decisions. We can learn from their example, and attempt to imitate their courage. But we cannot substitute their decisions for any decision that we have to make. We cannot slavishly say that because Socrates drank the hemlock, that's always the right choice to make; that because Caesar crossed the Rubicon, we don't have to agonize over the Rubicons in our own lives. Indeed, we can still argue today whether the decisions of Caesar or Socrates really were moral or correct. But to insist that those decisions in the past relieve us of the responsibility of making our own decisions is patently false; if that were true, it would limit our own freedom and thus our own human dignity.

Indeed, the very fact that we can find examples of ethical behavior in the past that still are models for our behavior today shows that the general quality of ethical decisions open to human beings does not develop with time; compare this to the way that, say, programmers today can make computers do things that were impossible to do in 1955. Paint and canvas technology today has improved over that available during the Renaissance; the same cannot be said for the quality of the artistry displayed by painters now, compared to then.

Scientific knowledge is cumulative. Other kinds of knowledge, such as moral knowledge, are not. This was a key point of Pope Benedict XVI's encyclical letter *Spe Salvi*, "Saved in Hope" (Benedict 2007). Answers to questions about right and wrong can never get "built into" human society the way that scientific answers get "built into" technology. As individuals we must have the freedom to make our own choices, including our own mistakes. And those mistakes are important.

If you were to encounter a race that has never sinned, how would you know that they had the freedom that is essential for the ability to be truly good? (That is one of the fatal theological errors in both *A Case of Conscience* and *The*

Sparrow, cited above.) But once you do allow for mistakes, you also need a way to remedy those mistakes. This is one of the features of "baptism" in its religious sense: it is an invitation for the redemption of the person who was baptized from the mistakes that haunt their past.

But since ethics is not cumulative, there is no reason to suppose that a more technically advanced race would necessarily have found answers to the moral problems that plague us, and thus no longer stand in need of redemption. Indeed, we should be very suspicious of any race that suggested it had such answers to prevent us from causing each other harm; such a system, even if it worked, would by its nature be totalitarian, denying us our freedom to choose right or wrong. If any such structure prevents me from doing evil, it prevents me from being human; instead, it turns me into a "clockwork orange" (as described in Anthony Burgess' classic science fiction novel from 1962, and the 1972 film based on it).

We have mentioned the redemptive nature of baptism; but that is not the only sense in which baptism is understood, and redemption is not the only function of baptism. Besides effecting a religious change within the baptized, baptism is also a rite of passage that signifies entrance into a closed group. Thus, "Would you baptize ET?" can be heard in different ways, with different emphasis. It can be heard either as asking if ET is someone to be baptized, or one could ask, who is to decide to let ET into our club?

The early Christians had to deal with the same ambiguity, when (as described in chapter 15 of the Acts of the Apostles) Paul argued with the other apostles about whether Gentiles (non-Jews) could be baptized. That's what Pope Francis was referring to in a recent homily; he used a colorful image of baptizing an ET to describe how the early Church reacted when gentiles approached the apostles and asked to be baptized. Imagine, he suggested, if "a Martian with a big nose and big ears came up and asked for baptism. What would you do?" (Naturally, the press decided that the Pope had just endorsed extraterrestrial baptisms . . .)

All religions have some sort of rite of passage in this sense, whether or not it is called "baptism." Once you've gone through the rite of passage, once you're in the club, you are a peer, an equal, and you have new rights and privileges. But life within a religion presumably is about more than just rights and privileges. So the question becomes more than a religious one, but one of authority; and this issue of authority will have consequences in the day-to-day interactions between various intelligent species. This is the topic as well of numerous science fiction stories, such as Michael Burstein's 2008 story "Sanctuary" about a non-human fleeing to a chapel on a space station run by terrestrials, seeking sanctuary from pursuit by the authorities of her planet.

This question of inclusion returns us to a basic issue in the question of "baptizing ET," which is our own assumed belief in the privileged role of ourselves (either as a person, or as a species), in the universe. The assumed tension is that finding other species might somehow challenge that privileged role. Some find this challenge a threat; others, a necessary balance against human pride.

Just as we are rightly wary about imposing our view of religion – for or against – onto any aliens, we must also be wary of imposing our own understanding of religion onto the way we do the science that might find those aliens: instead of baptizing aliens, we wind up "baptizing" science – imposing our own expectations on what we discover as a confirmation of either our own hopes or our own fears. We can see this occurring in the way that cosmology has been misused in religious arguments: some believers look at the latest Big Bang theory and say, "aha, the universe started with light, just like Genesis said!" while non-believers can take the possible origin of the universe as a quantum fluctuation of the gravity field and say, "aha, no need for God to start the universe!" Either way, it's a circular argument; you wind up concluding the very thing you assumed. This sort of fallacy is what happens when believers insist that finding extraterrestrial intelligence would confirm their belief in God, while the atheists insist just as strongly that such a discovery would prove all religions were meaningless. (Interestingly, none of these people are planning to change their faith, or lack of faith, in the face of the complete absence of evidence for ET up to now!)

In part, this difference arises from the fact that atheists and Christians have very different ideas of God. You have to have a clear concept of God in order to be sure you don't believe in it. But there are a lot of concepts of God which atheists reject, that most Christians would reject as well – such as the idea that God is merely centered on humanity and Earth.

In the same way, a real danger exists in assuming our own categories of religion when attempting to interpret or understand the social behavior of others. "Religion" plays so many different functions in our own society that go far beyond the level of theology: it is a marker of social class, of ethnic origin, of presumed moral behavior. It's common enough even in human interactions to make unwarranted assumptions and stereotypes across these lines – "all Hispanics are Catholic," or "all scientists are atheists," or "all persons who attend church regularly are deeply religious." In fact, reality is always far more complicated. A common flaw in many science fiction depictions of other species is that too often, while humanity is allowed its range of cultures and religions, all creatures from some other planet are assumed to be of the same religion, ethnicity, and culture.

The meaning of the question

So let us return to the question raised at the beginning of this chapter. Rather than expecting one could answer the question, "would you baptize an extraterrestrial?," what does it mean that this question itself is so commonly asked?

To some, the question is a shorthand way of demonstrating either the foolishness of believing in baptism, or the foolishness of believing in extraterrestrials. But as we have seen, this sort of attack is based on a fundamental misapprehension of the nature of transcendentals as compared to technology. A more technically advanced civilization, alien or otherwise, is by no means going to be more "ethically advanced." Ethics, truth, beauty, and other such transcendentals are not entities that can be quantified or accumulated. By their nature, requiring as they do free choices, they must be constantly sought but never "solved."

And yet, at the same time, our fundamental assumption (reflected in the stories we tell) is that other beings would be subject to the same propensity to failure and thus open to the same need for redemption. If that is so, then the need for something akin to "baptism" would be as universal among all intelligent species as the need for an energy source to fuel biological functions. The issue then is not, whether aliens would need it, but if they would want it from us.

By that same notion, naïve hopes or fears of encountering an alien intelligence as either the hope or the doom of humankind, neglects the inevitable complexity of any such encounter.

The fact is, we do not know at all how discovering extraterrestrial intelligence will change our religions or our conception of religion. In the absence of experiencing such a discovery, the best we can do is speculate. That is the sort of "what if" problem that science fiction is perhaps best qualified to approach.

Change in our understanding of the universe will always occur ... but questions of how we understand our place in this universe, and our relationship with whatever we identify as the source of truth, goodness, and love, will never cease to be mysteries for deeper contemplation. That remains true, whether or not we ever find extraterrestrials.

References

Benedict XVI. 2007. *Spe Salvi*. Accessed October 30, 2014. http://www.vatican.va/holy_father/benedict_xvi/encyclicals/documents/hf_ben-xvi_enc_20071130_spe-salvi_en.html.

Consolmagno, Guy J. 2012. "The new physics and the old metaphysics: an essay for the use of Christian teachers." *International Studies in Catholic Education* 4: 111–121.

Consolmagno, Guy and Paul Mueller. 2014. *Would You Baptize an Extraterrestrial?* New York, NY: Image Press.

Crowe, Michael. 1986. *The Extraterrestrial Life Debate 1750–1900: The Idea of a Plurality of Worlds from Kant to Lowell.* Cambridge: Cambridge University Press.

Jha, Alok. 2010. "Pope's astronomer says he would baptise an alien if it asked him." *The Guardian*, September 17. http://www.theguardian.com/science/2010/sep/17/pope-astronomer-baptise-aliens and http://www.dailymail.co.uk/sciencetech/article-1312922/Pope-astronomer-Guy-Consolmagno-Aliens-souls-living-stars.html)

Lewis, C. S. 1964. *The Discarded Image: An Introduction to Medieval and Renaissance Literature.* Cambridge: Cambridge University Press.

McMullin, Ernan. 2000. "Life and intelligence far from earth: formulating theological issues." In Steven J. Dick, ed., *Many Worlds: The New Universe, Extraterrestrial Life and the Theological Implications.* Philadelphia, PA: Templeton Foundation Press, pp. 151–175.

Nicholas of Cusa. 1985. *Nicholas of Cusa on Learned Ignorance: A Translation and Appraisal of De Docta Ignorantia*, trans. Jasper Hopkins. Minneapolis, MN: The Arthur Banning Press.

O'Meara, Thomas F. 2012. *Vast Universe.* Collegeville, MN: Liturgical Press.

Purfield, Brian. 2009. "The Letter to the Colossians: Jesus and the Universe." Accessed October 30, 2014. http://www.thinkingfaith.org/articles/20090623_1.htm

Walton, Jo. 2014. *What Makes This Book So Great.* New York, NY: Tor Books.

Wilkinson, David. 2013. *Science, Religion, and the Search for Extraterrestrial Intelligence.* Oxford: Oxford University Press.

Part IV Practical considerations

How should society prepare for discovery – and non-discovery?

Introduction

Over the past several decades sporadic attention has been given to the impact of discovering life beyond Earth, beginning with NASA's "Cultural Aspects of SETI" workshops in the early 1990s and extending to meetings sponsored by institutions as diverse as the Templeton Foundation, the Foundation for the Future, the American Association for the Advancement of Science, and the Royal Society of London (see Dick 2012, p. 917, for an overview and references). These meetings only partially dealt with practical steps to prepare for such a discovery. Here we enter the policy arena, and not just in a theoretical way. During Congressional hearings on astrobiology held in 2013 and 2014, members of Congress wanted to know "What should we do if life is found beyond Earth"? (US Congress 2013 and 2014). This is a policy question with no agreed-upon answers at this time. Discussions such as those found throughout this volume are essential background to decisions that will inevitably have to be made in the event of discovery. This section continues that discussion, but also attempts to tackle policy problems more directly, both from the point of view of approaches and practical steps.

If we are going to discuss policy at the interface of astrobiology and society, it would seem prudent to ask what lessons can be learned from the approaches, issues, and answers provided in previous endeavors such as the Human Genome Project, or in biology and society programs that exist at several universities around the world. In the opening chapter of this section, historian of science Jane Maienschein gives us the benefit of her experience as the long-time Director of the Biology and Society program at Arizona State University. She systematically lays out the issues and urges humanistic approaches to astrobiology. An important lesson learned is that scholars from the humanities, social sciences, and other areas cannot isolate themselves in their discussions and recommendations, but must interact in a meaningful way with both scientists and policy makers. Otherwise social science deliberations will remain an academic exercise divorced from political and scientific reality. Indeed, interactions across the sciences and social sciences in the astrobiological context bid fair to serve as a leading edge of what Harvard biologist E. O.

Wilson has called consilience, the unity of knowledge across the humanities, social sciences, and natural sciences (Wilson 1998; Finney 2000).

With this important caveat and call for consilience, the authors in this section examine policy considerations from a variety of perspectives. Margaret Race, a senior scientist at the SETI Institute who has led efforts to build an interdisciplinary community such as Maienschein describes (Race *et al.* 2012), analyzes in considerable detail what has been done to prepare for discovery in three search domains: SETI, extrasolar planets, and the Solar System. She further parses each of these search domains into three search phases, evaluating policy preparedness during the search, upon discovery, and post discovery. She finds all search domains generally well prepared during current searches, but identifies notable gaps especially in the Solar System arena with regard to policy both upon discovery and post discovery. This is an important finding since most astrobiologists consider the discovery of microbial life in the Solar System to be the scenario most likely to occur first. Former US State Department official Michael Michaud focuses further on the SETI domain. A long-time participant in drafting protocols in the event of a SETI discovery, Michaud outlines 16 concrete steps to prepare ourselves for discovering extraterrestrial intelligence. These range from the study of potential benefits and risks to raising consciousness among policy makers of the options, and learning to live with ambiguity.

In Chapter 19 John and Julian Traphagan crucially remind us once again that we must not be parochial in our considerations of the impact of extraterrestrial life. In particular we cannot be too Western-centric. Alternative worldviews do exist on Earth, and will certainly exist on other worlds. It therefore behooves us to study those alternative terrestrial worldviews in the context of their implications for astrobiology. The Traphagans point us to the obscure but important work of E. M. McAdamis (2011) on the capacity of world religions to contextualize the implications of finding life. The bottom line is that non-Western scholars must be drawn into the field if we are going to have a robust discussion. The Traphagans illustrate their point with a discussion of Buddhism, a worldview with 400 million adherents. They warn of the difficulties of generalizing any one religion, yet emphasize that Buddhists have a different way of seeing space and time in a philosophical (as opposed to a scientific) sense. And they conclude that, as a result, the world's Buddhists would have a very different reaction to the discovery of life. Surely the same is true of other religions, including the worldview of more than a billion Muslims.

Is it possible that life and intelligence do not exist beyond Earth – that the answer to the plaintiff question, "Are We Alone?," is "Yes"? It is indeed possible. Despite the scientific and philosophical arguments in the first two chapters in this volume, and the recognition that the Copernican and Darwinian preconceptions point toward a universe filled with life, even examined preconceptions are still preconceptions. The last two chapters of this section address a critical question not often enough raised in the context of astrobiology: what if no discovery is made? Linda Billings, an expert in communications and culture and a long-time participant-observer of the astrobiology and SETI communities, discusses how both the media and SETI "true believers" have conditioned the public to believe in an outcome that may in fact never occur. In addition, she demonstrates how the cultural environment and the human psyche shape both our representations of extraterrestrial life, and our potential reactions to its discovery. By arguing that "what we know about extraterrestrial life, of any sort, is nothing," she injects a healthy dose of skepticism into the volume. Astronomer Eric Chaisson takes this a step further, postulating what the absence of discovery – a "null signal" – may mean to humanity. Approaching the problem from his long-time work on ordered complex systems and energy consumption across a wide range of both the natural and technological world, Chaisson concludes that the message may be that advanced technological civilizations need to harness the energy of their parent star as soon as possible, or be "removed from the population of galactic civilizations." This turns the usual argument on its head: rather than finding salvation in extraterrestrials, the lack of extraterrestrials may be telling us something about what we need to do to ensure our own salvation as a species. While a sobering way to end the volume, this final chapter brings home the important point that the extraterrestrial life debate may have something important to offer even if life is never found beyond Earth – if not specifically in terms of energy alternatives, then certainly and more philosophically in illuminating our place in the universe.

References

Dick, S. J. 2012. "Critical Issues in the History, Philosophy, and Sociology of Astrobiology." *Astrobiology*, 12: 906–927.

Finney, B. 2000. "SETI, Consilience and the Unity of Knowledge." In *When SETI Succeeds: The Impact of High Information Contact*, ed. Allen Tough, 139–144. Bellevue, WA: Foundation for the Future.

McAdamis, E. M. 2011. "Astrosociology and the Capacity of Major World Religions to Contextualize the Possibility of Life Beyond Earth." *Physics Procedia*, 20: 338–352.

Race, M., K. Denning, C. M. Bertka, *et al.* 2012. "Astrobiology and Society: Building an Interdisciplinary Research Community." *Astrobiology*, 12: 958–965.

Wilson, E. O. 1998. *Consilience: The Unity of Knowledge*. New York, NY: Knopf.

16 Is there anything new about astrobiology and society?

JANE MAIENSCHEIN

At the intersections of biology and society, scholars have long explored ethical, legal, policy, economic, and other social issues, while also placing emerging science in the context of history and philosophy of science. One tradition has focused on the impact of scientific developments on society, reflecting on eugenics, recombinant DNA, reproductive technologies, human subjects experimentation, genetically modified foods, and other issues in largely reactive ways. Others are trying to anticipate where the science will be going and to outline issues that society is likely to face. Synthetic biology and technologies such as human reproductive cloning raise additional questions about whether we should forbid some science altogether. Stem-cell research or genetic engineering of food and people raise questions about appropriate regulatory responses. Experiments with pathogens and sequencing genes of dangerous organisms raise questions about control of knowledge. The National Science Foundation's Program on the Science of Science Policy explores issues of how science policy gets made and what factors influence the decisions.

Astrobiology falls into this complex world of biology and society, and here I ask: "Is there anything new under the Sun?" Or, more precisely, "Is there anything new under and beyond the Sun?" Have we already heard all the issues and are now just applying them to astrobiology in particular? Or are there special features of astrobiology that call for new thinking or raise new questions? Providing answers requires thinking about the presumed domain of astrobiology, then its implications, which in turn benefits from a look at the context of issues of biology and society more generally.

The domain of astrobiology

NASA defines astrobiology as "the study of the origin, evolution, distribution, and future of life in the universe." Further: "This multidisciplinary field encompasses the search for habitable environments in our Solar System and habitable planets outside our Solar System, the search for evidence of prebiotic chemistry and life on Mars and other bodies in our Solar System, laboratory and field research into the origins and early evolution of life on Earth, and studies of the potential for life to adapt to challenges on Earth and in space" (NASA 2014a). The "astro" is only part of the story, which includes the origins and evolution of all life.

Biology is the study of life, of living organisms and systems, their processes, and their interactions with the environment. In their excellent look at the history of astrobiology, Steven J. Dick and James E. Strick explain the evolution of approaches to astrobiology in particular and how it came to include the study of life on Earth. First there was exobiology, meant to focus on extraterrestrial life. Then, for a number of strategic reasons that they explain, NASA expanded to include origins and evolution more generally, with the initial 1998 articulated set of roadmap goals to (Dick and Strick 2005, p. 219):

1. Understand how life arose on Earth.
2. Determine the general principles governing the organization of matter into systems.
3. Explore how life evolves on the molecular, organism, and ecosystem level.
4. Determine how the terrestrial biosphere has coevolved with the Earth.
5. Establish limits for life in environments that provide analogues for conditions on other worlds.
6. Determine what makes a planet habitable and how common these worlds are in the universe.
7. Determine how to recognize the signature of life on other worlds.
8. Determine whether there is (or once was) life elsewhere in our Solar System, particular on Mars and Europa.
9. Determine how ecosystems respond to environmental changes on time scales relevant to human life on Earth.
10. Understand the response of terrestrial life to conditions in space or on other planets.

More recent versions of the roadmap have revised the goals somewhat, though the overall intent remains (Des Marais *et al.* 2008).

Goals 1–5 and 9 fall into the category of terrestrial biology and are not particularly "astro." Goals 6–8 turn outward, asking about the universe more generally – about habitable worlds, signature of life, and where life has existed. This is study about, though not study of, life that might exist extraterrestrially. Goal 10 takes terrestrial life into space and asks how it might fare. None asks about possible risks of bringing life forms from space to Earth – accidentally or on purpose.

Dick and Strick make clear that from the beginning of the space program concern about possible back contamination was nonetheless real. In 1960,

Norman Horowitz wrote a memo to Joshua Lederberg and invoked Christopher Columbus proposing to set sail, and noted that the fear of encountering risk and in particular any foreknowledge of syphilis would have kept Columbus from leaving. "Suppose, however, that they had known also of the tremendous benefits that were to flow from the discovery of the New World. Can there be any doubt what their decision would have been then?" (Dick and Strick 2005, 59). Yet worries grew stronger for some during the Cold War, with fears that enemies might try to use astro-knowledge or astro-materials as weapons. It seems a bit surprising, therefore, that protection is not listed as one of the astrobiology program goals. Presumably this is because of the existence of NASA's separate Office of Planetary Protection (Meltzer 2012). Yet surely that office is worth acknowledging. An eleventh goal might be in order; something like: "11. Understand the response of terrestrial life to life and materials from space."

Implications of astrobiology

For now, let's set aside the general goals of studying life on Earth, its evolution, and environment, which do not raise new questions in themselves. Instead, let's ask about three areas where astrobiology seeks to do something different, namely: (1) how to study and manage extraterrestrial life that ends up on Earth (either imported through back contamination or travelling here on its own); (2) how to study life elsewhere in its own environment; (3) how terrestrial life is affected by travel in space. These are sufficiently different sets of questions that it's worth looking at each in turn. I ask here about the science and how to address these questions. Then I'll turn to the social implications in the next section.

(1) How to study and manage extraterrestrial life that ends up on Earth?

While historians of the extraterrestrial life debate have documented the broad range of possible beings described in science fiction, these remain hypothetical. The issue here is: if life really arrived here from outside the Earth, what impact might it have on terrestrial life? Might it, for example, infect us with microbial "diseases," and how would we know? The possible existence of such life implies that we should plan for studying it and its biological impact. How would we know, what would we do, and what might we learn? Without thinking ahead about how the science should work, it is easy to fall into ad hoc reactions that fuddle one's thinking and lead to non-scientific hypotheticals that are not epistemologically well-grounded.

The second issue about managing extraterrestrial life is more complicated and, as noted earlier, seems not to be part of the NASA astrobiology program. Yet there are obviously real issues related to "planetary protection," as NASA recognized in setting up a special office for the job (Meltzer 2012). Although a great deal of sensible regulatory planning, guided by realistic scientific assessments of risks and costs and benefits, already exists, more will be needed to anticipate and respond to new discoveries.

(2) How to study life elsewhere in its own environment

If there is life in places other than Earth and we can study it with some version of biology, then how? Let's imagine that we want to do astrobiological fieldwork, which ecosystems ecologists argue is the best way to understand organisms and their environmental interactions. Doing fieldwork means respecting the field, and it's a little hard to know how to do that. Surely when we send rovers to Mars to march around and collect and stir up the environment, we are looking for what *we* want to find. And we are also changing the conditions even as we look; we have not yet developed any way to study without intervening. The same questions arise as with experimental biology on Earth: to what extent might alternative experimental work allow us to understand life in its natural context? How might we be able to do reliable biological inquiry on, or about, other planets or other bodies in the universe? The vigorous debates surrounding what it is that we see in looking at Mars or Moon rocks, for example, show just how many assumptions get made in working with unfamiliar materials with limited, familiar methods.

Biology of multicellular organisms typically includes study of reproduction, heredity, growth, differentiation, morphogenesis, development of functional systems, evolution, ecological interaction with the environment, and other processes related to each of these. In addition, some sort of cellular living units exist, and heredity is carried through information packed into units called chromosomes and genes.

Yet perhaps life does not always have to work this way. Rather than being based on carbon, for example, life could have a silicon base, as has been hypothesized repeatedly. But how much like our current life does it need to be in order to be considered "life?" In other words, what is the object of our biological investigation in space? Which functions of living systems are essential for life; might it be possible to generate more individuals through some sort of copying that would not be considered biological reproduction, for example? We need criteria to demarcate life from non-life, yet generating those has remained a major challenge in the history of exobiological studies and now of astrobiology (see Dick and Strick 2005 for many examples).

We also need criteria for deciding what to count as doing biology. If there is no reproduction or no heredity through information units such as chromosomes or genes, and if the systems are based on elements other than carbon and rely on molecules other than nucleic acids, then what kind of science do we use to study these systems? Is it still "biology" if the science relies on quite different methods and approaches? Therefore, the domain and the actual work of doing astrobiology are not entirely clear for both metaphysical and epistemological reasons. Nonetheless, let's make the assumptions that NASA seems to make and assume that we are going to work on something close enough to life to count as the same kind of thing, using something close enough to current biology to count as the same kind of science.

(3) How to understand how terrestrial life is affected by travel in space

As NASA's history website explains, before sending humans into space, several countries, including the United States, sent animals to test the effects of gravity reduction and other factors (NASA 2014b). Already in 1948 and 1949, American rocket scientists tested monkeys and mice, many of which did not survive the landing impact. The Soviet Union experimented with mice, rats, and rabbits. Tests suggested that conditions could support life, including human life, and the challenges came more with controlling landing.

As humans began to travel in space, researchers kept a close watch on their physiological and mental reactions, focusing on such concerns as impacts of cosmic radiation or effects of reduced gravity. Then came other issues in the field of space medicine, including concerns about muscle atrophy, digestion, and whether neural impairment might result from conditions in space. Those studies for the first decades were mostly reactive and focused on keeping astronauts apparently healthy and productive.

With time, launches began to include a wider range of biological experiments to determine the effects of the microgravity on living tissues and cells as well as whole organisms. Most recently, researchers have sent stem cells into space. Since the hopes for regenerative biology rely on our ability to control differentiation of cells, and since biological research has suggested since the late nineteenth century that gravitational forces can influence differentiation, stem-cell research seems an obvious candidate for experimentation in space. In fact, a NASA press release of December 6, 2013 announced that the Center for Advancement of Science in Space (CASIS) was making it possible to do stem-cell experiments on the international space station (NASA CASIS 2013). Experiments on mouse culture lines are currently underway to assess tissue loss in space conditions (NASA 2014c).

What issues arise for astrobiology and society?

If astrobiology is basically a form of biology, then we can start from issues of biology and society and build outward. Those issues include: how we understand assumptions and arguments in science, informed through history and philosophy of science; how we evaluate the field's impacts on society, including through ethics, law, policy, and other social lenses; and how we understand the relevant ecosystems, including through study of how living organismic systems interact with their environments. Let's look at the three sets of considerations laid out earlier with respect to these societal issues. In each case, the way that other fields have approached parallel problems suggests approaches for astrobiology.

(1) Societal impacts of extraterrestrial life that ends up on Earth

Steven Dick has shown that, historically, people have been fascinated by thinking about life elsewhere, imagining what might be possible and what that might mean. Such speculation throws light back on ourselves and on what we hope and what we fear. Dick has himself done some of the best thinking about where our imaginations can carry us. He shows that we can reason by analogy, drawing on history for help (Dick 2014). We can come to define ourselves more clearly in response to thinking about what is alien or "other" so that even if we do not find actual extraterrestrial life, the reflective process can be socially useful.

Then there are questions about what to do if extraterrestrial life were actually discovered. What if we find microbes or even more organized life forms? Presumably the usual ethical, legal, and social considerations hold. Standard biomedical ethical thinking, typically grounded on Thomas Beauchamp and James F. Childress's principles of respect for autonomy, non-maleficence, beneficence, and justice, would guide us to realize that we should respect this life (Beauchamp and Childress 2013). This would presumably also lead to invoking relevant animal care or human subjects protections. Would we want to respect the integrity of even microbial life or molecular components that might signal proto-life, on the reasoning that failing to do so violates ethical assumptions about what matters?

Perhaps we should think about extraterrestrial life as a special form of life. Genetically engineered organisms, chimeras, nano-enhanced organic parts, synthetically produced cells or organisms, cloning, and stem-cell research: all produce something new and raise new questions about how to understand the new, as well as how to control or manage it. They raise questions about how to contain risk and costs while enhancing benefits. In the United States, these questions have typically led to expert panels, often multiple expert panels, to weigh relevant factors. These panels then advise other organizations, and eventually either the issue goes away or some level of government does something. The United States has done nothing about stem-cell research at the federal level, much to the relief of many researchers. By the time all the deliberative panels had deliberated, new techniques had already begun to raise new and different questions. (National Academy of Sciences 2002a, 2002b, 2005).

In other cases, as for embryo research in the United Kingdom, the government called on a special committee, which in turn held a series of public consultations. The stimulus was the first "test-tube baby" born after an in vitro fertilization process. The result was the Human Fertilisation and Embryology Act (HFEC 1990), which has guided embryo research since its inception. Other governments and citizen groups have called for participatory democracy, inviting discussions of nanotechnology, genetically modified organisms, or other innovations. The goals have varied but have always assumed that if a wider public understands what is at issue, deliberation and resulting policy decisions will be wiser and more reflective. Perhaps astrobiology would benefit from public discussions of the science and its implications, as Steven Dick has suggested in organizing the public conference that led to this collection of essays.

Public discussion might address the complex of topics included as part of astrobiology and might make astrobiology seem less exotic. Whether the life being studied is truly alien, from outside Earth, or extremophiles in the deep ocean vents on Earth, perhaps the research involves something different and calls for unique approaches. Or perhaps not. Including a broader community in the discussions about what is going on and what is at issue might help make the research seem less alien. For example, the intense interest of visitors at Yellowstone listening to scientific lectures about extremophiles shows that people are fascinated by the possibilities and ask really sensible questions about costs and benefits. Perhaps it is only our history of feeling titillated by the possibilities of aliens that causes us to believe that astrobiology should cause public worry.

Issues of safety seem, more compellingly, to raise new questions. If we were to find new and different life forms that come from other environments than those on Earth, then we should not assume that we will know exactly how they will behave in this different environment. Perhaps we have an ethical obligation to build clearly and wisely thought-through regulatory protections against such "invasions" just in case? This would be true whether alien life forms arrived on Earth on their own or were brought back through space travel as back contamination. Questions about safety depend on an understanding of the science involved, and depend very much on expert explanations. This is true for such different topics as stem-cell research, nanotechnology, or invasive species: we rely on experts to tell us what is actually real rather than imagined. Then a larger community can help make decisions about what the risk assessments mean and what risks we are prepared as a society to take, since this is a social rather than a scientific question.

So far, this discussion has focused on our human perspective. Several of the papers here ask us to move beyond anthropocentrism. Why do we imagine life elsewhere as like us? And if it is not, how can we conceptualize it and respond to it? Life elsewhere might not be carbon-based, might not operate with DNA, and might be different in many other ways. It might, in other words, involve a different physics and metaphysics. But then, I would argue, to study it would probably not involve biology as such and would not be astro*biology*. We would need some other scientific field, beginning with different underlying assumptions and drawing on different methods. Whether this would require a different epistemological strategy is unclear, but epistemology does often follow metaphysics and suggests that we would need to think differently about how to carry out our study. In this case, we do not have good precedents upon which to draw, since we have assumed that we have one kind of science (and its epistemological approaches) and one kind of life (with the metaphysical assumptions about what life is).

There are also religious questions – both about what to believe and about how to feel. Those seem not to raise any really new questions. The question of whether to baptize an extraterrestrial (Chapter 15 in this volume) is not much about biology and remains a matter for conventional decision-making processes to work out. There isn't any fact of the matter, nor any clear set of guidelines except insofar as we make conventions and policies for such matters, separating scientific from social considerations. Eugenics had its biological roots, for example, but was driven by sets of social values and conventions that led to social impacts (Kevles 1985; Cold Spring Harbor Laboratory, n.d.).

Another factor concerns environmental impacts. If extraterrestrial life is actually living and close enough to earthly life to warrant biological study, then its environment matters. If we take these beings out of their "natural habitat" and move them or even allow them to travel to other locations, then they would be "invasive" species and thought by many environmentalists and conservationists to be undesirable. The same holds for terrestrial microbes found in the deep oceans and thought to be possibly similar to astro-life: we are potentially creating invasive species problems when we move them around. Again, assumptions abound. Again, we can start from our experience with invasive species, going back to early-twentieth-century worries about invading alien cherry trees from Japan (Pauly 1996; Chew and Laubichler 2003). Since studies of ecology and evolution show that context matters tremendously, taking an "astro" object and moving it could cause serious ecological questions. These include not just the usual concerns about impacts on the environment, but also questions about the impact on populations and therefore on evolution of Earth's biodiversity.

(2) Societal implications for studying life elsewhere in its own environment

Ecologists urge us to do field study of life in its own environment insofar as possible. Experimental work in the laboratory is fine for some purposes, but not for all aspects of complex natural systems. Research requires going to new places, whether to extreme environments of hydrothermal vents on Earth or more uniquely to space. Researchers might themselves travel to space, but we obviously have not yet developed human travel to those places thought most likely to prove habitable for life like Mars or Europa. Research is much more likely to be carried out by programmed machines and robots.

These machines will necessarily impact the field sites they are studying in their own ways. Perhaps they will contaminate sites with dust or microparticles. We have certainly done this in studies of the ocean floor or desert environments, with our eagerness to explore new worlds. Only gradually have we come to realize that we are impacting the nature we seek to study. Ecologists recently noticed, for example, that in studying why amphibians were dying off, they were themselves carrying pathogens and infecting populations that had previously been unaffected (Collins and Crump 2009).

With space travel, the vehicles will surely land on and impact the physical site. In effect, we are ourselves the invasive species in this case – acting remotely. Is satisfying our curiosity sufficient grounds for risking affecting the alien life we seek to study? Or can we, and ought we, to build in regulatory protections for any such research before we muck about? How do we think about how to weigh the risks, costs, and benefits? And who are the "we" who should do this weighing? These are all questions that have long been contemplated as part of NASA's planetary protection program (Meltzer 2012). Presumably, such reflection will benefit from scientists working with humanists and social scientists.

(3) Societal implications for terrestrial life that is affected by travel in space

Here we ask not about whether we might bring something alien back with us, which is question 1, nor about whether we might impact the field sites we wish to study, which is question 2, but about the impact on the humans or animals doing the study. Traditional considerations about interactions of biology and society suggest that we should think about several issues. We can draw parallels with study of nuclear development or radiation, in which scientists and citizens served as sometimes voluntary and at other times unwitting test subjects (Creager 2013). How can we test innovations without testing it on some*body* or some *thing*? With space travellers: if astronauts are, in effect, experimental subjects, do they receive appropriate information to be able to give informed consent? Do we have a clear standard of care for astronauts before, during, and after space travel? Given concerns about cosmic radiation, have we thought sufficiently about and imposed appropriate regulatory guidelines to protect germ lines so that future generations are not negatively impacted?

Then there are questions related to studies of other species, animals or microbes: are we influencing evolution in any way with space travel? If the germ line were affected in some significant ways, we could impact populations. Bioethicists and biologists both become worried at the idea of impacting the germ line, whether through genetic engineering or some other ways that we cannot predict. This seems very unlikely, but let's extend the consideration beyond just those few individuals actually travelling in space. In the laboratory, as researchers seek to create microgravity situations to discover the impact on stem-cell differentiation, for example, what impact might that have? Might selective breeding over generations lead to evolutionary change in microbial communities? If stem-cell (or other tissue and cell) researchers are successful in generating reliable results and move into commercial production, what effects might that have?

These are evolutionary questions that have ethical implications, for which our traditional thinking is applicable and informative. Space travel just happens to be one kind of condition to consider. Is there anything new under the Sun with astrobiology?

Are there any special issues for astrobiology and society?

The easy answer is probably not really. Not if we assume that astro-life and the study of astro-life are close enough to terrestrial biology. Yet we should consider that life on Earth and life beyond Earth might not be close enough to warrant the same kind of study, and also that the study itself will reveal significant differences. Perhaps extraterrestrial life forms will be so different that we will not know how to recognize them (except perhaps as being something different), how to study them, or what to do about them. If so, we have an entirely different context. As future-studies researchers like to point out, there are good ways to study the future. Yet, it takes work to lay the groundwork of assumptions as well as goals for study and values about what risks are worth taking.

Steven Dick's testimony to Congress on 4 December 2013 shows why my suggestion that there is nothing very new here may miss at least one important point. Dick challenges us to think differently. As Dick noted in his remarks, it is the imagination that matters. The very idea of astrobiology and the discoveries of those exploring life elsewhere in the universe have evoked "that sense of awe and wonder." Dick spoke as an historian with perspective on centuries of both hypothetical and empirically driven thinking about life beyond Earth, and on decades of space exploration that have led to discoveries of other planets and suggestive environmental conditions. As Dick put it, "Astrobiology raises fundamental questions and evokes a sense of awe and wonder as we realize perhaps there is something new under our sun, and the suns of other worlds" (Dick 2013).

Is he right and there is something new? And if there is some*thing* new, does that mean that we can know what it is or how to understand it? In 2012, Margaret Race and others looked at issues of astrobiology and society and asked what it would take to identify and address those issues. They asked what it would take to build an interdisciplinary research community for such study (Race *et al.* 2012). From a workshop and following efforts, they noted, as Dick's testimony had also suggested, that a large part of the impact of astrobiology on society is psychological.

The authors describe themselves as a working group for thinking about and developing a roadmap for the study of implications of astrobiology on society and of society on astrobiology. They point to the four implementation principles of NASA's Astrobiology Roadmap (Des Marais *et al.* 2008) and feel that much less progress has been made toward the third principle, namely recognizing "a broad societal interest in astrobiology's endeavors." They organized a workshop with their own five goals, to (Race *et al.* 2012):

(A) Explore the range and complexity of societal issues related to how life begins and evolves.
(B) Understand how astrobiology research relates to questions about the significance and meaning of life.

(C) Explore the relationships of humans with life and environments on Earth.
(D) Explore the potential relationships of humans with "other" worlds and types of life.
(E) Consider life's collective future – for humans and other life, on Earth and beyond.

To do all this, they maintain, involves consideration of religious, ethical, legal, cultural, and other concerns related to our current and future conditions. Philosophers and humanists should be part of addressing these issues, with the implication that they will also be informed about evolution and environmental sciences.

This effort has great attractions, including the potential for providing employment for humanists. Yet the process also requires that humanists and social scientists work closely with environmental and life scientists. It cannot be that the scientists do their science and hand the results over to others to assess the social impacts and implications. Too much of the Human Genome Project's ELSI (Ethical, Legal, and Social Implications) Program involved humanists talking to each other and thereafter not being heard by some of the scientists involved. Yet even when humanists and social scientists understand the details of the scientific work they are considering, they are often marginalized as working on "other" problems not central to the science. "Why do we need history when science is about the future?" is a common type of complaint, or "Why are we wasting money talking about bioethics when we scientists all intend to be ethical and don't need somebody telling us what to do?"

Race and others understand that, and call for interdisciplinary work including scientists. This is ideal. To have productive discussions, it cannot be that the humanists pronounce on what the social issues seem to be. The questions need to arise from mutual discussion. For example, scientists may well not think there are serious religious issues – or not anything new raised by astrobiological science or even its hoped-for discoveries. Or they may feel that ethical questions are the same as for other fields, and ask why we need more discussions about the topic. It will be important to discover what the questions are through mutual exploration, informed by contributions from all participants. Offering a list of issues already presumed to be the important ones may well put off otherwise willing collaborators. What could truly make astrobiology new under the Sun would be for humanists, social scientists, and leading scientists to work closely and collaboratively together with mutual respect on shared research problems and methods.

References

Beauchamp, Tom L. and James F. Childress. 2013. *Principles of Biomedical Ethics*. 7th edition. New York, NY: Oxford University Press.

Chew, Matthew K. and Manfred D. Laubichler. 2003. "Natural Enemies: Metaphor or Misconception?" *Science*, 301: 52–53.

Cold Spring Harbor Laboratory (CSHL). "Eugenics Archive." Online at http://eugenicsarchive.org/eugenics/list3.pl.

Collins, James P. and Martha L. Crump. 2009. *Extinction in Our Times. Global Amphibian Decline*. Oxford: Oxford University Press.

Creager, Angela N. H. 2013. *Life Atomic: A History of Radioisotopes in Science and Medicine*. Chicago, IL: University of Chicago Press.

Des Marais, David J., Joseph A. Nuth III, Louis J. Allamandola, *et al.* 2008. "The NASA Astrobiology Roadmap." *Astrobiology*, 8: 715–730. Online at https://astrobiology.nasa.gov/media/medialibrary/2013/09/AB_roadmap_2008.pdf.

Dick, Steven J. 1998. *Life on Other Worlds: The 20th-Century Extraterrestrial Life Debate*. Cambridge: Cambridge University Press.

Dick, Steven J. 2013. Testimony before the Committee on Science, Space, and Technology, US House of Representatives. December 4. Online at http://science.house.gov/hearing/full-committee-hearing-astrobiology-search-biosignatures-our-solar-system-and-beyond.

Dick, Steven J. 2014. "Analogy and the Societal Implications of Astrobiology." *Astropolitics: The International Journal of Space Politics & Policy*, 12: 210–230.

Dick, Steven J. and James E. Strick. 2005. *The Living Universe. NASA and the Development of Astrobiology*. New Brunswick, NJ: Rutgers University Press. Quoting from the NASA Astrobiology Roadmap released 6 January 1999. Online at http://astrobiology.nasa.gov/roadmap/.

HFEA 1990. "Human Fertilisation and Embryology Act 1990." Online at http://www.legislation.gov.uk/ukpga/1990/37/contents.

Kevles, Daniel J. 1985. *In the Name of Eugenics: Genetics and the Uses of Human Heredity*. New York, NY: Knopf.

Meltzer, Michael. 2012. *When Biospheres Collide: A History of NASA's Planetary Protection Programs*. Washington, DC: NASA.

NASA. 2014a. "Astrobiology." Online at http://science.nasa.gov/planetary-science/astrobiology/

NASA. 2014b. "A Brief History of Animals in Space." Online at http://history.nasa.gov/animals.html.

NASA. 2014c. Online at (http://www.nasa.gov/mission_pages/station/research/experiments/851.html)

NASA CASIS. 2013. "NASA, CASIS Make Space Station Accessible for Stem Cell Research." Online at http://www.nasa.gov/press/2013/december/nasa-casis-make-space-station-accessible-for-stem-cell-research/-.U-TqTUj-LXc.

National Academy of Sciences. 2002a. *Scientific and Medical Aspects of Human Reproductive Cloning*. Washington, DC: National Academies Press.

National Academy of Sciences. 2002b. *Stem Cells and the Future of Reproductive Medicine*. Washington: National Academies Press.

National Academy of Sciences. 2005. *Guidelines for Human Embryonic Stem Cell Research*. Washington: National Academies Press.

Pauly, Philip J. 1996. "The Beauty and Menace of the Japanese Cherry Trees: Conflicting Visions of American Ecological Independence." *Isis*, 87: 51–73.

Race, Margaret, Kathryn Denning, Constance M. Bertka, *et al.* 2012. "Astrobiology and Society: Building an Interdisciplinary Research Community." *Astrobiology*, 12: 958–965.

17 Preparing for the discovery of extraterrestrial life: are we ready?

Considering potential risks, impacts, and plans

MARGARET S. RACE

Science fiction novels and Hollywood movies have explored the subject of alien encounters for decades, but they provide little in the way of practical guidance on how humankind should respond to a verified discovery of extraterrestrial (ET) life. Today we are in the midst of unprecedented advances in our understanding of the potential for life in space. Without exaggeration, the prospects for finding ET life seem more likely all the time, raising questions of how well we are prepared to respond to a discovery. Do we Earthlings have the necessary plans and preparations for responding to potential risks and impacts of different discovery scenarios? Would it matter what kind of life is discovered first? Who would be involved in decision making on behalf of humankind?

This chapter takes a systematic approach to evaluating in detail how current decision-making processes and policies are prepared to deal with future discoveries and the associated risks and consequences of interacting with different types of ET life. It examines the three main search types (SETI, extrasolar planets, and Solar System searches), at three different search phases (during searches, upon discovery, and post discovery) and assesses comparative preparedness for systematic deliberations involving multiple stakeholders, scientific and otherwise. The evaluation borrows heavily from approaches used by the hazard-management and risk-analysis communities, which have extensive experience and research involving threats of many types, whether natural or man-made, and predictable, deliberate, or accidental (Alexander 2000; Eisner *et al.* 2012; Tierney 2014). In addition to providing a detailed comparison of current search efforts, this risk-centered approach also serves to highlight particular topics or procedural steps that may need more attention in order to develop practical and coordinated implementation plans for responding to any future discovery of ET life, whenever and wherever it may occur.[1]

[1] This analysis is necessarily qualitative in nature. The unpredictability of discovery scenarios makes it inappropriate to use quantitative analyses like those associated with most other studies of hazards and risks. Even so, qualitative approaches can be quite useful in guiding the consideration of future risks and preparedness (see MacKenzie 2014).

Putting searches in context

Before discussing the similarities and differences between current searches, it is important to recognize that there is no such thing as planning *a response* to the discovery of ET life, but rather *multiple responses*, built upon the characteristics of the different search efforts under way. Each anticipates different scenarios and types of life (intelligent, complex, or simple) as well as potential impacts, risks, and timeframes. In addition, each raises assorted questions that will require inputs from different perspectives – scientific and otherwise – regardless of who makes the first discovery. We may not be able to predict when, where, or how the first discovery of ET life will occur, but it is appropriate to think in advance about what might happen upon discovery, as well as afterwards.

While some may consider these concerns far-fetched, others suggest it is appropriate, even necessary, to think far in advance about them. Each year, the World Economic Forum (WEF) undertakes an annual survey of over 1,000 experts from industry, government, academia, and civil society to identify leading global risks and their potential for significant or cascading effects. While focusing mainly on traditional categories like economic, environmental, societal, geopolitical, and technological risks, the survey also seeks to identify trends or "X factors" that could be emerging "game-changers" with unanticipated impacts. These X factors are deemed serious issues grounded in the latest scientific findings and representing "... broad and vaguely understood issues that could be hatching grounds for potential future risks (or opportunities)." The 2013 WEF report included the "discovery of alien life" as one of five X Factors that decision makers should be aware of (World Economic Forum 2013).[2] In fact, the report suggested that "In 10 years' time we may have evidence not only that Earth is not unique but also that life exists elsewhere in the universe."

Because the discovery of alien life could happen at any time and has assorted unknown consequences, it warrants special examination to help anticipate future challenges, adopt more pro-active approaches, and avoid being caught by surprise or forced into fully reactive mode in the future. No doubt a discovery would be a major news story with intense interest (see Billings, Chapter 20), but it would not likely change the world immediately. The WEF report asserts that the largest near-term impact would likely be on science itself – and possibly a push for greater funding for robotic or human missions to study the life forms *in situ* (World Economic Forum 2013). Over the long

[2] Other X factors included on the 2013 included the risks of runaway climate change, cognitive enhancement of humans in daily life and military situations, rogue deployment of geoengineering, and the societal costs of people living longer.

term, the psychological and philosophical implications of the discovery could also be profound, as speculation about the existence of alien life, particularly intelligent beings, may challenge many assumptions that underpin philosophy, religion, ethics, and world views. Additionally, there are known direct risks to consider as well, such as the potential for cross-contamination between planets and uncertainties about the nature of possible microbial ET life. Thus, it is appropriate to be prepared – to anticipate, minimize, or prevent undesired physical and/or societal consequences, knowing that regardless of the discovery scenario, it will raise a mix of concerns and represent a major paradigm shift about the very nature of life, our position in the universe, and even humankind's self-image.

Overview of current searches

Today's searches for ET life are confined largely to the realm of scientists and technical experts whose methods collect assorted types of data and evidence in vastly separated locations in space. While some outside the space community may be aware of ET search efforts, there is likely a wide public disparity in basic understanding about the searches and assumptions involved – the types of life that may be found; their impacts, significance, and meaning; what, if anything, should be done in response to a discovery; and by whom. Thus, before evaluating overall preparedness for future discoveries, it is necessary to understand the basic facts associated with the different searches – from the nature of evidence, type of life, and likely discovery scenarios, to the locations searched, methods used, and time frames involved. In addition, we need to prepare for rational, practical discussions about societal responses by including details on what is assumed and involved for different searches; what risks, consequences, mitigation, impacts, and meanings arise for whom, when, and where; and what institutions and groups of decision makers will be involved in deliberations on behalf of humankind.

Currently, there are three general categories of searches involving the international scientific community and space-faring nations: (1) the search for extraterrestrial intelligence (SETI); (2) searches for extrasolar planets (exoplanets); and (3) astrobiological searches in the Solar System. Each category is described briefly in the sections below, and in more detail elsewhere.[3] Table 17.1 provides a matrix summarizing the basic characteristics of the different search types currently underway, outlining their locations, methods,

[3] See, e.g., www.seti.org; http://planetquest.jpl.nasa.gov; and www.astrobiology.nasa.gov. Any detections made via alien-initiated actions are not covered by current planning efforts.

Table 17.1 Overview of current searches: comparative features, potential risks and impacts, and institutional leadership *during* searches

Search type →	**ETI**	**Extrasolar planets**	**Our Solar System**		
			Space missions		**On Earth**
			Robotic	**Human**	**Lab and field**
Location and search distance	In galaxy (tens to thousands of light years)	In galaxy (tens to thousands of light years)	Planets, moons, small bodies (e.g., Europa, Enceladus, Titan) (light minutes away)	Mars and its moons (light minutes away)	Earth (real time)
Methods	Electromagnetic signals (e.g. radio, light) Passive SETI and METI (?)	Telescopes; remote sensing and Earth-based analyses	Multiple mission types (with science instruments and analyses)	Multiple mission types (with science instruments and analyses)	Diverse science instruments and analyses
Anticipated evidence	Signals, message (coded, uncoded, translated)	Terrestrial planets; habitable locations; atmospheric signatures and biomarkers	*in situ* ET life or biomarkers; fossils, remnants?; sample analysis	*in situ* ET life or biomarkers; sample analysis *in situ* or sample return	Meteoritic or extremophile life; simulations; sample return
Type of life	Unknown biology; technological; Presumed intelligent and complex	Uncertain; Presumed biological? Simple or complex?	Microbial (simple?)	Microbial (simple?)	Microbial (simple?)

Status of life	Uncertain if still in existence	Uncertain if still exists? Still habitable?	Real time	Real time	Real time
Risks and potential impacts during search	SETI: no direct impacts (psychological, societal concerns) METI: uncertain?	None	Forward contamination (and back contamination if sample return?)	Forward and back contamination; EHS?	Earth EHS concerns; back contamination if sample return
Institutional leadership	IAA; SETI Permanent Committee	Space agencies; science peer review; science working group?	COSPAR; Planetary Protection Panel; space agencies; science peer review	COSPAR; Planetary Protection Panel; space agencies; science peer review Earth EHS regulators?	COSPAR; Planetary Protection Panel; space agencies; science peer review; Earth EHS regulators?

type of evidence and type of life anticipated, and the search distances and time lags involved. In addition, it indicates the types of risks and potential impacts that are considered during search processes, as well as the institutions that are involved in deliberations and decision making about on-going searches.

The current search for extraterrestrial intelligence (SETI)

The search for intelligent ET life is accomplished using a combination of telescopes and signal-processing technology on Earth to detect electromagnetic signals or messages sent across interstellar distances by presumed extraterrestrial civilizations elsewhere in the Milky Way Galaxy. Astronomers distinguish two types of searches for ET intelligence: passive SETI and active SETI. Passive searches, designed to receive but not transmit signals, account for the majority of searches under way since the 1960s. More recently, the idea of active SETI has been proposed. Also known as METI (messaging to ET intelligence), such searches involve the deliberate transmission of *de novo* signals into space in the hope they will be intercepted by alien civilizations that would presumably send targeted messages back to Earth. The objective of both search types is to seek and detect signals from distances of tens to hundreds of light years away, which would represent the technological manifestation of intelligence, but cannot provide additional information about the nature of the beings who sent the message. Because the signals would have travelled many light years from their source, there is no way to know if the civilization still exists when astronomers receive a signal. In addition, the signal may be encrypted or not readily understandable, so it may take a while to translate messages, analogous to the Rosetta stone process.

Institutional leadership for SETI activities is provided through the International Academy of Astronautics (IAA), a non-governmental organization founded in 1960. IAA works closely with the International Astronautical Federation and national and international space agencies in promoting the development of astronautics for peaceful purposes and expanding the frontiers of space. Since the 1970s, the IAA's SETI Committee has been involved in deliberations about international issues and activities related to SETI, including the social consequences of a discovery.[4]

Current searches for extrasolar planets

Searches for extrasolar planets represent the newest type of search in our galactic neighborhood, with the first detections of exoplanets (around pulsars)

[4] See www.iaaweb.org.

in 1992.[5] By mid 2015, a combined total of over 5,000 candidate and confirmed extrasolar planets had been discovered by astronomers worldwide using a variety of space telescopes (e.g. Kepler, Hubble, Spitzer) and search methods, including transits, radial velocity, imaging, and microlensing (Seager 2013).[6]

Exoplanet searches focus on locations that are tens to thousands of light years distant, and thus cannot provide details about the existence or nature of ET life directly. Rather, the discoveries help build a taxonomy of solar systems, paying particular interest to the potential habitability of planets orbiting Sun-like stars at distances where temperatures and conditions are suitable for liquid water to exist. While astronomers cannot know whether the planets are inhabited, they can collect spectroscopic data on atmospheric biosignatures and potential habitability through evidence of chemical or biogenic processes known to be associated with life as we know it on Earth. While there is extensive coordination and sharing of exoplanet data and discovery information by scientists and space agencies worldwide, there are no officially designated organizations to address policy or legal issues associated with the searches – and none are anticipated. Indeed, over the past two decades, information about repeated exoplanet discoveries has been treated as news and communicated broadly to scientists and the general public without concerns about whether or how humans might interact with these distant locations *in situ* or indirectly.

Current astrobiology research/exploration

Astrobiology searches for extraterrestrial life in the Solar System involve the widest variety of mission types, methods, locations, and activities, tracing back to the earliest days of the space era. They build upon diverse activities involving everything from Earth-based laboratory experiments of primordial conditions; to meteorite studies; robotic spacecraft on flyby, orbital, or landed missions to the Moon and beyond; remote sensing of distant space environments; analog studies of extreme environments on Earth; sample return missions; human missions in low Earth orbit; and some day perhaps, human missions to Mars and other celestial bodies. Astrobiological research and exploration integrate science methods of all types to look for evidence of either *in situ* ET life (extant or dormant) or biomarkers of any type, including remnants, fossils, atmospheric or metabolic signatures, geological/geochemical evidence, and more. Because searches in the Solar System are relatively nearby with only minor communication lag times, any discovery would be considered current, real-

[5] Although discovery of alien life is not the main objective of their searches, their methods and data may provide demonstrable evidence of ET life.

[6] See http://planetquest.jpl.nasa.gov and exoplanetarchive.ipac.caltech.edu.

time evidence of ET life. Currently, all expectations are that ET life will be microbial, possibly alive at the time of collection, and with the prospect for additional study to verify if it is related to life as we know it or is different biochemically and perhaps representative of a second genesis of life. The discovery of even a single verified example of independently evolved microbial ET life in the Solar System would increase the odds that we live in a biological universe (Dick 1996).

Unlike the searches for ET intelligence or extrasolar planets, astrobiological searches in the Solar System raise distinct concerns about assorted real-time, direct risks and potential impacts on Earth and in space. Depending on where research or exploration activities are conducted, and what kinds of methods are used, a variety of policy and legal constraints may arise. For example, scientists and space agencies face diverse Earth-based environmental, health, and safety (EHS) regulations depending on their respective home countries. In addition, astrobiological search missions raise questions about potential biological cross contamination similar to impacts of microbes, epidemics, and possibly invasive species on Earth. Some day, there may also be prospects for beneficial uses of ET microorganisms, just as there are for Earth microbes, raising varied non-science concerns including questions about claims, ownership, patents, and damages. In general, space missions and search activities are implemented by launching agencies or science institutions, and conducted in compliance with the Outer Space Treaty (United Nations, 1967), which has over 100 signatory nations. Leadership on science considerations is provided by the Committee on Space Research (COSPAR), a non-governmental organization having permanent observer status with the United Nations Committee on the Peaceful Uses of Outer Space (COPUOS). COSPAR planetary protection policies to prevent harmful contamination during exploration (COSPAR 2011) are set in consultation with the national academies of the signatory nations – and implemented through planetary protection offices or relevant administrative units within space agencies.[7]

Overview of existing policy/regulatory frameworks

Just as the types of searches differ from each other, so too do their specific written policies and regulatory frameworks. This section begins with an overview of the particular policies applicable to different searches at different times, and then evaluates the information for the adequacy of plans and policies

[7] For detailed information on COSPAR international policies, see https://cosparhq.cnes.fr/sites/default/files/pppolicy.pdf.

applicable *during* searches, *upon* discovery (in the short term of weeks to years) and *post* discovery (years to decades).

Based on experiences in the risk and hazards community, it is well known that decision making under conditions of uncertainty is inadequately described by traditional models of rational choice (Eisner *et al.* 2012; Slovic 2000). Thus, when making plans for follow-on activities it is important to acknowledge and anticipate the likelihood of interactions between the scientific facts and other features, including human factors, institutional involvement, and cultural/international perspectives. In many ways, the different discovery scenarios raise questions similar to those experienced in debates about new and emerging technologies (e.g. synthetic biology, nanotechnology, artificial life, drones, etc.), which are likewise generally ahead of full understanding of their implications. While disciplinary experts may view the discovery of ET life as exciting and even scientifically inevitable, those who are less knowledgeable or unfamiliar with searches may perceive the situation as fraught with impacts and disruptive consequences for individuals, groups, and even societal stability. Recommendations for subsequent actions or responses by experts may raise assorted questions based on different risk perceptions, including concerns about scientific unknowns or inaccuracies; uncertainty about mitigation options; unfamiliar governance or oversight frameworks; unclear decision-making processes; mistrust of involved experts; and inadequate involvement of non-expert stakeholders in deliberations.

In order to assess the adequacy of current processes and guidelines for ET discovery, it is helpful to articulate what specific steps and deliberations will occur upon discovery and afterward, what incremental examinations or actions are prescribed, who is involved, and what institutions or officials will oversee deliberations. By comparing the specific plans associated with different searches, it is possible to identify gaps or inadequacy in preparedness and prioritize steps that may need attention. Table 17.2 provides an overview of the current policies and frameworks that apply to the different search types at different times. Tables 17.3 and 17.4 focus on the detailed steps anticipated *upon discovery* and *post discovery* indicating who is involved and what written protocols or sanctioned processes (if any) will apply.

It is clear that the different searches approach their activities and potential discoveries in different ways. Nonetheless, it should also be said that all activities are first and foremost conducted using longstanding traditions of responsible scientific research and exploration, regardless of which scientists or space agencies are involved. As such, they integrate foundational notions and ethical practices of proper science, including self-governance, information

Table 17.2 Existing policy/legal frameworks and institutions applicable to ET searches, *during*, *upon*, and *post* discovery

Search type →	SETI, METI	Extrasolar planets	Our solar system		
			Space missions		On Earth
			Robotic	Human	Lab & Field
Policy/legal during searches	Passive: IAA SETI principles; METI: under debate; proposed IAA San Marino two-level scale?	None	COSPAR planetary protection (PP) policy (forward and back contamination controls); agency notify COSPAR and UN COPUOS? Policy revisions as needed	COSPAR PP policy (forward and back contamination controls); agency PP directives Earth EHS laws/regulations upon return to be determined (TBD)	Diverse Earth EHS laws/ regulations; COSPAR PP policy; backward contamination controls and laboratory protocols for returned samples
Policy/legal upon discovery	Both SETI and METI: IAA SETI post-detection, (PD) protocols; notify UN COPUOS and General Assembly (GA)	Peer review	COSPAR PP policy; forward and back contamination controls; agency notify COSPAR and UN COPUOS?; policy revisions TBD	COSPAR and agency robotic and human PP policy (forward and back contamination controls); EHS regulations TBD; agency notify COSPAR and UN COPUOS?; policy revisions TBD	COSPAR PP (backward contamination controls and laboratory protocols); EHS science advisory committee TBD; agency notify COSPAR and UN COPUOS?; policy revisions TBD
Policy/legal post discovery	IAA SETI PD protocols; suggest consult with UN GA and COPUOS (not official); details TBD	None	COSPAR PP policy; agency notify UN COPUOS; policy revisions TBD	COSPAR human PP policy; agency notify UN COPUOS; policy revisions TBD	COSPAR PP policy; back contamination and laboratory protocols and EHS regulations; policy revisions TBD?

Table 17.3 Protocols upon discovery: existing plans and procedural steps

Search type →	SETI	Extrasolar planets	In our solar system		
			Space missions		On Earth
			Robotic	Human	Lab and field
Leadership and oversight role?	IAA	None Agency and science peer review	COSPAR, space agency, science community?	COSPAR, space agency, science community; EHS regulations	COSPAR, space and Earth agencies, science community; EHS regulations
Written discovery policy?	IAA SETI principles and post-detection protocols	None	None specified	None specified	Mars Sample Return Protocol; EHS regulations (keep contained/ re-evaluate)
Interpretation and consultation?	IAA Permanent SETI Committee	Routine; peer review	Considered routine?; agency and science peer review	None specified; peer review?; science advisory committee? EHS TBD	Routine peer review and Mars Sample Return Science Advisory Committee?
Notifications	UN COPUOS and UN GA (recommended); IAU telegram	Routine	Not specified; routine agency/science (notify COSPAR and UN COPUOS?)	Not specified; space agency; EHS regulations; COSPAR and UN COPUOS?	Not specified; routine agency/ science review?; notify COSPAR and UN COPUOS?
Public communication	SETI protocols via mass media	Routine	Routine: agency to mass media?	Agency to mass media?	Agency to mass media?

Table 17.4 Preparedness for post-discovery actions involving ET life (deliberate next steps)

Search type →	SETI	Extrasolar planets	Solar system		
			Space missions		On Earth
			Robotic	Human	Laboratory and field
Written policy or protocol ?	SETI principles (unofficial – not adopted by UN)	None	None	None	Update laboratory protocols and EHS considerations?
Clear governance and decision-making process identified	Science review; IAA; and UN proposed	No	No	No	No
Non-science stakeholders involved?	IAA urges international and societal involvement	No	No	No	No
"No action" situations addressed?	Not addressed	No	No	No	No

sharing, transparency, and peer-review practices, in addition to appropriate regulatory requirements, as described below.

SETI policy framework during searches

From the earliest radio-telescopic searches, the SETI community recognized the importance of developing coordinated plans for their search activities. In 1985, the IAA SETI Committee began discussions on the development of appropriate principles and protocols for signal detection, verification, validation, announcement, and response scenarios (Tarter and Michaud 1990; Michaud 2007, 358–376). In 1989, the IAA endorsed a set of guiding SETI principles, officially known as the "Declaration of Principles Concerning Activities Following the Detection of Extraterrestrial Intelligence," informally known as the SETI protocols. These principles, also adopted by the International Institute of Space Law and other non-government organizations, stipulated explicitly that openness and transparency should apply to handling of signals and associated data during searches (Tarter and Michaud, pp. 153–154, online at http://avsport.org/IAA/protocol.htm). Over the decades as participation has grown to include ever more professional and amateur astronomers worldwide, these principles have continued to apply, with updates adopted in 2010. Even searches involving citizen-volunteers – like SETI-at-Home (which allows individuals to register and donate their computer time to the task of analyzing incoming data) – provide registrants with information on the basic SETI principles and consultative approach that should be used, particularly when presumed signals are detected.[8]

Because passive SETI activities (the majority of current searches) have no known direct risks or concerns beyond those of other astronomical pursuits, and because "evidence" for a discovery would be entirely indirect in the form of incoming electromagnetic signals, there are no anticipated direct risks or impacts during the search phase itself aside from possible psychological, cultural, or societal concerns, nor any applicable legal/regulatory issues or penalties for non-compliance during passive search activities.

In recent years, growing interest in adopting active-SETI methods (METI) has raised new concerns. Some critics, including Stephen Hawking in his book *A Brief History of Time*, have argued that active messaging is foolhardy because it may alert superior ET intelligence to our existence and technological inferiority. The presumption of risk from METI draws upon historical analogs, arguing that pro-active methods somehow increase the likelihood of harmful consequences – much like contacts long ago between technologically superior

[8] http://setiathome.berkeley.edu/forum_thread.php?id=76730

and less-developed civilizations on Earth. The IAA SETI Permanent Study Group discussed policy approaches that might be used for METI searches and adopted the San Marino scale (Almár and Shuch 2007) as a way to evaluate the significance of deliberate transmission based on a combination of signal intensity and information content. In the meantime, no formal policy for METI has been set within the astronomical community. Denning (2011) suggests it is actually a social rather than a scientific question about global citizenship and participation, similar to governance and policy debates in other areas of science and technology. Even now the METI debate remains unresolved drawing attention among scientists as well as the public (Tarter and Black 2015.) A group of prominent scientists and citizens has begun circulating a petition to halt pro-active messaging until a worldwide scientific, political, and humanitarian discussion can occur (Siemion *et al.* 2015).[9]

SETI – upon discovery

The policy and regulatory framework applicable to SETI *upon discovery* is again organized under the leadership of IAA and the SETI Permanent Committee, through its Post Detection Science and Technology Subcommittee. Specific ET intelligence detection protocols were first adopted as part of the Declaration of Principles in 1989, and later revised and updated in 2010.[10] They amount to a set of written guidelines for astronomers who receive a tentative credible signal. The protocols stipulate communication to the International Astronomical Union (IAU) to notify the astronomical community upon initial detection of a presumed signal in order to alert and involve them in verification and confirmation. In addition, the protocols address the importance of preserving the frequencies and signal data, and early notification about the discovery to relevant governments and the United Nations (UN), including the UN COPUOS and General Assembly (GA). The procedural objectives are to avoid false alarms through careful expert data analysis prior to public discovery announcements. Based on the protocols, announcement of the discovery should be coordinated by the UN or appropriate international organization, with the original discoverer having the honor of participating. The announcement will be communicated via the mass media as well as through professional publications.

Although the protocols are quite forward looking, they may need to be further amended or updated. As noted by Schenkel (1997), the current protocols do not address alternative scenarios such as how to respond to an ET visit

[9] An overview of SETI and METI and IAA activities may be found at http://iaaseti.org/protocol.htm

[10] Available at http://avsport.org/IAA/protocol.htm.

or the discovery of artifacts. In addition, they do not address situations that may arise in response to today's 24/7 communication realities, including misinformation or leaks disseminated via the Internet or social networks beyond the control of scientific or governmental institutions. Finally, although the IAA has consistently recommended the involvement of the UN, there has been no official adoption of the SETI detection protocols, and no international committee has been charged with developing tentative plans that involve more than SETI scientists.

SETI post discovery

Certain features of the IAA SETI protocols address what to do *after* information about the discovery is announced worldwide. According to the protocols, there should be no response to a message or signal without consultation and deliberation about the planned responses or activities. Again, the protocols indicate that post-discovery deliberations should be undertaken through the auspices of an international governance organization – presumably the UN COPUOS and GA – and involve multiple stakeholders, including non-scientists. The SETI post-detection plans amount to a self-governance agreement among scientists to pause and seek international input including broad societal representation before return messages or other activities are undertaken. As with the initial discovery protocol, there is no official agreement beyond IAA, nor any penalties or enforcement mechanisms for those who undertake response actions on their own. Because lag times for round-trip communications will likely be more than a generation's lifespan, the Post Detection SETI Committee also suggested the need to establish a long-term infrastructure within UN COPUOS to respond if and when messages are received in the distant future.

Framework for extrasolar planet searches

For exoplanet searches, which have no particular risks or impacts during operations, there are no special regulatory or institutional frameworks or policies for the discovery of new planets, or their designation as terrestrial and potentially habitable by virtue of biogenic signatures. All work is done within the context of space agencies and scientific communities, and involves science working groups and peer review in a process of science self-governance.

The existence and significance of extrasolar planets was a scientifically debatable notion until the first discoveries were made in 1992 (around pulsars) and 1995 (around sun-like stars). Since that time, thousands of discoveries have been reviewed, validated, and announced, with many others designated as "candidate" planets pending further verification, which may take a long time depending on their orbital durations in relation to their star. All detection and interpretation activities as well as public announcements are done via international science working groups, with transparency and collaboration, and no need for legal or regulatory oversight. No policy issues have arisen thus far with these announcements. Like SETI, the time lag for receipt of the signal data from exoplanets means there is no way to verify whether or not the planetary conditions and potential biogenic signatures exist at present. So far, most of the discovered exoplanets are within hundreds of light years. The chances of planetary conditions changing are not as likely as the possibility of civilizations evolving over that time scale, since astronomical and biological evolution occur much more slowly than cultural evolution. With the passage of time and improvement of technological methods for examining nearby stars, astronomers have become increasingly optimistic about the likelihood of finding evidence of ET life beyond Earth (Seager 2013).

Since the only post-discovery activities for extrasolar planet searches are those involved in searches for new planets or re-examination of previously discovered planets, there are no foreseeable legal or regulatory complications likely for follow-on activities. Likewise, since the involved distances and time lags are so great, there are no current societal concerns about *in situ* or other direct studies with potential impacts on Earth, nor questions about whether our actions could have impacts upon the alien environments or possible life there.

Framework for astrobiology searches

The thousands of robotic and human Solar System missions undertaken since the dawn of the Space Age have contributed in various ways towards exploration for ET life on the Moon and other celestial bodies. In addition, research on Earth – including studies of meteorites, laboratory experiments of Earth's primordial conditions, analog studies of microbes and extreme habitats, and a range of other topics – can all be considered part of astrobiological research and searches for life. Policies and constraints applicable during exploration of the Solar System are rooted in the UN Outer Space Treaty of 1967, which addresses planetary protection in Article IX:

> ... parties to the Treaty shall pursue studies of outer space including the Moon and other celestial bodies, and conduct exploration of them so as to avoid their harmful contamination and also adverse changes in the environment of Earth resulting from the introduction of extraterrestrial matter and, where necessary, shall adopt appropriate measures for this purpose ...

COSPAR is assigned responsibility to develop, maintain, and promulgate knowledge, policy, and plans to prevent the harmful effects of such contamination, and to inform the international community through the UN's COPUOS. Specifically, the COSPAR planetary protection policy seeks to avoid harmful cross contamination by controlling both forward and backward contamination, regardless if the mission is designed to search for ET life or not.

Current COSPAR planetary protection policy is based on a categorical approach depending on the nature of the mission, where it will go, and what activities are planned (robotic or human; one-way or round-trip; designed as a flyby, orbiter, lander, or subsurface probe). In general, if a particular mission might encounter habitable locations during its lifetime, more rigorous cleanliness levels and controls will be required.[11]

Launching nations may modify or update their recommended control measures and requirements consistent with COSPAR policies. In the United States NASA's implementation guidelines, methods, and other requirements are stipulated in detail in management directives and other documents.[12] Plans for proposed US launched missions beyond Earth orbit must be reviewed for planetary protection compliance by the NASA planetary protection officer prior to launch. To date all one-way and round-trip robotic missions beyond Earth orbit are reviewed to determine whether and how planetary protection policies apply to them (Meltzer 2011). In general, robotic missions to Mars – either one way or proposed sample return – have the most stringent planetary protection control requirements. Future missions to other potentially habitable locations like Europa and Enceladus will also require strict controls.

Human missions beyond Earth orbit are also subject to COSPAR planetary protection policy. The earliest implementation of planetary protection controls occurred during the Apollo lunar missions over four decades ago. A comprehensive protocol to protect astronauts, samples, and Earth was developed and overseen by an interagency committee on back contamination (ICBC). Rigorous planetary protection controls including quarantine, containment, and testing were included on Apollo 11 through 14 (Meltzer 2011, 211–246). After Apollo 14, it was determined that return of lunar materials and handling of samples in laboratories on Earth posed no significant risks to crew, technical personnel, or the environment. Thus, there is no longer the need for extreme planetary protection controls during future round-trip lunar visits. Because of the increased potential for finding ET life on Mars, future human missions must comply with COSPAR policy updated in 2008. NASA and the

[11] See https://cosparhq.cnes.fr/sites/default/files/pppolicy.pdf
[12] See documents link on www.planetaryprotection.nasa.gov

international planetary protection community are currently working to develop detailed procedural instructions and implementation requirements for planetary protection controls for future crewed Mars missions. Future human planetary protection requirements will address forward and backward contamination concerns associated with habitats, operations, life support, extravehicular activities (EVAs), *in situ* resource utilization (ISRU), crew health and quarantine, instruments, and hardware (NASA 2015).

Astrobiology searches upon discovery

Because there are diverse mission types and celestial bodies involved, there could be multiple scenarios for discovery of ET microbial life. These include discovery by robotic probes *in situ*, by crews during future human missions, or inside containment laboratories on Earth. To date, the Mars Sample Return Protocol (Rummel *et al.* 2002) is the only case that discusses how to handle a discovery of presumed microbial ET life in a sample-receiving facility on Earth. Otherwise, there are currently no specific written guidelines about what to do if simple or complex ET life is discovered on or from bodies beyond Earth.

Based on other examples of significant planetary science discoveries, it is likely that the detection of presumed ET microbial life would be subjected to intense peer review and verification by mission scientists and others.[13] Likewise, the initial notifications and public announcements presumably would be handled similarly to other major science discoveries, and may not require specific written communication protocols. However, if ET life were discovered *in situ*, either robotically or by a crewed mission, there might be a need for international consultation before any samples or materials were returned to Earth.

Astrobiology post discovery

If and when microbial or complex ET life is found elsewhere in the Solar System, presumably extensive international deliberations will be needed to re-address questions about direct risks and associated handling concerns, as well as the guidelines for follow-on exploration activities. Currently, there is no written plan for what to do about future science exploration missions beyond Earth after verification that ET life exists elsewhere. There is also no framework or plan for considering the many indirect implications of "other" life or a verified second

[13] The current approach for verifying a microbial ET discovery in returned samples involves avoiding risks of exposure via a conservative approach using maximum laboratory biosafety, high-level containment, and quarantine. Pristine samples may be released for study beyond the primary receiving facility only *after* a rigorous battery of tests demonstrates there are no replicating entities or biohazards in returned materials.

genesis. Without provisions for integrating broader societal input into deliberations it is unclear how humankind will address a variety of legal, policy, and management issues ahead. To what extent should philosophical and theological perspectives about indigenous ET life impact future plans and activities – whether undertaken by the scientific, commercial or private sector? Who should decide questions about future claims, ownership, exploitation, property rights, liability, or impacts upon truly ET life? Should deliberate terraforming be restricted in any ways? A host of novel issues will undoubtedly arise that go well beyond current planetary protection and space exploration policies.

Findings and implications

To summarize, all three search types are generally well prepared with rational and practical plans applicable *during current searches* for ET life. As new questions or issues have arisen, the existing frameworks have been adaptable for revising plans and addressing questions about what constitutes responsible searches for ET life.[14] When it comes to the scenarios *upon discovery*, both searches for ET intelligence and extrasolar planets are adequately prepared to respond to credible detections. With no known direct impacts involved and very long lag times for interacting with presumed ET life, the consequences of detection will not likely have any real-time effects upon continuing search activities. In contrast, while astrobiologists are prepared in general for interpreting scientific aspects of possible discoveries, there are a number of gaps in their plans. Although COSPAR and space agencies eagerly anticipate the discovery of ET life, they have not yet deliberated about short-term responses or plans following a credible discovery. This gap in preparedness is particularly important in considering how a discovery might impact science exploration or missions already under way or planned. (For example, should there be stricter limitations on sample returns, future human missions or possible environmental disturbances on planets with ET life?) In addition, the existing framework is based entirely upon planetary protection constraints during exploration, with no major deliberations about when and how controls might apply to non-science activities such as "uses" of outer space or resource exploitation that could impact indigenous ET life or environments. Looking

[14] Among examples of this responsiveness are passive versus active SETI, planetary protection controls for future human missions to Mars, and interpretation of exoplanet discoveries regarding potential habitability and biogenic signatures. In general, the existing stepwise, deliberative processes have taken approximately 5–10 years or more to fully explore new issues and reach suitable international resolutions.

ahead, potential impacts on ET life from actions of private, commercial, or other non-government entities have also not been considered.

Finally, *post discovery* plans of the three search types are quite different in their levels of preparedness. Searches for extrasolar planets do not have, or need, any *post-discovery* plans aside from ongoing peer review and science deliberations. Due to the distances and time lags involved, all post-discovery activities will continue to have no direct impacts or consequences for a long time.

Astronomers involved in SETI searches have thought the longest about possible discoveries and protocols, but their post-discovery plans amount to a stop-and-discuss approach. They recommend no follow-on responses without international deliberations involving multiple stakeholders including diverse scientific and societal interests. To date, no post-detection protocols or guidelines have been officially incorporated into UN plans or committee structure, meaning that actual deliberations would start anew post discovery. Even so, with no direct impacts and long time lags involved, an incremental deliberative process is probably not unreasonable.

The least prepared post-discovery planning involves astrobiological searches of various types. Direct, real-time, and regulated risks have been acknowledged regarding possible interactions with microbial ET life. Even so, there have been no general or specific discussions to date about whether or how the existence of ET life might impact long-term activities in space. Presumably, the discovery of ET life in the Solar System would prompt re-evaluation of planetary protection policies and discussion about potential impacts of future visits to and uses of specific locations beyond Earth. If the deliberations are undertaken in incremental steps like most international policy discussions, considerable time will likely elapse until consensus is reached, with delays having the potential for unintended consequences for science missions and other space activities.

People have long suggested that the discovery of extraterrestrial life would be as historically significant as the Copernican revolution, with similar prospects for unanticipated scientific and societal consequences over time. An important difference is that a discovery today would be made in the context of increasingly fast-paced technological progress driven by enthusiastic, knowledgeable advocates interested in taking the next steps, which may precede full awareness of the overall consequences and impacts. Even though we live in a world with rapid communication about breaking news and widespread dissemination of science advances, there is no international deliberative process in place that includes multiple stakeholders and a broad perspective on whether and how to proceed – or not – and in what time period.

As the World Economic Forum suggests, if discovery of ET life is indeed an X factor of concern, it is important to identify risks, map and understand them as comprehensively as possible, and evaluate their implications proactively. Based on the analysis in the previous sections, it is clear that all current searches involve responsible science activities with rational and practical approaches during search efforts. All are guided by international science groups whose open deliberations have routinely led to policy revisions and updated plans in the face of new findings. Yet, there have been no coordinated discussions about protocols or plans for the discovery of ET life. Based on the complexity of the different search and discovery scenarios, it may be advisable to proactively and jointly have the three ET search communities consider the assorted science and societal questions together. In addition, it would be helpful if researchers from varied non-science disciplines could be encouraged to extend their focus about ET life more broadly across the spectrum of possibilities (simple, complex, and intelligent), rather than continuing to focus heavily on just a single type (ET intelligence) and its implications.

As future framework revisions are considered, it is also useful to acknowledge two implicit perspectives that form the backdrop of policies for all search efforts, and against which future deliberations and revisions will presumably be undertaken. First, all the current and proposed policy processes implicitly incorporate notions that science experts have priority in interpreting the significance and meaning of new discoveries and thus will be leaders in policy formulation. At the very least, it will be important to recognize where and how non-science perspectives should fit into future deliberations and how to balance the decision-making frameworks and processes appropriately. Second, it is appropriate to question whether future implementation policies in response to discovery of ET life will build upon our long-held anthropocentric perspectives about life and values, or somehow adapt to accommodate notions of "other" life (possibly dissimilar from Earth life – whether simple, complex, or intelligent) and their values. Because the discovery of ET life may involve major paradigm shifts in both scientific and societal domains (see the next chapter), the very assumptions that are so important during search efforts may represent policy complications upon discovery and afterwards, depending on the scenario(s) encountered.

Finally, it may also be advisable to seek an official recognition of the need for comprehensive discovery protocols and plans within a single subcommittee of UN COPUOS or other appropriate international institution. At this point, all search types are essentially in reactive mode in the event of discovery. Once the discovery is verified by science experts, there is no official designated review group that is familiar with the full range of issues and ready to begin deliberations about next steps. If the intention is that deliberations and decision making should respond in a reasonable time and reflect multiple, international stakeholders (not just the science and space community), now is the time to think about what issues might arise upon discovery and afterwards – for all of humankind.

Already, there is widespread recognition that the anticipated questions will go beyond scientific definitions of ET life and consideration of alternative biochemistries. How would the verified knowledge of ET life in the Solar System affect the management of celestial bodies, either for continued exploration or for "use" and exploitation as well? Would the verification of a "second genesis" of life impact legal, ethical, cultural, or philosophical perspectives and guidelines? How would "other life" be dealt with when deliberating about property rights, patenting, claims, or resource extraction? Perhaps COSPAR should undertake preliminary discussions similar to those held in the SETI community, exploring tentative plans for the next steps in deliberating the implications of a verified discovery of non-intelligent ET life.

Some might argue that the discovery of ET life is so far off as to not warrant special attention. Perhaps it will have few or no impacts on day-to-day life. More likely, while scientists and space explorers urge continued searches for ET life near and far, people worldwide will understand that we have crossed a significant threshold with implications for life of all types. Addressing the many questions will involve a stepwise path forward and consideration of diverse risks and perspectives. Now is the time to focus on gaps in legal and policy frameworks, and to think about where and how the international coordination and policy deliberations should occur. As importantly, it will be critical to consider how cultural meaning and societal concerns will be integrated into the public dialog, along with relevant technical and scientific considerations.

References

Alexander, D. 2000. *Confronting Catastrophe*. Harpenden: Terra Publishing.

Almar, I. and H. P. Shuch. 2007. "The San Marino Scale: a New Analytical Tool for Assessing Transmission Risk." *Acta Astronautica* 60, 57–59. For an online calculator for the Rio scale, devised by the Permanent SETI Committee of the IAA. See http://avsport.org/IAA/riocalc.htm.

COSPAR. 2011. COSPAR Planetary Protection Policy, [20 Oct 2002 as amended to 24 Mar 2011]. See https://cosparhq.cnes.fr/sites/default/files/pppolicy.pdf.

Denning, K. 2011. "Unpacking the Great Transmission Debate." In D. Vakoch, ed. *Communication with Extraterrestrial Intelligence*. New York, NY: SUNY Press, pp. 237–252.

Dick, S. J. 1996. *The Biological Universe: The Twentieth Century Extraterrestrial Life Debate and the Limits of Science.* Cambridge: Cambridge University Press.

Eisner, J. R., A. Bostrom, I. Burton, *et al.* 2012. "Risk Interpretation and Action: A conceptual framework for responses to Natural Hazards." *International Journal of Disaster Risk Reduction* 1, 5–16.

MacKenzie, C. A. 2014. "Summarizing Risk Using Risk Measures and Risk Indices." *Risk Analysis* 34, 2143–2162.

Meltzer, M. 2011. *When Biospheres Collide: A History of NASA's Planetary Protection Programs.* Washington, DC: NASA.

Michaud, M. A. G. 2007. *Contact with Alien Civilizations: Our Hopes and Fears about Encountering Extraterrestrials*, New York, NY: Copernicus Books.

NASA. 2015. Workshop on Planetary Protection Knowledge Gaps for Human Extraterrestrial Missions. See http://www.nasa.gov/sites/default/files/files/LPI-PP-Workshop-Announcement_2015_01-08v7.pdf

Rummel, J. D., M. S. Race, D. L. DeVincenzi, *et al.* (eds.), 2002. *A Draft Test Protocol For Detecting Possible Biohazards in Martian Samples Returned to Earth.* Washington, DC: NASA.

Schenkel, P. 1997. "Legal Frameworks for Two Contact Scenarios." *J. British Interplanetary Society* 50, 258–262.

Seager, S. 2013. "Exoplanet Habitability." *Science* 340, 577–580.

Siemion, A. *et al.*, 2015. "Regarding Messaging to Extraterrestrial Intelligence (METI) / Active Searches for Extraterrestrial Intelligence (Active SETI)" [Anti Active-SETI petition]. UC Berkeley. See https://setiathome.berkeley.edu/meti_statement_0.html.

Slovic, P. 2000. *Perception of Risk.* London: Earthscan.

Tarter, J. and M. A. G. Michaud. 1990. "SETI Post Detection Protocol." *Acta Astronautica* 21, 69–174.

Tarter, J. and D. Black, 2015. "Active SETI: Is it Time to Start Transmitting to the Cosmos?" Conference Session at AAAS Annual Meeting 2015, and follow-on Workshop at SETI Institute, Mountain View CA, Feb 2015.

Tierney, K. 2014. *The Social Roots of Risk: Producing Disasters, Promoting Resilience.* Stanford, CA: Stanford Business Books.

United Nations. 1967. "Treaty on Principles Governing the Activities of States in the Exploration and Use of Outer Space, including the Moon and Other Celestial Bodies." See http://www.unoosa.org/oosa/SpaceLaw/outerspt.html.

World Economic Forum. 2013. "X Factors – Discovery of Alien Life." In Lee Howell, ed., *Global Risks 2013 Eighth Edition*, Section 5. See http://reports.weforum.org/global-risks-2013.

18 Searching for extraterrestrial intelligence: preparing for an expected paradigm break

MICHAEL A. G. MICHAUD

Fortune favors the prepared mind.

LOUIS PASTEUR[1]

Finding many planets in orbit around other stars has provoked new interest in the long-running debate about the existence of extraterrestrial intelligence. Now we are asked to think about what rational, practical steps we can take to prepare for discovering evidence of sapient aliens, an event whose timing and exact nature are unpredictable. Our preparations must be based on probabilities, analogies, and disciplined speculations rather than on confirmed evidence.

Many of those interested in this debate expect that what they see as the final blow to anthropocentrism – discovering extraterrestrial intelligence – is just a matter of time and effort. If that discovery occurs, it might lead to a change in the way we see our status and our position in the universe. Such a discovery might not be the kind of paradigm break that Thomas Kuhn discussed in his famous book, *The Structure of Scientific Revolutions* (Kuhn 1970). Kuhn was writing about breaking *scientific* paradigms: replacing one set of physical laws with another. The new paradigms were not just different; they were better. The anticipated discovery of extraterrestrial intelligence is part of a different kind of paradigm break. Contact with an alien civilization would involve much more than science, raising philosophical and societal questions where laws are less certain or non-existent.

Evidence, scenarios, choices

Standards of evidence

A century ago, in the wake of Percival Lowell and the controversy over the canals of Mars, many humans believed that the red planet was inhabited by an alien civilization (Crowe 1986). Subsequent observations discredited that idea;

[1] Louis Pasteur, Lecture, University of Lille, 7 December, 1854, quoted in Roger A. MacGowan and Frederick I. Ordway III. 1966. *Intelligence in the Universe*. Englewood Cliffs, NJ: Prentice-Hall, p. 359.

the theory failed the empirical test. Beginning in 1947, thousands of people reported seeing exotic craft in our atmosphere, even on the Earth's surface (Jacobs 1975; Peebles 1994). While large numbers of our fellow humans believed that those objects were visitors from other worlds, most analysts concluded that nearly all UFOs actually were IFOs – identifiable flying objects. As astronomer Seth Shostak pointed out, there still are no artifacts to examine (Shostak 1998, 135). Others have suggested that the sightings most difficult to explain might be unknown phenomena that are unrelated to alien visitors. Again, the empirical standard has not been met.

Many people long believed that planets must exist around other stars, though we had no confirmed evidence before the 1990s. Some astronomers were skeptical, even dismissive (Sage 2014). The year 1995 brought us the first report confirming the existence of a planet orbiting a normal star other than the Sun. Since then, thousands of other planets have been discovered through astronomical observations. The belief in a plurality of worlds has met the empirical standard, though not the belief in a plurality of *inhabited* worlds. In 1996, scientists reported that a rock blasted off the Martian surface long ago contained fossil evidence that life once existed on Mars (McKay *et al.* 1996). Other researchers sharply criticized that finding, leaving the issue unresolved. The empirical standard may or may not have been met. Some have speculated that the "Wow" signal detected by a radio telescope in 1977 was a transmission from an extraterrestrial civilization (Gray 2014). That signal never was found again; its source is unknown. A one-time observation is not enough to meet normal scientific standards of empirical proof.

This history suggests caution in accepting claims about the discovery of extraterrestrial intelligence. There will be resistance to such a paradigm change, particularly if the evidence is fleeting or ambiguous. The Mars rock controversy might be a preview of what will happen.

What would constitute strong evidence? Shostak recently stated that there are three types of possible evidence for the existence of extraterrestrial life: finding alien life in our Solar System, hearing signals from far away, or detecting traces of biological activity in exoplanet atmospheres (Shostak, Chapter 1 in this volume). Other forms of evidence have been proposed, including astroengineering and alien artifacts in our Solar System.

We cannot precisely predict the nature of the evidence for our first confirmed discovery of alien intelligence. We cannot be sure that we are searching for the right evidence, or that we have the right means to detect it. Astronomer Thomas Kuiper proposed that detection will be the result of an accumulation of phenomena that are hard to explain (McDonough 1987, 218). Arthur Clarke famously said that any sufficiently advanced technology would be indistinguishable from magic (Clarke 1999, 399). What if a very advanced technology is indistinguishable from nature? Searching may lead to a false negative based on the limitations of our equipment or our assumptions. Absence of evidence is not evidence of absence.

How can we judge the significance of a detection? The scientific community that studies near Earth objects gave us a model after false alarms about asteroid threats. They developed the Torino scale, which ranks possible impactors according to several criteria (NASA 2014). SETI researchers proposed a similar scale for assessing evidence of extraterrestrial intelligence, including artifacts. The factors include the class of phenomenon (ranging from traces of astroengineering to an Earth-specific message), the means of discovery, and the distance. The assessment of significance would weigh probable consequences with the credibility of the claim. This is known as the Rio scale (Almár and Tarter 2011).

Scenarios of discovery and visions of the alien

The implications of finding evidence of extraterrestrial intelligence would depend heavily on the circumstances of that discovery (see Chapters 3 and 4 in this volume). While writers of science fiction and science speculation have suggested many possible scenarios for human encounters with sapient aliens, film and television presentations have imprinted two extreme models on many human minds. At one extreme is the classic SETI scenario, in which radio astronomers detect a faint signal generated by another technological species located many light years away. According to the orthodox view, the signal will be either a beacon intended to attract the attention of other technological civilizations or a message targeted on us. The remoteness of the aliens would imply that they will be no threat to us, and that the major outcome of contact will be an exchange of information.

At the other extreme is a direct contact scenario in which alien spacecraft come to the Earth. In this vision, we meet the aliens face to face. As depicted in science fiction films and television programs, the extraterrestrials could be as harmless and benevolent as the alien botanist *E.T.* and the childlike innocents of *Close Encounters of the Third Kind*, or as vicious and malevolent as the marauding invaders of *Independence Day* and *V*.

These extremes leave out many possible alternatives. What if a discovery is a one-time event, never repeated? What if we detect remote evidence of alien technology that is not a deliberate signal, such as industrial pollutants in an exoplanet atmosphere? Detecting such evidence would be a form of contact, even if there were no communication.

The implications of contact also will be influenced by the visions of aliens in our heads, most of them from science fiction. Anthropomorphized notions of culture have generated assumptions about what a non-human intelligence would be like; those visions change over time (Traphagan, Chapter 8 this volume). Before astronomers began finding thousands of planets around other stars, our model of planetary systems was based on the one example we knew – our own Solar System. Now we know that our case is not typical of those we find elsewhere (Sage 2014). Is that also true of biology, intelligence, and behavior? Our models of aliens and their societies, often based on human examples, may prove to be exceptions to galactic general rules. We cannot assume that all extraterrestrials would conform to any standard model; it is more likely that separate alien societies would evolve differently.

Choices of actions

Unlike non-intelligent life, intelligent beings can have intentions. They can make choices and take actions. We cannot assume that those choices and actions will be the ones we prefer. If one species sends a message, would the other one choose to reply? If one of the civilizations has the technology for interstellar flight by uninhabited machines, would it choose to send a probe towards the other? The orthodox SETI scenario assumes that both civilizations would welcome contact and pursue it further through radio communications. What if the alien belief system has never encompassed the existence of other intelligences? What if one civilization is suffering from causes of fear such as war? Human responses to contact might vary across a wide spectrum (Billingham *et al.* 1999). We and our alien counterparts might be more welcoming at some times than at others.

We have no evidence of what motivates extraterrestrials. We have no basis for assuming that another technological civilization would welcome contact with us, nor do we have any basis for assuming that such a society would be hostile to us. We should be wary of both utopian and apocalyptic visions of what contact might bring. We have no way of knowing the intentions of extraterrestrials until they take some sort of action we can detect. The great mystery of contact may be their intent.

Assuming alien motives may be particularly questionable if we encounter postbiological societies. Highly intelligent machines may not share our emotions, our ethics, or our respect for biological life forms. Our assumptions about alien behavior have not passed the empirical test. Their reaction to contact with us is unpredictable.

We have only two ways of analyzing the possible consequences of contact: analogy with ourselves, and probability based on what we know of human history and behavior. Our history gives us many examples of contacts between very different societies (Dick, Chapter 3 in this volume). The results have ranged from relatively benign to disastrous. The bottom line is that we cannot assume any particular outcome from contact. Given our ignorance, a certain degree of prudence is in order.

Organizing ourselves for contact

How can we prepare for such a diverse menu of possibilities? Years ago, the astrobiology community established procedures for avoiding contamination of alien biology that might be discovered through planetary exploration, and measures for protecting humankind from potentially dangerous alien life that might be returned to the Earth (Race, Chapter 17, this volume). In the 1980s, members of the SETI community developed a Declaration of Principles (better known as the first SETI protocol) that established procedures for handling the detection of an alien signal or other evidence of an extraterrestrial civilization. The basic principles are three: confirm the detection; announce a confirmed detection to the world; do not send a message to the detected intelligence until an international process of consultation has taken place (Michaud 1998; Michaud 2007, 359–363). Science fiction author Stephen Baxter argued that this protocol establishes precedents in several ways. It is based on discussion and consensus building. It emphasizes the need for scientific responsibility and a transparent process. It requires sharing discovery information with all of humankind. And it adds a caution: don't take actions that might imperil humankind without a prior discussion (Baxter 2013).

Others have proposed protocols for the detection of extraterrestrial artifacts in our Solar System. That may not be necessary. While the existing declaration originally was driven by the radio-signal scenario, it includes the phrase "or other evidence of extraterrestrial intelligence." A few have proposed protocols for visits to the Earth by inhabited spacecraft. Most people involved with these debates regard an alien landing on Earth as highly unlikely.

The Association of Space Explorers gave us a model for organizing international action. Raising the issue of asteroid impacts on the Earth within the United Nations, they persuaded key national governments to support principles for action: establish, by means of a clear legal structure, who would be entitled to speak with authority; require informing the relevant entities of the threat; use transparent, internationally acceptable mechanisms for decision-making; establish the right to respond to and mitigate the threat; accept the obligation to do these things (Association of Space Explorers 2008). Preparing for planetary defense could be a rehearsal for preparing for contact. But getting the United Nations to pay attention may be problematic. In 2000, the

International Academy of Astronautics (IAA) briefed the United Nations Committee on the Peaceful Uses of Outer Space (COPUOS) about proposed principles for handling contact. The committee report noted the IAA's submission and filed it without further action.

Another form of preparation is to address the issue of active SETI or messaging, in which some of our fellow humans send out powerful signals for the specific purpose of attracting attention from alien technological civilizations. While most of the signals sent so far lacked the power to reach even nearby stars, the strongest ones have generated controversy (Benford *et al.* 2014; Billingham and Benford 2014; Zaitsev 2014). A civilization more technologically advanced than ours might be capable of sending machines across interstellar space.

The content of the message is secondary to the issue of attracting attention to ourselves. Whatever the consequences of such transmissions may be, our descendants will not be able to opt out of them. We might expose our species to risks we cannot calculate. As one scientist wrote about climate change, uncertainty about the unknown unknowns of what might go wrong is coupled with essentially unlimited liability (Weitzman 2010).

Some have called for a process of international consultation to determine whether such signals should be sent, extending the policy enunciated in the Declaration of Principles that no reply to an incoming signal should be sent until such a consultation has taken place. Others strongly disagree. One approach would be to set quantitative thresholds for proposed signals, such as the normal power and duration of pulses from military and planetary radars. Above that level, transmissions would require approval. Those who fund, control, or regulate the largest radars and transmitting radio telescopes could require observance of that threshold.

Astronomer Ivan Almár proposed what he called the San Marino scale, intended to quantify the potential hazard of transmitting powerful messages into space. The main factors were the signal strength in relation to Earth's normal background radiation and the characteristics of the transmission such as information content, direction, and duration (Almár and Shuch 2007).

The debate about active SETI is an opportunity to consider what is in the best interests of the human species. It also is a way of dealing with a basic question: who speaks for the Earth? Should we speak with one voice or many?

Where do we go from here?

We can take concrete steps to prepare ourselves for discovering extraterrestrial intelligence. Here are my recommendations:

1. Establish a contact literature base. Many authors have written articles, journal papers, and books about the search for extraterrestrial intelligence and the implications of finding it. Compiling those works into a searchable data base would be a big help to interested people. Such a bibliographical compilation was started long ago for another speculative topic: interstellar flight. The late Robert Forward and others listed hundreds of articles and papers, showing that a real literature existed as early as the 1970s (Mallove and Forward 1972). Science fiction author Steven Baxter already has given us a brief summary of the contact literature of science fiction (Baxter 2009). It would be useful, though challenging, to prepare something similar for non-fiction.

2. Reach outside narrow disciplines. Reading and conference-going in any particular discipline will not prepare individuals for the discovery of extraterrestrial intelligence. Contact is an inherently multidisciplinary subject. Interested academics should think beyond their disciplines, educating themselves about others that discuss extraterrestrial intelligence and contact. Anthropologist Ben Finney thought that the search could help bridge intellectual gulfs within humankind, particularly between physical scientists and social scientists (Finney 1992). Many of the issues to be studied already have been identified, including ways to get social scientists more involved in SETI. (Billingham 1998; Harrison *et al.* 2000; Michaud 2007, 329–330). Happily, more work by social scientists is appearing in scholarly volumes, including this one (Vakoch 2011, 2014). More extensive work might require funding. This broadening must include learning from the histories of first contacts between unequal human civilizations. Historians could write an agenda-free survey of those events and their outcomes. None of those examples will be perfect analogs of contact with extraterrestrial intelligence.

3. Globalize the debate. This field has been dominated by white English-speakers, with some important contributions from continental Europe. It is time to bring in non-Western perspectives, particularly from Asia. China, India, and Japan are poorly represented in the contact literature, at least as published in Western languages.

4. Support broadening the search for life and intelligence. Humankind will be deploying major new astronomical capabilities during the coming decade such as the James Webb Space Telescope and proposed space telescopes designed to search for planets. Ground-based mega-telescopes like the Thirty Meter Telescope in Hawaii and the Giant Magellan Telescope and the European Very Large Telescope in Chile may help. New spectroscopic instruments may be able to detect chemical signatures of biology in an extrasolar planet's atmosphere, and possibly the waste products of industrial activity. SETI astronomers may be able to get observing time on the Square Kilometer

Array of radio telescopes, due to begin construction in 2016. SETI experts have recommended several steps for technologically broadening the search for evidence of extraterrestrial intelligence, particularly by searching in non-radio wave lengths (Ekers *et al.* 2002). A broadened search may have a higher probability of success than active SETI. Future funding for SETI is uncertain. People who want to expand the search should look for allies with overlapping interests, seeking to arrange mutual support. Most obvious are astronomers searching for and characterizing extrasolar planets. Others include people whose work on interstellar probes is reported on the Centauri Dreams blog.

5. Fill in and expand the Drake equation. The later factors in the Drake equation – originally a device for estimating the number of communicating civilizations – lack reliable numbers (Chick 2014). Producing more accurate estimates for the probability of life and intelligence may require further discoveries. As Marino pointed out, we have no agreed definition of intelligence. (Marino, Chapter 6 in this volume). In effect, the original equation identified intelligence by signals that could be made only by intelligently created technology. However, radio eras may be relatively brief in our civilization and others. It would help to have other markers of intelligent life. Several authors have argued that additional factors should be added to the equation. Particularly relevant is interstellar flight. What percentage of technological civilizations explore or expand their presence beyond their original biospheres? How far would they go? No single calculation of the probability of discovery covers all of the possibilities – nor does any single estimate of the consequences.

6. Design alternate worlds. Non-astronomers interested in the search should absorb some of the most basic findings about extrasolar planets where life and intelligence might evolve. (see, e.g., planetquest.jpl.nasa.gov/). Those worlds are a basic context for the contact question. In our minds and our computers, we could create alternate worlds where intelligence may evolve under different circumstances and may be shaped by different histories.

7. Practice the message. Several people have presented reasons for drafting a message to extraterrestrial intelligence in advance (Tarter *et al.* 1992; Vakoch 1998; Michaud 2007, 371). The practice involved in drafting detailed messages could leave us better prepared for certain scenarios of contact. Such drafting is not necessarily tied to prompt transmission.

8. Rehearse the reaction. We could practice reacting to different scenarios of contact, projecting what we and intelligent aliens might do. Simulations have been used at the CONTACT conferences where teams role-played the first human–alien contact. Another model is the role-playing that military services use for training and planning. We might reach beyond war games to inter-civilizational games. These rehearsals should include the drafting of public statements to be made by governmental authorities after a discovery. The public would want to learn not only what governments know, but also what governments plan to do.

9. Spell out potential benefits and risks. We need a more systematic and objective calculation of the potential benefits and risks of contact. We should reflect on the full range of possible outcomes, not just those we prefer. One approach is to apply the Rio scale to a wide variety of possible scenarios for our first encounter. Shostak (2002) has given us an introductory example.

10. Prepare a plan for international action. Dwight Eisenhower reportedly said that, "Plans are worthless, but planning is everything." His planning process examined multiple contingencies and defined policy goals so that he could "do the normal thing when everybody else is going nuts" (Nichols 2011). Following the asteroid example, we could prepare a plan for international action that might be formally introduced to the United Nations COPUOS as an agenda item. Like the Association of Space Explorers, we could propose a decision-making process separate from COPUOS, since COPUOS itself would be inadequate in the immediate aftermath of a discovery. This would require the sponsorship of at least one national government.

11. Fly to the stars, in your mind. Finding extraterrestrial life and intelligence ultimately may require interstellar probes to explore nearby star/planet systems. While this is beyond our present capabilities, we should recall that communicating across interstellar distances seemed impossible until the Cocconi–Morrison paper pointed out that technological advance had made it feasible (Cocconi and Morrison 1959). Before dismissing interstellar flight on the basis of its cost to us, we could try to measure its feasibility by its cost to a more advanced technological civilization.

12. Form the night watch. Proposals have been made for some sort of permanent committee of distinguished persons to monitor and advise on the contact issue. Such a permanent group would be resented by many of those outside of it, and its members would age. The contact issue does not belong to any particular generation. We cannot accept the opinion of any person or group as the final word; there are no authorities on extraterrestrial intelligence until it is found. If a group is formed, it should have no permanent chairperson and no permanent membership.

13. Raise consciousness among policy makers. Governments will be forced to respond to contact. Yet we don't know what decision makers will do in a discovery situation (Billings, Chapter 20, this volume). Few senior officials in any country will take preparations seriously until a detection is confirmed. Before then, we can get their attention only by periodic targeted briefings on the implications of contact. Those briefings should include multiple scenarios, and the full range of possible consequences. Such briefings would be a never-ending process; turnover assures that government institutions have short memories. While it may be impossible to restrain media speculation immediately after a discovery, authorities can help with a media strategy of correcting false information and putting that event in context. It would be useful to look at the successes and failures of government announcements about other major phenomena. Hurricane experts, much better informed than we are about extraterrestrial intelligence, struggled to find the right balance in announcements about superstorm Sandy (Miles 2014).

14. Recognize the limitations of prediction. We humans are notoriously inaccurate when we predict the future more than a few years ahead. Entire books have been published recording utterly wrong predictions made by people who were considered authorities in their fields (Cerf and Navasky 1998). We do not know which scenario of contact we will face. At best, our planning will be only partly successful.

15. Learn to live with ambiguity. Until we have facts about extraterrestrial intelligence, we must live with uncertainty and ambiguity. Physical scientists often are impatient with those qualities. Yet uncertainty and ambiguity are familiar territory for foreign policy makers. As Henry Kissinger explained, decisions to meet the most consequential issues must be taken before it is possible to know what the outcome will be (Kissinger 2014, 371–374).

16. Be patient. Finding evidence of extraterrestrial intelligence will be more difficult than most searchers had hoped. It may require rigorous and repetitive search and data analysis that last beyond individual lifetimes. It may require a broader strategy, and a willingness to look in new places. It may require technical means not yet available to us. This is not a reason to give up. It is a reason to be more clever, and more determined. Fortunately, the interest shown by some younger scholars helps to compensate for jaded elders.

Conclusion

What was once an exotic small-scale scientific enterprise has become a vast, multidisciplinary thought experiment about the nature and behavior of intelligence, both on and beyond the Earth. That experiment draws on many different sorts of knowledge and on many unproven assumptions. Debates about extraterrestrial intelligence have been plagued by overstated assertions and dismissive denials. We know too little to be intellectually arrogant. We should seize opportunities to introduce greater objectivity, and treasure the facts we find. The search for extraterrestrials is a test of human intelligence.

References

Almár, Ivan and Paul Shuch. 2007. "The San Marino Scale: a New Analytical Tool for Assessing Transmission Risk," *Acta Astronautica* 60, 57–59. Additional information on the San Marino scale may be found at http://avsport.org/IAA/smiscale.htm

Almár, Ivan and Jill Tarter. 2011. "The Discovery of ETI as a High Consequence, Low-Probability Event," *Acta Astronautica* 68, 358–361. Online at http://avsport.org/IAA/abst2000/rio2000.pdf

Association of Space Explorers. 2008. "Asteroid Threats: A Call for a Global Response." Report issued on 25 September. Online at www.space-explorers.org/committees /NEO/ASE_NEO-Final-Report_excerpt.pdf.

Baxter, Stephen. 2009. "Imagining the Alien," *Journal of the British Interplanetary Society* 62, 132–138.

Baxter, Stephen. 2013. "Project Icarus: Interstellar Spaceprobes and Encounters with Extraterrestrial Intelligence," *Journal of the British Interplanetary Society* 64, 51–60.

Benford, James, Dominic Benford and Gregory Benford. 2011. "Building and Searching for Cost-Optimized Interstellar Beacons," in *Communication with Extraterrestrial Intelligence*, ed. Douglas A. Vakoch. Albany, NY: State University of New York Press, 279–306.

Billingham, John. 1998. "Cultural Aspects of the Search for Extraterrestrial Intelligence," *Acta Astronautica* 42, 711–719.

Billingham, John, Roger Heyns, David Milne, *et al.*, eds. 1999. *Social Implications of the Detection of an Extraterrestrial Civilization.* Mountain View, CA: SETI Press.

Billingham, John and James Benford. 2014. "Costs and Difficulties of Interstellar 'Messaging' and the Need for International Debate on Potential Risks," *Journal of the British Interplanetary Society* 67, 17–23.

Cerf, Christopher and Victor S. Navasky. 1998. *The Experts Speak.* New York, NY: Villard.

Chick, Garry. 2014. "Biocultural Prerequisites for the Development of Interstellar Communication," in Douglas A. Vakoch, editor. *Archaeology, Anthropology, and Interstellar Communication.* Washington, DC: NASA, 203–226.

Clarke, Arthur C. 1999. *Greetings, Carbon-Based Bipeds!* New York, NY: Saint Martin's Press.

Cocconi, Giuseppe and Philip Morrison. 1959. "Searching for Interstellar Communications," *Nature* 184, 844–846.

Crowe, Michael J. 1986. *The Extraterrestrial Life Debate, 1750–1900. The Idea of a Plurality of Worlds from Kant to Lowell.* Cambridge: Cambridge University Press.

Ekers, Ronald D., D. Kent Cullers, John Billingham, and Louis K. Scheffer, eds. 2002. *SETI 2020: A Roadmap for the Search for Extraterrestrial Intelligence.* Mountain View, CA: SETI Press.

Finney, Ben. 1992. "SETI and the Two Terrestrial Cultures," *Acta Astronautica* 26, 263–265.

Gray, Robert H. 2014. *The Elusive WOW: Searching for Extraterrestrial Intelligence.* Chicago, IL: Palmer Square Press.

Harrison, Albert A., John Billingham, Steven J. Dick, *et al.* 2000. "The Role of Social Sciences in SETI," in Allen Tough, editor, *When SETI Succeeds*, Bellevue, WA: Foundation for the Future, 71–86.

Jacobs, David. 1975. *The UFO Controversy in America.* Bloomington, IN: Indiana University Press.

Kissinger, Henry. 2014. *World Order.* New York, NY: Penguin Press, 371–374.

Kuhn, Thomas S. 1970. *The Structure of Scientific Revolutions*, 2nd edition. Chicago, IL: University of Chicago Press.

Mallove, Eugene F. and Robert L. Forward. 1972. "Bibliography of Interstellar Travel and Communication," *Hughes Research Laboratories Research Report* 460.

McDonough, Thomas R. 1987. *The Search for Extraterrestrial Intelligence.* New York, NY: Wiley.

McKay, David, Gibson Jr., Everett K. Thomas-Keprta, Kathie L., *et al.* 1996. "Search for Past Life on Mars: Possible Relic Biogenic Activity in Martian Meteorite ALH84001," *Science* 273, 924–930.

Michaud, Michael A. G. 1998. "Organizing Ourselves for Contact," *Analog*, January, 51–63.

Michaud, Michael A. G. 2007. *Contact with Alien Civilizations.* New York, NY: Copernicus (Springer).

Miles, Kathryn. 2014. *Superstorm.* New York, NY: Dutton.

NASA. 2014. "The Torino Impact Hazard Scale." Online at http://neo.jpl.nasa.gov/torino_scale.html. Accessed December 4, 2014.

Nichols, David A. 2011. "Eisenhower Showed Mideast Mastery," *Los Angeles Times*, March 15.

Peebles, Curtis. 1994. *Watch the Skies! A Chronicle of the Flying Saucer Myth.* Washington, DC: Smithsonian Institution Press.

Sage, Leslie. 2014. "Introduction to special section on exoplanets," *Nature* 513, 327.

Shostak, Seth. 1998. *Sharing the Universe.* Berkeley, CA: Berkeley Hills Books.

Shostak, Seth. 2002. "The Rio Scale Applied to Fictional SETI Detections," paper presented at the International Astronautical Congress. Online at http://avsport.org/IAA/abst2002/rio2002.pdf.

Tarter, Jill, John Billingham, and Michael Michaud. 1992. "A Reply from Earth?" *Acta Astronautica* 26, 295–297.

Vakoch, Douglas A. 1998. "The Dialogic Model: Representing Human Diversity in Messages to Extraterrestrials," *Acta Astronautica* 42, 705–710.

Vakoch, Douglas A., editor. 2011. *Communication with Extraterrestrial Intelligence*. Albany, NY: SUNY Press.

Vakoch, Douglas A., editor. 2014. *Archaeology, Anthropology, and Interstellar Communication*. Washington, DC: NASA. Available online at http://history.nasa.gov/what.html

Weitzman, Martin L. 2010. "Insurance for a Warming Planet," *Nature* 467, 784–785.

Zaitsev, Alexander. 2014. "Calling ET, or Not Even Answering the Phone?," *Journal of the British Interplanetary Society* 67, 30–32.

19 SETI in non-Western perspective

JOHN W. TRAPHAGAN AND JULIAN W. TRAPHAGAN

One of the more significant weaknesses in the search for extraterrestrial intelligence (SETI) research has been the limited contribution from scholars working outside of the Western cultural context including North America and Europe. While there has been interest among astronomers in countries like Japan and South Korea in the search, involvement has been limited particularly when it comes to more speculative research and writing about communication with extraterrestrial intelligence and the possible nature of civilizations inhabiting exoplanets. As argued in Chapter 8 of this volume, there is an ethnocentric bias in much writing about extraterrestrial life that is shaped by both the Western intellectual and religious traditions of Social Darwinism and Christianity, and this ethnocentrism significantly shapes the ways we think about what constitutes a civilization and how cultures and civilizations evolve over time.

In this chapter, we are interested in giving some thought to the potential influence contributions from a non-Western perspective might make in expanding our thinking about extraterrestrial intelligence. This has both pragmatic and theoretical consequences. From a theoretical perspective, alternative worldviews (from the Christian perspective) – such as what we see in Buddhism or Taoism – have the potential to shape our thinking about the nature of progress and change. Pragmatically, this should widen our scope of imagination as we contemplate the possibilities and difficulties we may encounter should contact occur.

Before moving on, we want to emphasize that this chapter is quite speculative. Our perspective here is intended only as an example of how a non-Western perspective *might* influence our thinking about SETI. In many respects, this chapter is a call for SETI scholars to work on finding new ways to draw non-Western scholars into the discourse. It was interesting that during the Library of Congress symposium that formed the basis for this volume, there were participants from the Vatican and those who worked in areas of Western theology and ethics, but there were no thinkers involved who work from a Buddhist, Muslim, or Hindu perspective (just to name a few possibilities). Indeed, as McAdamis (2011: 339) has argued, "[w]hether a result of ethnocentrism, or of the global influence of Western philosophy, most research engaging astrobiology's relationship with religion has tended to disproportionately focus on Christian theology." McAdamis' argument is largely demographic – the central point seems to be that since there are a lot of people in the world who aren't Christians, we should take their perspective into account. We agree, but would go a step further and argue that in some ways, the speculative writing on the nature of extraterrestrial intelligence is an indirect product of Christian theology and Christian beliefs about time and progress and the anthropocentric qualities evident in SETI research are an outgrowth of the strong anthropocentrism of Christianity. Thus, in order to help scientists move away from this tacit influence, non-Western perspectives, whether religious or non-religious, have the potential to bring much to the discussion and challenge assumptions floating around in many areas of SETI research.[1]

For the remainder of this chapter, our aim is to offer some speculative thoughts on how Buddhist perspectives might contribute to SETI and our thinking about extraterrestrial civilizations. To begin, we need to devote some space to presenting, in a very limited way, some key elements of the Buddhist worldview.

An example in Buddhism

Writing about Buddhism, particularly in such a brief chapter, is difficult. There are many potential pitfalls that arise when trying to generalize about a religio-philosophical framework that involves nearly 400 million adherents across multiple cultural contexts and that has two major, and distinctive, traditions (Mahayana and Theravada). These are broken into numerous sects with somewhat varied philosophical concepts about the efficacy of mediation, the nature of enlightenment, and the afterlife, to name a few key points. One of the more significant problems in the West when it comes to Buddhism and other forms of Eastern religions and philosophies has been the plethora of popular books published displaying a rather naïve or even just plain misguided understanding of either pragmatic, political, or philosophical aspects of traditions such as Buddhism and Taoism in particular – Frijof Capra's *The Tao of Physics* (2010) is a fine example of a rather naïve understanding of the topic, but one that has had a profound influence on Western ideas about Eastern philosophy and religion as it might relate to concepts found in physics. Another is Eugen Herrigel's *Zen in the Art of Archery*

[1] It is worth noting that the organizer of the symposium, Steven Dick, did in fact try to bring in a scholar working on Buddhism, but he was unable to participate.

(1999), which explores some of the philosophical ideas about the world that the author attributes to the influence of Zen in Japanese archery. These books are good reads and certainly raise interesting ideas about life, the universe, and everything. But they shouldn't necessarily be taken as adequate representations of Eastern philosophy or religion and, in fact, typically are highly romanticized and filled with inaccuracies or misunderstandings (Yamada 2009).

When it comes to Japan and Buddhism, one of the most troubling problems is a tendency to conflate philosophical Buddhism (the Buddhism of texts such as the Lotus Sutra) and practiced Buddhism. A good example of this can be found in McAdamis' (2011: 344) article on what he terms "astrosociology" and world religions, in which he argues that "Buddhism has no place for the supernatural." This misunderstanding is common among Western scholars who focus only on the philosophical, textual side of Buddhism. Some scholars have argued that while they may not be gods, per se, the Buddha and buddhas fulfill the definition of a counter-intuitive agent and, thus, function in a way similar to the gods of Western and other religions and have features that seem supernatural or superhuman. The attainment of enlightenment (nirvana) also involves a super (meaning detached) understanding of the world (Pyysiäinen 2003: 148).

Regardless of the philosophical issues related to the supernatural in Buddhism, when considering the practice of Buddhism among, for example, most Japanese, there is much less to debate. Buddhism at the pragmatic or ritual level can involve spirits or ghosts that have the potential to be dangerous to the living and in the minds of some the dead can return to the living in dreams – an event that is variously interpreted as being either simply a dream or the actual return of the deceased to warn the living of potential danger (Connor and Traphagan 2014, Traphagan 2003). And most of Buddhist ritual in Japan is devoted to care of the dead who, despite the notion of reincarnation being present, are in some ways viewed as being permanently dead in the form of ancestral spirits (Traphagan 2013).

That said, there is little doubt the worldview which emerges throughout East Asia, and which is deeply influenced by Buddhism, Taoism, and Confucianism (which is not a religion in Japan, but which can operate as a religion in China and Korea), generate a different way of seeing both time and space, just to name two important aspects of that worldview (Raud 2004). Obviously, there is not sufficient space in a chapter like this to explore the complexities of Buddhist philosophy, so we want to focus on two very specific ideas that we think might influence SETI. First, let's take a look at the notion of time.

Time and Buddhism

Whether we discuss Western or Eastern philosophical viewpoints, the idea of time is actually quite complex; we somewhat unfairly simplify both perspectives for the purpose of raising the issue here (cf. van Fraassen 1970). In the Abrahamic tradition, time operates as a linear-historical progression aimed at a particular end to history and then a goal beyond history in the form of a utopic post-apocalypse reality. As King (1968: 219) writes, "For each moment is inherently unique in its contribution to the final eternity, never repeating what has gone before – despite such a world-weary lament as that of Ecclesiastes."

The Buddhist (and Hindu) approach to time is not linear, but cyclical and as a result tends to relativize historical time. "All time, even eternities of it, finally passes. Whether we consider a second or an eon, one is as fleeting as the other, depending entirely upon our choice of time-scales. So also all times are repeated endlessly. Thus, nowhere in time – past, present, or future – is there any time or time-conditioned structure that is unique, and hence holy" (King 1968: 218). There is not a final state of blissful happiness, nor is there any salvation from the world of suffering here that leads to a permanent life in a better world. Instead, a fundamental observation of Buddhism is the impermanence of the world and, as a result, the impermanence of all being. Time isn't going anywhere – there is no telos towards which the arrow of time is directed. Instead, time is just a series of endless cycles – the only way out is through enlightenment, which is in the end the eradication of self that allows one to escape the wheel of samsara or the endless cycles of birth and rebirth. In most schools of Buddhism, the way we experience time is seen as being an illusion – the movement from past to present to future that we appear to experience is not real, but a product of our desires and tendencies to cling to things in the world. Japanese Zen master Dogen writes, "Time is not separate from you, and as you are present, time does not go away. As time is not marked by coming and going, the moment you climbed the mountains is the time-being right now. If time keeps coming and going, you are the time-being right now" (in Tanahashi 1999: 71). Through the concept of *uji* Dogen equates being and time as the constantly changing activities that unfold in the ongoing becoming that is the present.

Although not all forms of Buddhism take this approach, in Zen there is a broad rejection of religious approaches that dichotomize existence in a way that loses sight of the immediate present, such as Christianity with its focus on a historical flow leading from a primordial beginning such as Genesis to a future both for individuals (heaven or hell) and for humanity as a whole (King 1968: 222). Within Zen, historical, sequential time is without existential significance, "[i]ts own primordial-time expressions, such as 'What was your original face before you were born?,' have absolutely no time-sequential significance. They represent only a fully immanental sense of the human being's identity with the eternal world-process, in which there is neither before nor after but only Now" (King 1968: 222). The only reality – and this is not just a philosophical exercise – from the Zen perspective is the here and now and the aim of Zen – the point of Zen meditation – is to develop the capacity to experience the world completely in the present. In other words, time is an illusion; in reality there is nothing but now.

One way to think about this is that each worldview sees time as a function of reality, but in different ways. What we mean by this is that the effects that time are perceived as having on an individual vary depending on the person's religious or philosophical position. In Buddhism, an event in time is like throwing a rock into a still pond, causing a ripple effect throughout the rest of the continuum of aquatic space. While in Christianity, an event in time is like throwing a rock into a river: it may create a brief ripple, but the course of the water will always redirect into its original path.

Buddhism and the nature of the universe

The second point I want to explore is Buddhist, specifically Mahayana, notions about the nature of the universe itself. One example of Buddhist views of reality can be found in the Hua-yen or Flower Garland school of Buddhism and the writings of Fa-Tseng, one of the school's patriarchs who lived from 643–712 AD. Fa-Tseng writes about the Jewel Net of Indra, a lattice with jewels at each node (quoted in Liu 1982: 65).

> Due to their brightness and transparence, [the jewels] reflect each other. In each of the jewels, the images of all the other jewels are [completely] reflected. This is the case with any one of the jewels, and will remain forever so. Now, if we take a jewel in the southwestern direction and examine it, [we can see] that this one jewel can reflect simultaneously the images of all other jewels at once ... Since each of the jewels simultaneously reflects the images of all other jewels at once, it follows that the jewel in the southwestern direction also reflects all the images of the jewels in each of the other jewels [at once] ... Thus the images multiply infinitely and all these multiple infinite images are bright and clear inside this single jewel. The rest of the jewels can be understood in the same manner.

The essence of this notion of reality is that the universe consists not of a bunch of discrete parts, but of inter-penetrating nodes that contain and reflect all of the other nodes – all things in the world are interrelated and interdependent. As a result, to cause harm to one point in the network implies the causing of harm to all other nodes, and to cause harm to one being is to cause harm to all beings.

This metaphorical representation of the world leads to a kind of leveling of value assigned to living, or at least sentient, beings and provides a foundation for moral ideas in Buddhism related to the karmic cycle of samsara, or the cycle of birth and rebirth in which the quality (good or bad) of one's rebirth is a product of the thoughts and behaviors (good or bad) carried out in a given life. This brings us back to the notion of time, because there is no end point here – the cycle goes on forever, unless one manages to attain enlightenment, which means escaping the cycle through the attainment of no-self (the elimination of selfish desires).

Alan Watts (1966: 73) a British philosopher who had experience living in Japan and wrote extensively on Eastern philosophical ideas for Western audiences, summed up the way in which Buddhism interprets life very clearly:

> . . . every organism is a process: thus the organism is not other than its actions. To put it clumsily: it is what it does. More precisely, the organism, including its behavior, is a process which is to be understood only in relation to the larger and longer process of its environment. For what we mean by "understanding" or "comprehension" is seeing how parts fit into a whole, and then realizing that they don't *compose* the whole, as one assembles a jigsaw puzzle, but that the whole is a pattern, a complex wiggliness, which has no separate parts. Parts are fictions of language, of the calculus of looking at the world through a net which *seems* to chop it up into bits. Parts exist only for purposes of figuring and describing, and as we figure the world out we become confused if we do not remember this all the time.

Implications for SETI

This rather cursory and selective discussion of Buddhist philosophy should be considered with caution. As noted above, Buddhism has a long history and has crossed many different cultural contexts. There are notions of interdependence and unity of reality in Buddhism, but there are also notions of hierarchy in the ideas related to the karmic cycle of rebirths. On the one hand, Buddhist morality is impersonal with no god indicating right or wrong; on the other hand, in practice spiritual entities like ancestors can and do watch over and respond to the actions of those who are alive (Traphagan 2004). Buddhism, like Christianity, is complex and it is very difficult to make generalizations.

That said, we can point out that Buddhist ideas about time and the nature of reality are quite different from those found in Abrahamic religions like Christianity. The Zen concept of time as cyclical or even illusory is clearly quite different from the Western idea of time as a progression from past/present/future and this type of worldview needs to be considered when we think about the nature of extraterrestrial intelligence. If we apply this observation to SETI research, we might ask: What does cultural evolution mean within a worldview like that of Zen? Oddly enough, the Zen idea is much closer to the notion of biological evolution as simply change, rather than directional change, than is the Abrahamic idea of time as linear with a clearly defined telos or endpoint to history. Working from an assumption that time is not directional, can lead to a much broader way of thinking about the possible scope and nature of extraterrestrial intelligence, because there is no assumed path through which a civilization might evolve – there is no sense that there might be an infancy, adolescence, or maturity to any given civilization. In fact, one might be inclined to think in terms of cycles, where civilizations rise and fall (note these are very biased terms that adhere to notions of up and down, as opposed simply to ideas of change) and there one might not imagine any given civilization as being more advanced than any other, at least from social and moral perspectives.

The intriguing idea, suggested by scholars like Steven Dick (2003), that cultural evolution might inevitably lead to a postbiological civilization, can be challenged from this perspective. Dick uses what he calls the "intelligence principle" in which the driving force of cultural evolution is the "maintenance, improvement, and perpetuation of knowledge and intelligence" and that intelligence will inevitably "improve" if it is able to do so. As a result, intelligent extraterrestrial civilizations that are much older than our own will have evolved – improved – to the point that they may have advanced beyond biology to artificial intelligence in which they inhabit a postbiological world. This assumes a teleological concept of time and evolution, one that is not reflected in the Buddhist notion of cyclical time. While I don't think there is anything in Buddhism that precludes the emergence of a postbiological universe, biased concepts such as improvement and advancement would not be included in a Buddhist formulation of the possible futures of technological civilizations. Those civilizations would be different, because they are simply different iterations of the ongoing cycle; but they would not represent advancement, and certainly not moral advancement.[2]

When it comes to imagining the moral and social aspects of an extraterrestrial intelligence, perhaps drawing on Buddhist ideas related to interdependence and the universe as a single thing, rather than as a conglomeration of

[2] Dick does not argue that these postbiological civilizations would be morally superior to our own, but others do occasionally argue for the idea of moral advancement as a corollary to technological advancement when it comes to cultural evolution. See Chapter 8 in this volume.

independent things (which is actually much more like the observations of modern science than those of Abrahamic religions), might encourage us to believe that extraterrestrial intelligence would be unlikely to pose a threat – if there are extraterrestrial Buddhists, then one would assume that they might see us as reflections of themselves, assuming they see us as sentient. Frankly, we are not entirely sure how extraterrestrial intelligence and extraterrestrial civilizations might be imagined from Buddhist, Hindu, or Taoist perspectives; however, we do think we need to find ways to factor these perspectives into our discussions. In its current form, SETI continues to be heavily influenced by the tradition of Abrahamic religion and more general Western notions related to the nature of time and the nature of the universe as a whole. Incorporating broader perspectives from non-Western traditions has the potential of expanding our capacity to imagine and think about the nature of extraterrestrial intelligence as well as the implications of contact for life here on Earth.

This raises our final point. Quite simply, the assumption that contact with an intelligent extraterrestrial life form would be a profound event in human history is a product of Western ideas about the significance of life in general and human life in particular. The Buddhist worldview also values life deeply, but because it lacks the anthropocentrism of Western religious ideas, the notion that intelligent life would exist elsewhere in the universe might, rather than being a stunning revelation that, in fact, we are not alone, simply bring a shrug of the shoulders and a sort of "well, duh" reaction. Of course, there is life elsewhere; only the self-centered conceit of humans would assume that we might be alone or might even be important. Christianity, in particular, is very much structured around the idea that humans are supremely important in the order of creation; Buddhism and Taoism aren't, although they do claim that all life is valuable. Where that leads those influenced by these philosophical and religious traditions is an open question, but we doubt it will produce the same sense of a profound moment in human history that many in the SETI community assume will be the universal response. Whatever that response might be, clearly there is a deep need for bringing the voices and insights of those from non-Western philosophical and religious traditions into the SETI discourse.

References

Capra, Fritjof. 2010. *The Tao of Physics: An Exploration of the Parallels between Modern Physics and Eastern Mysticism*. Boston, MA: Shambhala Books.

Connor, Blaine P. and John W. Traphagan. 2014. "Negotiating the Afterlife: Emplacement as Process in Contemporary Japan." *Asian Anthropology* 13: 1–17.

Dick, Steven J. 2003. "Cultural Evolution, the Postbiological Universe and SETI." *International Journal of Astrobiology* 2: 65–74, reprinted as "Bringing Culture to Cosmos: The Postbiological Universe," in S. J. Dick and M. Lupisella (eds.), *Cosmos and Culture: Cultural Evolution in a Cosmic Context*. Washington, DC: NASA, 2009, pp. 463–488.

Herrigel, Eugen. 1999 [1953]. *Zen in the Art of Archery*. New York, NY: Vintage Books.

King, Winston L. 1968. "Time Transcendence-Acceptance in Zen Buddhism." *Journal of the American Academy of Religion* 36: 217–228.

Liu, Ming-wood. 1982. "The Harmonious Universe of Fa-tsang and Leibniz: A Comparative Study." *Philosophy East and West* 32: 61–76.

McAdamis, E. M. 2011. "Astrosociology and the Capacity of Major World Religions to Contextualize the Possibility of Life Beyond Earth." *Physics Procedia* 20: 338–352.

Pyysiäinen, Ilkka. 2003. "Buddhism, Religion, and the Concept of 'God'." *Numen* 50: 147–171.

Raud, Rein. 2004. "'Place' and 'Being-Time': Spatiotemporal Concepts in the Thought of Nishida Kitarō and Dōgen Kigen." *Philosophy East and West* 54: 29–51.

Tanahashi, Kazuaki. 1999. *Enlightenment Unfolds: The Essential Teachings of Zen Master Dōgen*. Boston, MA: Shambhala Books.

Traphagan, John W. 2003. "Older Women as Caregivers and Ancestral Protection in Rural Japan." *Ethnology* 42: 127–139.

Traphagan, John W. 2004. *The Practice of Concern: Ritual, Well-Being, and Aging in Rural Japan*. Durham, NC: Carolina Academic Press.

Traphagan, John W. 2013. "Rituelle Modulation, Liminalität und die Nembutsu-Praxis im Bäuerlichen Japan." *Paragrana: Internationale Zeitschrift für Historische Anthropologie* 22: 197–213.

van Fraassen, Bas C. 1970. *An Introduction to the Philosophy of Time and Space*. New York, NY: Random House.

Watts, Alan. 1966. *The Book on the Taboo Against Knowing Who You Are*. New York, NY: Random House.

Yamada, Shoji. 2009. *Shots in the Dark: Japan, Zen and the West*. Chicago, IL: University of Chicago Press.

20 The allure of alien life

Public and media framings of extraterrestrial life

LINDA BILLINGS

In this chapter I will explore the modern history of public conceptions and perceptions of extraterrestrial life and speculate on how people might respond to its discovery. Although most astrobiologists assume that "first contact" with extraterrestrial life will be the discovery of microbial life beyond Earth, in public discourse, and especially in popular culture, "first contact" tends to be characterized as contact with extraterrestrial intelligence. I will consider popular representations of extraterrestrial life – from single-celled to intelligent – in their cultural context. How does the cultural environment affect these representations? How does the political economy of the mass media industry shape these representations? How does the human psyche influence these representations? The theoretical framework for this analysis is, more or less, neo-Marxist, incorporating elements of psychoanalytic and ideological critique. The story of the search for evidence of extraterrestrial intelligent life has taken on a standard form, a litany repeated over and over again in scholarly and popular accounts, with little critical analysis (e.g. Davies 2010; Ekers *et al.* 2002; Vakoch 2014). In this chapter I will offer some critique and then draw on this critique to offer thoughts about possible responses to the discovery of extraterrestrial life.

Cultural environment and representations of extraterrestrial life

The public discourse about alien life takes place in a complex and ever-changing cultural environment. Communication theorist James Carey (1992, 44, 65) described culture "as a set of practices, a mode of human activity, a process whereby reality is created, maintained and transformed .... What is called the study of culture can also be called the study of communications." From this perspective, communication is a ritual enacted to maintain culture over time, a symbolic process of creating, maintaining, and transforming social reality. Science, literature, and film are among the many symbol systems constructed communicatively to "express and convey our knowledge of and attitudes toward reality" (Carey 1992, 30). Culture is the context in which power arises and operates, and the mass media are an integral element of culture and a site where power arises and operates.

In popular discourse the vast, largely uncharted territory that stretches from our knowledge of single-celled life as we know it to our knowledge of complex intelligent life as we know it tends to collapse. In science fiction as well as in popular discourse, the most common conception of extraterrestrial life is intelligent life – something we humans seem to believe we'll be able to recognize.

One obvious reason for this focus is that agency is necessary for drama to unfold, and agency requires motivation – that is, thinking, or intelligence. Another common explanation, well stated by astrophysicist Jean Heidmann (1997, xiii), is that fictional extraterrestrials "are generally an idealization of what humanity would like to be, or else a caricature of what we fear it might become" (Figure 20.1). As physicist and science fiction author Gregory

Figure 20.1 Some fictional accounts of human encounters with intelligent extraterrestrials depict aliens as highly advanced beings offering to help solve Earth's problems. Credit: Aaron Gronstal, reproduced by permission.

Benford (1987, 13–14) has noted, "There is probably no more fundamental theme in science fiction than the alien .... Aliens have been used as stand-in symbols for bad humans, or as trusty native guides, as foils for expansionist empires, and so on."

Alien as "Other"

It is beyond the scope of this chapter to map popular conceptions of aliens against the politics of difference over the past couple of decades. My observation is that the image of the alien in popular culture today still tends to conventional stereotype – monstrous, violent, warlike. Why is this the case? In his book *Orientalism*, a seminal text in the field of post-colonial studies, Edward Said begins, if somewhat obliquely, to answer this question. Said (1978, 5) explained how Western/European/US white people have come to understand people who look different from us. It is political – that is, we do it in pursuit of political interests. "Ideas, cultures, and histories cannot seriously be understood or studied without their ... configurations of power also being studied .... The relationship between the Occident and the Orient is a relationship of power, of domination, of varying degrees of a complex hegemony." This way of thinking about "the Other," which has endured as a way of thinking from the mid-nineteenth century to the present, positions "us" against all "those" non-Europeans and maintains the West in a superior position. Semiologist Roland Barthes addressed the "otherizing" of difference as well. Explaining what he called "bourgeois mythology," Barthes (*Mythologies*, 1972, as quoted in Hebdige 2006, 157) characterized "the petit-bourgeois person" as someone who is "unable to imagine the Other ... the Other is a scandal which threatens his existence."

The fictional extraterrestrial being is a popular way of depicting the Other in American culture and also a favorite subject of critical analysis, especially in film studies. I will touch only briefly on a rich body of science fiction film criticism that is relevant to the subject of this chapter.

Science fiction author Ursula K. Leguin (1975, 208–209) had this to say about the alien Other in science fiction, circa the 1970s: it is "the being that is different from yourself .... the sexual alien, the social alien, and the cultural alien, and finally the racial alien," all subject to domination by "all those Galactic Empires, taken straight from the British Empire of 1880." Old-school pulp fiction established this framing of the alien Other, she observed: "the only good alien [was] a dead alien – whether he is an Aldebaranian Mantis-Man or a German dentist" (Leguin 1975, 209).

Science fiction film theorist Annette Kuhn (1990, 10) has captured the flavor of critical analysis of the Other in the 1980s, writing that the "overt contents of science fiction texts are reflections of social trends and attitudes of the time." Such texts "relate to the social order through the mediation of ideologies, society's representation of itself in and for itself," she has observed. They "voice cultural repressions in 'unconscious' textual processes which, like dreams require interpretation in order to divulge meanings."

By the 1990s critics were charting "the new cultural politics of difference," as Cornel West (1993, 577) characterized the trend:

> Distinctive features of the new cultural politics of difference are to trash the monolithic and homogeneous in the name of diversity, multiplicity, and heterogeneity; to reject the abstract, general, and universal in light of the concrete, specific, and particular; and to historicize, contextualize, and pluralize by highlighting the contingent, provisional, variable, tentative, shifting, and changing .... These gestures are not new in the history of criticism or art, yet what makes them novel ... is how and what constitutes difference, the weight and gravity it is given in representation, and the way in which highlighting issues like exterminism, empire, race, class, gender, sexual orientation, age, nation, nature, and region at this historical moment acknowledges some discontinuity and disruption from previous forms of cultural critique.

West's contemporary, bell hooks (2006, 366), offered her views as well:

> Cultural taboos around sexuality and desire are transgressed ... as the media bombards folks with a message of difference ... bringing to the surface all those "nasty" unconscious fantasies and longings about contact with the Other embedded in the secret (not so secret) deep structure of white supremacy. In many ways, it's a contemporary revival of interest in the "primitive," with a distinctly postmodern slant.

Psychoanalytic perspectives

In considering the "alien" in film and television, it is worth thinking about the range of meanings and uses that image producers may embed in these texts, intentionally or not, and the meanings and uses that audiences may derive from them. Audience research is difficult to do, as individual responses are unique. In lieu of that, psychoanalytic and ideological critique (see Kuhn 1990) offer some interesting perspectives on the meanings and uses of the alien Other.

Critic Carl Malmgren (1993, 15) has observed that alien encounters in science fiction allow audiences to explore "who we are ... but also what we might become." In stories of confrontations between humans and aliens, "alien-encounter SF broaches the question of Self and the Other. The reader recuperates this fiction by comparing human and alien entities, measuring the Self by examining the Other."

In short, stories about aliens are stories about us.

Critic Margaret Tarratt (2003, 347) has described hostile movie aliens as "monsters from the id," expressions of repressed anger and desire. In science fiction films, "battles with sinister monsters or extraterrestrial forces are an externalization of the civilized person's conflict with his or her primitive subconscious or id." For critic John Rieder (1982, 26), "The alien in science fiction is the projection of the Other ... the depiction of a historically-determined figure who is an 'outsider'" in Western capitalist society. The science fiction alien "incarnates certain characteristics of the bourgeois mind itself – it exposes certain secret, autonomous urges and drives of the bourgeois subject in dealing with his own interior experience. It reflects the self-alienation by which the body becomes the fundamental *property* of the individual."

Perspectives from ideological critique

Turning to ideological analysis, it is worth considering how the political economy of the mass media industry shapes popular representations of alien life. Why does the global media industry keep feeding audiences stories about aliens? The simple answer is because they make money. Popular movies and television shows are mass-produced for a global audience, so producers tend not to stray too far from tried-and-true formulas for marketable content. Science fiction media content about alien life is usually about something else – horror, war, fantasy, doomsday – with aliens standing in for monsters, enemies, and evil. They are, in short, formulaic entertainment. In science fiction film and television, stories about doomsday/Armageddon/apocalypse appear to be especially popular. These stories are typically Christian-flavored morality tales. Even media content that is purportedly about the scientific search for extraterrestrial life tends to quickly veer into speculation about alien intelligence, alien technology, and alien belligerence.

How do these stories about aliens prepare us for the discovery of extraterrestrial life? They don't. But they do contribute to the maintenance of social order. Mass media are a vital structural element of culture. The media play a major role in constructing and affirming social norms, reinforcing the culture–power matrix and enabling and justifying the exercise of power. The process of constructing some sort of "commonsense," a basis for depicting and interpreting meanings in texts, is one of the most important ideological functions that the media perform, as communication scholar David Morley (1992) has explained. It is the way the media "translate" politics, economics, and science for the public.

Communication theorist George Gerbner (2000, 3) documented how "formula-driven assembly-line produced programs increasingly dominate the airways" and how "the formulas themselves reflect the structure of power that produces them and function to preserve and enhance that structure of power." Gerbner's cultivation theory explains how dominant modes of cultural production – mass media megacorporations that own the means of production, marketing, and distribution of radio, television, movies, books, Internet content, you name it – "generate messages and representations that nourish and sustain the ideologies, perspectives, and practices of the institutions and cultural contexts from which they arise" (Morgan 2002, p. 13).

So, as Annette Kuhn (2003) has asked, what does science fiction film do in culture? That is, what is its cultural instrumentality – its function(s), purpose(s), meaning(s)? How are science fiction film representations – their systems of meaning – produced and with what consequences? The process by which representations shape our understanding of the world we live in is a process of ideology. "Meaning is never neutral," as Kuhn (2003, 53–54) noted, "but always caught up in relations of power," and "ideology is invisible, naturalized." Considered from this perspective, science fiction films about aliens, "under the cover of entertainment," place audiences within a framework of ideology and contribute to the shaping of our views of the world.

What critic Vivan Sobchak (1987, 304) observed 25 years ago holds today, perhaps even more so: popular, mainstream science fiction movies – let's call them the products of Hollywood – "display an ideological conservatism and conventionality" that is "as much a function of the conditions of their economic production as they are a function of the political climate" in which they are produced and consumed. These movies are, first and foremost, commodities, and typically highly profitable ones (Figure 20.2)

Take, for example, the *Star Wars* movie series, which began in 1983. The *Star Wars* films "celebrate[d] the technology of destruction from the point of view of those employing it as victors," writes critic Michael Stern (2003, 67). By means of special effects, the "naturalization of artifice reenacts the ideology of the film[s] as a whole in [their] celebration of the scope of war, weapons, and large-scale technology in daily life . . .. [S]pecial effects accomplish the political work of legitimizing current structures of domination." In contemporary American culture, bigotry, racism, sexism, and xenophobia are considered socially unacceptable, yet they all remain operative. Thus, "bad" extraterrestrial intelligent life – and on television and in the movies, extraterrestrial intelligence is usually "bad" – is feared, fought, and conquered over and over

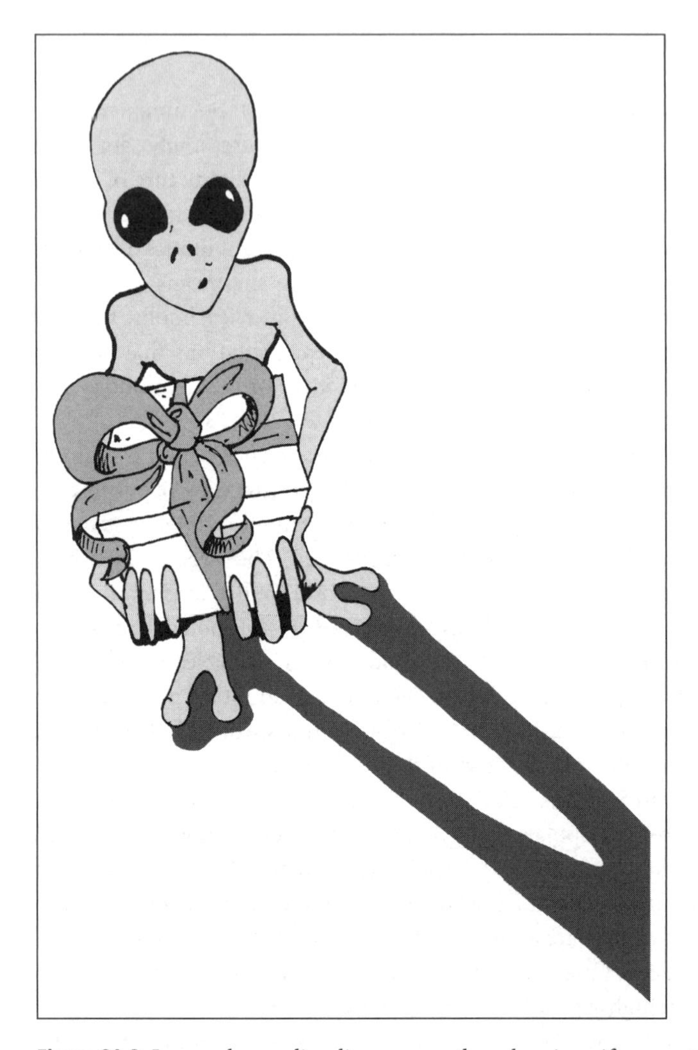

Figure 20.2 In popular media, aliens – even those bearing gifts – are typically depicted as enemies of humanity. Credit: Aaron Gronstal, reproduced by permission.

again in the mass media, providing an outlet for our fear and loathing. It is no longer socially acceptable to root for cowboys killing Indians – so the media urge audiences to cheer star warriors as they wipe out "bad" aliens and save the world.

Perhaps the most pernicious ideology that science fiction film and television enforce is the belief in the concept of "race" – which arguably has no biological basis (Hochman 2013) – and thus the self-fulfilling prophecy of racial (or even species) hierarchy. "Resemblances between aliens and humans preserve the subordination of 'other worlds, other cultures, other spheres' to the world, culture, and 'speciality' of White American culture," says Sobchack (1987, 297). She continues, "We can see this new American 'humanism' literally expand into and colonize outer space, making it safe for democracy, multinational capitalism, and the Rolling Stones."

Through the 1990s, science fiction film and television offered several iterations of what may be the most terrifying sort of alien Other: the collective alien – intelligent, technological, not quite conscious, and devoid of individuality, psyche, or soul. Its goals are takeover ("assimilation") and annihilation. Consider the alien swarms of *Independence Day* (1996); the collective, parasitic, telepathic Hive of *Dark Skies* (Fox TV, 1996), the caste-conscious alien "newcomers" of *Alien Nation: The Enemy Within* (Fox TV, 1996); and the very scary Borg (*Star Trek: Next Generation*, 1990).

Consider the $75 million Hollywood blockbuster *Independence Day*, released July 3, 1996. By August 4 the movie had grossed more than $240 million. "ID4" reportedly was the highest grossing film of 1996 and is the seventh highest grossing film to date. The aliens of ID4 are described as "locusts," roaming the galaxy, mass murdering, and appropriating territories. But they ultimately lose in battle against human (especially American) bravery and ingenuity and a "ready" global war machine. In ID4, the alien "antichrists" initiate a global apocalypse. Responding to direction from the president of the United States, the people of Earth band together to fight off the aliens, finally blowing them to smithereens – with nuclear weapons. In ID4, a conservative Christian Armageddon also unfolds: a separated husband and wife reunite, and they survive; a stripper shows that all she really wants to be is a wife and mother and nurse, and she survives; a careerist mother who leaves her child at home for a job and refuses to come home when her husband asks dies; a gay male character is killed; an entire crowd of "New Agers" is killed. The apocalypse of ID4 purges the Earth of "inferior" people and saves those who do and believe the right things – especially those who are willing to take up weapons of mass destruction. Superficially a doomsday tale, ID4 also propagated mainstream American political ideology.

Turning from Hollywood-blockbuster depictions of alien life to journalistic reporting on the science of the search for evidence of extraterrestrial microbial life, the profit-driven culture of the media megacorporation holds the same sway over content production. Writing of "journalism and its troubles," sociologist Herbert Gans (2003, 22) has observed how the news media have become "commercial media" in recent decades, how journalists have been "disempowered by changes taking place in the news industry," such as the consolidation of media ownership, the shift from news to infotainment, and the demolition of the traditional wall between editorial and publishing (p. 22).[1]

News Corporation (now operating as 21st Century Fox – producer of *The X-Files* in film and on television – and News Corp.) is only one example of the media megacorporation with a broad and deep reach. News Corp. reported $34 billion in revenues for 2012.[2]

Also consider Discovery Communications, which bills itself as "the world's #1 nonfiction media company reaching more than 2 billion cumulative subscribers in 220 countries and territories."[3] Discovery Communications earned $1.1 billion in net income on revenues of $5.5 billion in fiscal year 2013. Among the corporation's holdings is the Science Channel, the self-described "home for alien programming" on television.

In 2014 the Science Channel aired Season 3 of its television series *Alien Encounters*, which mixes science with fiction to speculate on "the impact of aliens on humanity." [4] Preceding Season 3 was the Science Channel's "Are We Alone? Week." A press release[5] announcing the week of "alien" programming stated: "Seventy four percent of Americans believe in the existence of aliens – and 15 million believe that they have actually made contact with extraterrestrials."

According to Debbie Myers, general manager of the Science Channel, "So many people are obsessed with the existence of aliens .... ARE WE ALONE? ignites their imaginations with bold new questions, and engages current research happening in the field of extraterrestrial life. It's programming that asks questions and makes you think. We hope ARE WE ALONE? advances the conversation even further." (In 2013, the Science Channel programmed an "Are We Alone? Month," including the airing of Season 2 of "Alien Encounters."[6]) Science Channel also has aired *Aliens: The Definitive Guide*, publicized as a program that "showcases the brilliant scientists who are currently grappling with extraordinary questions about alien life including: what

[1] See also Picard (2014).
[2] See News Corp.'s 2012 annual report to the Securities and Exchange Commission.
[3] http://ir.corporate.discovery.com/phoenix.zhtml?c=222412&p=irol-irhome.
[4] "Science Channel explores the impact of aliens on humanity in an all-new season of *Alien Encounters*," 2014. May 15. http://press.discovery.com/us/sci/press-releases/2014/alien-encounters-season-3-3080-3080/.
[5] "ARE WE ALONE? WEEK Makes Contact This March on Science Channel," 2014. February 18. http://press.discovery.com/us/sci/press-releases/2014/are-we-alone-week-2014-2961-2961/.
[6] "Science Channel Joins the Search for Extraterrestrial Life This March with *Are We Alone?* Month, 2013. February 20. http://press.discovery.com/us/sci/press-releases/2013/science-channel-joins-search-extraterrestrial-2303/.

will aliens really look like; how will they sound; what might their worlds look like; and, of course, will they come in peace?"[7]

Finally, consider SyFy (formerly the Sci Fi Channel), a division of NBC Universal, producer of the *Harry Potter* film series. NBC Universal is owned by Comcast, which reported $65 billion in revenues for 2013. A decade ago, Sci Fi waged a massive campaign to publicize a mini-series called *Taken* (2002), about alien abductions. Sci Fi framed its advertising as an effort to inform the public about the reality of alien visitations and abductions and convince government officials to take UFOs seriously (Billings 2005). Sci Fi described *Taken* as fiction based on true stories.

It is no wonder that public discourse about alien life is quite muddled, often contradictory, and sometimes misleading.

SETI as ideology

Moving from fictional conceptions of extraterrestrial intelligent life to the scientific search for extraterrestrial intelligence (SETI), it is worth considering how assumptions and beliefs influence science as well as fiction. Reporting on a 1961 meeting of scientists interested in SETI, then-*New York Times* science editor Walter Sullivan (1966, 15) wrote, "The participants shared a strong feeling that [extraterrestrial] civilizations exist." A feeling was all they had to hang on to, as evidence was lacking. It still is.

For more than 50 years, a small community of scientists, scholars, and others – which I will call the SETI faithful[8] – has sustained and propagated a belief in the existence of extraterrestrial intelligent life and the inevitability – now even the imminence – of contact with it. For more than 50 years, the mass media – in news, infotainment, and entertainment – have helped the faithful to propagate their beliefs. The SETI faithful, with the aid of the media, have used the so-called Drake "equation" to construct a mythology, a sort of origins myth, about extraterrestrial intelligent life in the universe and inevitable, perhaps imminent, human contact with it (Billings 2010).

Many of the SETI faithful are members of the International Academy of Astronautics' (IAA's) SETI Permanent Committee. Paul Davies, chair of the IAA's SETI Post-Detection Subcommittee, has been described in *The*

[7] See also http://press.discovery.com/us/sci/press-releases/2013/science-channel-joins-search-extraterrestrial-2303/.

[8] I have been following the scholarly and popular discourse on SETI since 1983. I worked as a contractor with NASA's SETI program from 1988 until it was cancelled in 1993. Among the individuals whose SETI advocacy I am familiar with and whom I deem "SETI faithful" are John Billingham (deceased), Paul Davies, Steven Dick, Frank Drake, Ben Finney, Al Harrison, Claudio Maccone, Michael Michaud, Seth Shostak, Jill Tarter, and Douglas Vakoch.

Guardian as "the man who'll welcome the aliens" and "the scientist with an awesome responsibility" (Ronson, 2010). In his book *The Eerie Silence: Renewing Our Search for Alien Intelligence*, Davies (2010) advocates expanding the search for evidence of extraterrestrial intelligent life. Davies readily admits that SETI is "speculative to a degree far beyond that of conventional science" (xii). Yet he argues that the space community should be more rather than less enthusiastic about searching for and ultimately finding evidence of extraterrestrial intelligence. He closes his book with a "three-hats answer" to the question, "Are we alone?" As a scientist, he says, "I have yet to see a convincing theoretical argument" for widespread intelligent life in the universe. As a philosopher, he wonders what the universe is for, if not for intelligent life. As a "dreamer," he wants to believe that the universe hosts widespread intelligent life, though he claims his scientist-self keeps him shy of outright belief (Davies 2010, 207).

Many of the SETI faithful,[9] including Davies, participated in a 1999 SETI workshop organized by the Foundation for the Future. A report on the workshop, entitled "When SETI Succeeds" (Tough 2000) offers further insight into the ideology of SETI. The title alone reflects a belief – not posing a scientific question of if, but making an ideological assertion of when. The workshop focused on "the long-term implications of a potential dialogue with extraterrestrial civilization," such as "the impact of any practical information and advice that we receive; the impact of new insights, understanding, and knowledge about major questions that go far beyond ordinary, practical, day-to-day matters; [and] the impact of a transformation in our view of ourselves and our place in the universe" (Tough 2000, 33). The report reflects a considerable amount of wishful thinking, even a hope for extraterrestrial salvation (a hope not at all uncommon among the SETI faithful and often reflected in television and film encounters with extraterrestrial intelligence).

For example, following the assertion that extraterrestrial civilizations "will be far older than we" come these "implications": "Their longevity is inconsistent with organized monotheistic religions typical of Earth. Such religions are responsible for the longest lasting warfare and destruction we have witnessed. There may be something like a universal religion that integrates the existence of intelligent life forms with a fully developed cosmology. There will be no sects, as religious warfare would be the result, and that is inconsistent with longevity .... They may tell us how it is possible to transition from the 'My God vs. Your God' conflicts of ... Earth to a more stable universal religion/understanding" (Tough

[9] See Note 7. All but Drake and Shostak were participants in the Foundation for the Future workshop.

2000, 45). The report goes on to say, "We would most likely be dealing with beings who are far in advance of us not only technologically and scientifically, but also in the general sense, spiritually" (Tough 2000, 51).

Workshop organizer Allen Tough (who died in 2012) was a leader of the IAA's SETI Committee and the founder of a project called "Invitation to ETI," whose mission "is to establish communication with any form of extraterrestrial intelligence able to monitor our World Wide Web."[10] Here is the invitation: "Greetings to extraterrestrial intelligence! If you originated in some other place in the universe, we welcome you here. And we invite you to establish communication with us and with all of humanity. We enthusiastically look forward to that dialogue. This invitation comes to you from a diverse group of approximately 100 individuals. We come from various parts of our planet. Almost all of us are related to science – as researchers, engineers, artists, writers, benefactors, or graduate students. Many of us are involved in [SETI] . . . ."[11] The "Invitation to ETI" group includes many of the SETI faithful I've come to know.

A 549-page SETI "roadmap" published by the SETI Institute in 2002 (Ekers *et al.* 2002, 2) relies on the SETI belief system. In it these questions are posed: "Are we the only creatures to view the universe with understanding? In a cosmos filled with billions of galaxies containing trillions of stars, is it possible that Earth, a world of inconsequential size and ordinary position, is alone in housing life that can discern the natural order?" (My own answer to the first question is, who knows? And what is "understanding?" Is it possible we are it? Yes. Is it probable? It depends on what you believe.) "SETI has important implications for the future of society on Earth," this roadmap dutifully notes. "The actual discovery of a signal will lead to sociological, intellectual and practical changes in our civilization" (Ekers *et al.* 2002, 9). These claims, couched as certainties, are, at best, speculations. To accept them as certainties requires an embrace of the ideology of SETI. The SETI belief system is reminiscent of messianism, millenarianism, anarcho-communism, and utopianism (see Cohn 1970) – ideologies that fostered hope for social change, salvation, a new way of living.

To sum up, what we now know about extraterrestrial life, of any sort, is nothing. Yet optimism about contact with extraterrestrial intelligent life prevails.

Conclusion

A decade ago, astrobiologist David Grinspoon noted in *Lonely Planets* (Grinspoon 2004, xxix) that in researching the book he had been "repeatedly

[10] http://ieti.org/admin/mission.htm. [11] http://ieti.org/hello/index.html.

been struck by the great similarity between our current ideas about alien life and those that were expressed decades and even centuries ago." I'm similarly struck. Grinspoon pointed out then what I am pointing out again here, 10 years later: what we know about extraterrestrial life, of any sort, is nothing.

As to how the discovery of extraterrestrial microbial life in our Solar System will affect our self-image, I would guess not a whole lot. Humans have a hard enough time understanding, communicating with, and relating to other complex terrestrial life forms, from plants to birds to other humans. Most people do not understand, communicate with, or relate to microbes.

As to how contact with extraterrestrial intelligent life will affect our self-image, our behavior, our culture – while speculation runs wild, no one knows. Predictions about response are simply guesses – some educated, some not so much. Perspectives on contact are subjective – dependent on what one chooses to believe and not believe (optimistic or pessimistic estimates, the goodness or badness of humanity, the nature of human cultures). My opinion – my belief – is that such contact is nowhere near as likely as the SETI faithful purport it to be. And it is important to note that, if it should ever occur, response would be contingent on context. We cannot predict when or where or how such contact will occur – because we don't know IF it will occur – so speculations about response are of limited use in preparing for contact.

"It's hard enough to understand certain assumptions of the Samoans, the Balinese or the Americans," as one science fiction critic (Durgnat 1971, 252, cited in Sobchak 1987, 93) has observed, "and all but impossible to empathize into the perceptions and drives of, say, a boa constrictor. How much more difficult then to identify with the notions of, say, the immortal twelve-sensed telepathic polymorphoids whose natural habitat is the ammonia clouds of Galaxy X7?"

The SETI faithful believe that science and mathematics are universal. But no one knows. They believe that biology is universal. But no one knows. What is the "landscape" – or shall I say the "universe-scape"?" – of intelligence? We have no idea (but see Chapter 6). SETI ideology asserts that the sort of technology developed by humans is a universal quality of intelligence. But no one knows. As Gregory Benford (1987, 17) has observed, "Scientists often say that communication with aliens could proceed because . . . we both inhabit the same physical universe and [because] the laws of physics are universal. I'm not so sure . . . . Language can't simply refer to an agreed-upon real world, because we don't know if the alien agrees about reality."

Speaking at the University of Oregon in 2013, the Dalai Lama offered some advice on preparing for contact with extraterrestrial intelligence. The fact that I am a Buddhist and the Dalai Lama, he said, creates a distance between "me" and "you." "As soon as you look at another as something different from you . . . then this sort of uneasiness comes." The human tendency to be wary of strangers, outsiders, "creates more anxiety, more fear," he said. "If we receive visitors from outer space, respect them, look at them as sentient beings."[12]

In the closing chapter of his book *First Contact* – entitled "The day after contact" – Marc Kaufman (2011, 187–188) offers three possible answers to the question of whether we humans are alone in the universe: (1) we are alone in the universe, (2) only Earth has complex life, or (3) "life exists beyond Earth and, in some instances, has become complex and most likely includes what we would consider intelligence." I would like to offer a fourth option, a more general statement than the three above: life exists elsewhere. This option is my chosen belief. As to where this life might be and what it might be like, I await data.

By playing into the media industry's appetite for cheap entertainment, by ignoring the gap between the origin and evolution of single-celled life and the evolution and nature of complex walking-and-talking life, by propagating the idea that scientists understand what intelligence is, the SETI faithful and their helpers in the media may be entertaining their audiences. However, they are not promoting scientific literacy or critical thinking. Scientists, journalists, Hollywood, media megacorporations – all parties involved in propagating the mythology that the universe must be swarming with intelligent life that we can recognize and communicate with – are missing an opportunity to broaden public dialogue about what it means to be human, what it means to be intelligent, what it means to be complex and evolving, and what it means to live on a planet in a solar system in a galaxy in a universe.

References

Benford, Gregory. 1987. "Effing the ineffable." In *Aliens: The Anthropology of Science Fiction*, edited by George E. Slusser and Eric S. Rabkin, 13–25. Carbondale: Southern Illinois University Press.

Billings, Linda. 2012. "Astrobiology in culture: the search for extraterrestrial life as 'science'," *Astrobiology* 12: 966–975.

Billings, Linda. 2010. "Search for extraterrestrial intelligence." In *Encyclopedia of Science and Technology Communication*, edited by S. H. Priest, 785–787. Thousand Oaks, CA: Sage.

[12] http://youtu.be/zKTu6VBhU2A

Billings, Linda. 2005. "Sex! Aliens! Harvard! Rhetorical boundary-work in the media (A case study of role of journalists in the social construction of scientific authority)," Ph.D. dissertation, Indiana University.

Carey, James. 1992. *Communication as Culture: Essays on Media and Society*. New York, NY: Routledge.

Cohn, Norman. 1970. *The Pursuit of the Millennium: Revolutionary Millenarians and Mystical Anarchists of the Middle Ages*, revised and expanded edition. New York, NY: Oxford University Press.

Davies, Paul. 2010. *The Eerie Silence: Renewing our Search for Alien Intelligence*. New York, NY: Houghton Mifflin Harcourt.

Durgnat, Raymond. 1971. *Films and Feelings*. Cambridge, MA: MIT Press.

Ekers, Ronald D., D. Kent Cullers, John Billingham, and Louis K. Scheffer, editors. 2002. *SETI 2020: A Roadmap for the Search for Extraterrestrial Intelligence*. Mountain View, CA: SETI Press.

Gans, Herbert J. 2003. *Democracy and the Media*. New York, NY: Oxford University Press.

Gerbner, George. 2000. "Cultivation analysis: an overview." *Communicator*, October–December: 3–12.

Grinspoon, David. 2004. *Lonely Planets: The Natural Philosophy of Alien Life*. New York, NY: Harper Collins.

Hebdige, Dick. 2006. "Subculture: the meaning of style." In *Media and Cultural Studies: Key Works*, revised edition, edited by Meenakshi Gigi Durham and Douglas M. Kellner, 144–162. Malden, MA: Blackwell Publishing.

Heidmann, Jean. 1997. *Extraterrestrial Intelligence*. Cambridge: Cambridge University Press.

Hochman, Adam. 2013. "Racial discrimination: how not to do it." *Studies in the History and Philosophy of Science Part C* 44: 278–286.

hooks, bell. 2006. "Eating the other: desire and resistance." In *Media and Cultural Studies: Key Works*, revised edition, edited by Meenakshi Gigi Durham and Douglas M. Kellner, 366–380. Malden, MA: Blackwell Publishing.

Kaufman, Marc. 2011. *First Contact: Scientific Breakthroughs in the Hunt for Life Beyond Earth*. New York, NY: Simon and Schuster.

Kuhn, Annette, editor. 1990, 2003. *Alien Zone: Cultural Theory and Contemporary Science Fiction Cinema*. London: Verso.

Leguin, Ursula K. 1975. "American SF and the Other." *Science Fiction Studies* 2: 208–210.

Malmgren, Carl D. 1993. "Self and Other in SF: alien encounter." *Science Fiction Studies* 20: 15–33.

Morgan, Michael, editor. 2002. *Against the Mainstream: The Selected Works of George Gerbner*. New York, NY: Peter Lang.
Morley, David. 1992. *Television, Audiences, and Cultural Studies*. New York, NY: Routledge.
Picard, Robert G. 2014. "Twilight or dawn of journalism? Evidence from the changing news ecosystem." *Journalism Studies* 15: 500–510.
Rieder, John. 1982. "Embracing the alien: science fiction in mass culture," *Science Fiction Studies* 26: 26–37.
Ronson, Jon. 2010. "First contact: the man who'll welcome the aliens." *The Guardian*, 5 March. Accessed June 5, 2014. http://www.theguardian.com/global/2010/mar/06/paul-davies-aliens-welcome-jon-ronson.
Said, Edward. 1978. *Orientalism*. New York, NY: Pantheon.
Sobchak, Vivian. 1987. *Screening Space: The American Science Fiction Film*, 2nd edition, New Brunswick, NJ: Rutgers University Press.
Stern, Michael. 2003. "Making culture into nature." In *Alien Zone: Cultural Theory and Contemporary Science Fiction Cinema*, edited by Annette Kuhn, 66–72. New York: Verso.
Sullivan, Walter. 1966. *We Are Not Alone*. New York, NY: New American Library.
Tarratt, Margaret. 2003. "Monsters from the id." In *Film Genre Reader III*, edited by Barry Keith Grant, 346–365. Austin, TX: University of Texas Press.
Tough, D. Allen, editor 2000. *When SETI Succeeds: The Impact of High-Information Contact*. Bellevue, WA: Foundation for the Future.
Vakoch, Douglas A., ed. 2014. *Archaeology, Anthropology, and Interstellar Communication*. Washington, DC: NASA.
West, Cornel. 1993. "The new cultural politics of difference." In *Social Theory: The Multicultural and Classic Readings*, edited by Charles Lemert, 577–589. Boulder, CO: Westview Press.

21 Internalizing null extraterrestrial "signals"

An astrobiological app for a technological society

ERIC J. CHAISSON

One of the beneficial outcomes of searching for life in the universe is that it grants greater awareness of our own problems here on Earth. Lack of contact with alien beings to date might actually comprise a null "signal" pointing humankind towards a viable future. Astrobiology has surprising practical applications to human society; within the larger cosmological context of cosmic evolution, astrobiology clarifies the energetic essence of complex systems throughout the universe, including technological intelligence that is intimately dependent on energy and likely will be for as long as it endures. The "message" contained in the "signal" with which today's society needs to cope is reasonably this: only solar energy can power our civilization going forward without soiling the environment with increased heat yet robustly driving the economy with increased per capita energy usage. The null "signals" from extraterrestrials also offer a rational solution to the Fermi paradox as a principle of cosmic selection likely limits galactic civilizations in time as well as in space: those advanced life forms anywhere in the universe that wisely adopt, and quickly too, the energy of their parent star probably survive, and those that don't, don't.

The context

A few years ago, I had the pleasure of attending the 50th anniversary of Project Ozma – the first dedicated search for extraterrestrial intelligence (SETI) conducted by Frank Drake in 1960. The celebratory gathering was held at the National Radio Astronomy Observatory in Green Bank, West Virginia, where that initial search was attempted and where I had in the intervening years operated dozens of radio-frequency experiments of my own, including a few unauthorized reality-checks for signs of otherworldly life. Although I never detected there any signal implying contact, I often wondered why not. Astronomers are commissioned by society to keep our eyes on the sky, yet we have never found any unambiguous, confirmed evidence for life beyond Earth. Are alien civilizations out there but not advanced enough to betray their presence? Or are they so advanced they are actively hiding from us? Perhaps they just don't exist at all, thereby ensuring that we are alone in the observable universe.

I have always imagined the SETI search parameter space to resemble a vast chandelier containing some billion light bulbs, each representing a star with a habitable planet orbiting it in the Milky Way Galaxy. I cannot recall who first conceived this analogy; it probably wasn't me although I've used it for decades in my classes and writings to frame a viable response to the famous Fermi question about extraterrestrial intelligence – namely, where are they? My reply includes a serious timing issue. Conceivably, over the history of our galaxy, virtually all the bulbs in this extraordinary light fixture eventually illuminate when myriad technologically competent civilizations emerge as evolution aimlessly yet sufficiently twists the many bulbs for each of them to glow – but perhaps only a few light up at any given time. That is, such a chandelier might never fully shine brilliantly since only a few bulbs simultaneously brighten – maybe only one (or none) illumes at any moment, such as a dim and dirty bulb now signifying us toward the edge of the chandelier. All of which reverie suggests that advanced civilizations might not endure – their technological longevity is brief.

Many SETI enthusiasts lament the fact that no signal has thus far been heard, seen, or otherwise recognized. SETI proponents are quick to note that today's search strategies resemble hardly more than a romp in a haystack – and that a lack of signal means little as we have so far only briefly sampled the total search domains of space, time, and wavelength. Among them, I, too, have often mused that only after searching for perhaps a thousand years and still not having detected anything might the observed silence be significant. Even so, more than five decades of scanning the skies for extraterrestrial intelligence ought to be sufficient to infer something about the prevalence of smart, long-lived aliens. This is especially so given the recent rash of exciting discoveries of habitable exoplanets, which many colleagues argue increases the probability of making contact, yet by contrast makes lack of contact even more troubling given that we now know the universe to be rich in favorable extraterrestrial abodes. For a universe that seems bio-friendly, absence of evidence for extra-terrestrial intelligence to date implies either that biological intelligence is everywhere slow to evolve as was the case on Earth, or that galactic civilizations don't survive long after achieving technical capability to intercommunicate. While remaining a staunch SETI supporter, I nonetheless surmise that, at any one time, there are likely very few "needles in the cosmic haystack."

Might we learn something useful from the lack of positive extraterrestrial intelligence detection? Is it possible that the absence of an incoming literal signal represents an important figurative signal, which conveys ways and means that our own "bulb in the chandelier" might stay lit longer?

The signal

At the Green Bank anniversary meeting, I was assigned to a panel that most participants carefully avoided. I was asked to critically assess long-standing assumptions regarding SETI in front of esteemed SETI pioneers who had made those same assumptions for decades, and to speculate about what might happen if extraterrestrial intelligence was found – much as I had a decade earlier without much endorsement (Chaisson 2000). So I trotted out my favorite chandelier analogy and was surprised to hear myself saying that the evident silence could well be telling us that it's time to get our own planetary house in order.

I was even more startled when this inward-focused notion resonated with a couple of administrative assistants who had helped organize the meeting and were eavesdropping on a few of the panels. These clerical workers as well as a few engineers who had been quietly sitting in on the meeting surrounded me after the session, wanting to know more about how the outward-focused work of the observatory could conceivably aid humankind at a time when global social problems seem to be mounting. The discussion continued well on into the evening when some cooks and servers joined us after dinner in the laboratory cafeteria. These non-science folks were not interested in contacting extraterrestrial intelligence as much as learning how the search itself might identify ways that our own civilization on Earth could benefit.

I found this off-line, after-hours conversation quite sobering. While the distinguished celebrity scientists adjourned for drinks in an upstairs lounge, several observatory support staff had corralled me for a spirited discussion that I never recall having had as a member for nearly two decades of NASA's SETI Science Working Group. More than ever before, I realized that the lack of extraterrestrial intelligence signals might actually amount to a "signal" itself – indeed one that might both contain a message and have an impact for us on Earth. In short, the negative findings thus far for other intelligent life forms might well be alerting us to get our own worldly act together and thereby enhance society's technological longevity as much as possible – to maximize the "*L*" factor at the end of the famous Drake equation.

Could the eerie silence from alien worlds actually help to improve *this* world? Might the null "signal" contain a "message" that our civilization needs to learn to cope with global problems that could harm, reverse, or even extinguish life on Earth?

The message

To appreciate any embedded "message" within the null "signal," consider how astrobiology might help us identify practical ways to aid human society at key milestones of our own recent evolutionary trajectory. Harshly stated though no less true, evolution is all about winners and losers – and regarding life on Earth, the ~ 99% of all species that are extinct comprise the losers; life's development is not much different, as for example in any typical forested area hardly 20 trees result from 1,000 seedlings, most of which die while unable to acquire enough energy. So might most extraterrestrial intelligences be losers. The broader context of cosmic evolution – a synthesis of change writ large among radiation, matter, and life throughout the history of the universe – can potentially guide us to become a rare winner, hence to increase *L* and survive longer. With that in mind, I recently published a study that I never thought I would write – or even ever be able to write – exploring how the cosmology of cosmic evolution might have numerous practical applications to human society, including global warming, smart machines, world economics, and cancer research (Chaisson 2014a).

The grand scenario of cosmic evolution strives to provide a comprehensive worldview of the rise of complex, material systems in the known universe, from big bang to humankind (Chaisson 2001, 2014b; Dick and Lupisella 2009; Vidal 2014). Also sometimes termed "universal history," "big history," "epic of evolution," or simply "origins," this interdisciplinary worldview combines complexity science and modern cosmology in order to grant humankind a sense of place in the cosmos. Cosmic evolution also places the study of life in the universe into an even larger perspective, providing for astrobiology an effective unifying theme across three preeminent sequential timeframes – Radiation Era → Matter Era → Life Era – the last of these beginning not when life originates but rather when it becomes technologically sentient (Chaisson 1987, 2003).

Basically the idea is this: energy aids the rise of complexity among all material systems, living and non-living. Energy flow, in particular, through open, organized, non-equilibrium systems seems to be a key facilitator in all aspects of evolution, and may actually drive it, albeit meanderingly, while mixing chance causes with necessary effects. It's a relatively simple idea, dating back to Heraclitus some 25 centuries ago, that is now constantly being tested with new experiments and observations, all in accord with the modern scientific method as scientists of different disciplines go about their business exploring the origin and evolution of myriad diverse systems spanning the universe. Particularly notable is the rise in complexity of many varied systems over the course of time, indeed dramatically so (with some exceptions) within the past half-billion years since the Cambrian geological period. Remarkably, this energy-centered concept, supported by vast amounts of quantitative data, details how a wealth of complexity emerged, flourished, and grew, despite the entropy-based second law of thermodynamics that implies things ought to be breaking down.

Theory, experiment, observation, and computer modeling together affirm that islands of ordered complexity – mainly galaxies, stars, planets, life, and society – are numerically more than balanced by great seas of increasing disorder elsewhere in the larger environments beyond those systems. Emergent complex systems encountered in the cosmic-evolutionary scenario agree quantitatively with the valued principles of thermodynamics, including its celebrated second law. Furthermore, an underlying principle, general law, or ongoing process might well aid in the creation, organization, and maintenance of all complex systems everywhere and everywhen, from the early universe to the present.

Neither absolute energy alone, however, nor merely energy flow are sufficient to explain the origin and evolution of the multitude of systems observed in Nature. Life on Earth is surely more complex than any star or galaxy, yet the latter utilize much more total energy than anything now alive on our planet. Accordingly, I have found that by normalizing energy flows in complex systems by their inherent mass, a more uniform analysis and effective comparison is achieved between and among virtually every kind of system known in Nature. Mass-normalized energy flow, characterized by the term "energy rate density," is a leading candidate underlying a ubiquitous universal process capable of building structures, evolving systems, and creating complexity throughout the cosmos.

Figure 21.1 plots energy rate density literally "on the same page" for many complex systems, whose numerical values for major systems are compiled in the five bubble inserts. This single graph shows this same quantity consistently and uniformly describing physical, biological, and cultural evolution for an extremely wide range of complex systems spanning ~ 20 orders of magnitude in spatial dimension and nearly as many in time. For those who prefer words devoid of numbers, a simple "translation" of the figure's rising curves suggests a ranked order of increasingly complex systems across all of time to date:

- mature galaxies are more complex than their dwarf predecessors
- red-giant stars are more complex than their main-sequence counterparts
- eukaryotes are more complex than prokaryotes
- plants are more complex than protists

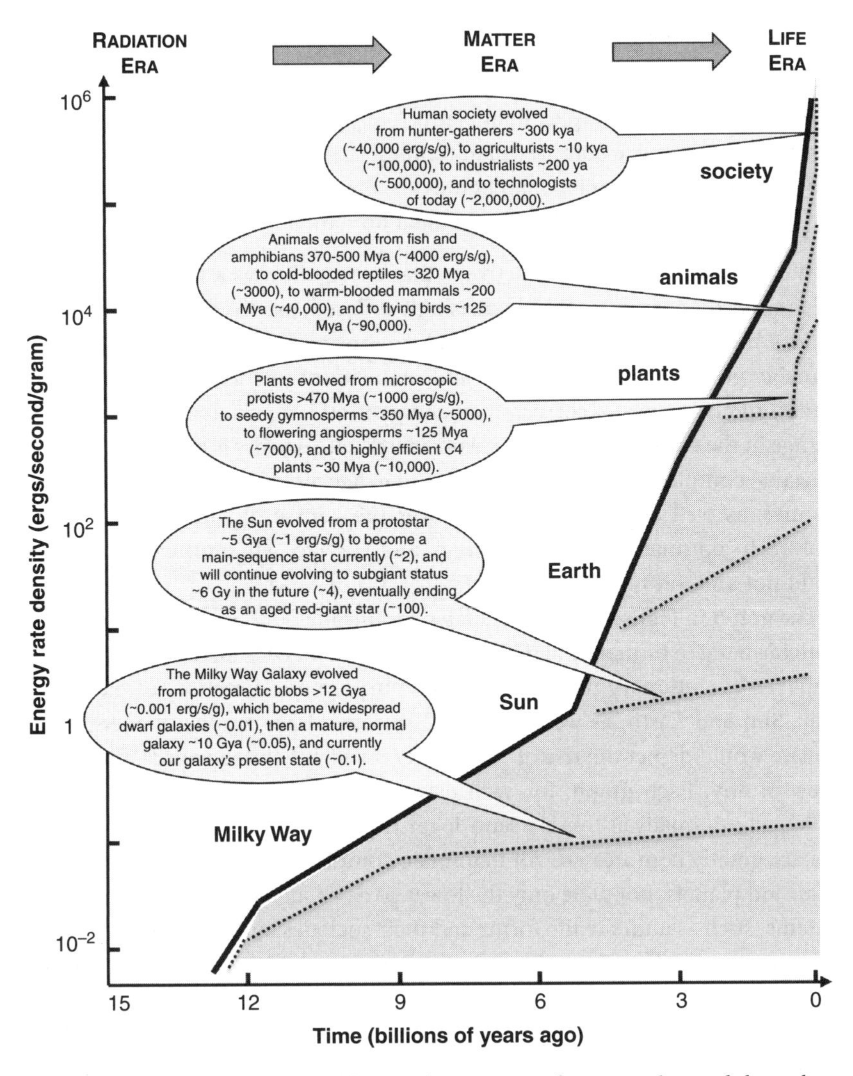

Figure 21.1 Energy rate densities for a wide spectrum of systems observed throughout Nature display a clear trend across ~14 billion years as simple primordial matter changed into increasingly intricate, complex systems. The solid black curve implies an exponential rise in system complexity as cultural evolution (steepest slope at upper right) acts faster than biological evolution (moderate slope in middle part of curve), which in turn surpasses physical evolution (smallest slope at lower left). The shaded area includes a huge ensemble of energy rate densities as many varied types of complex systems continued changing and complexifying since their origin; the several dotted black curves delineate notable evolutionary paths traversed by the major systems labeled. The energy-rate-density values and historical dates plotted here are estimates for specific systems along the evolutionary path from big bang to humankind, namely, our galaxy, star, planet, life, and society, as compiled in the bubble inserts (Chaisson 2014b). Similar graphs likely pertain to extraterrestrial life forms, as all complex systems fundamentally hark back to the early Radiation Era, evolve throughout the Matter Era, and potentially enter the Life Era (left to right across top). Credit: Eric Chaisson.

- animals are more complex than plants
- mammals are more complex than reptiles
- brains are more complex than bodies
- society is more complex than individual humans.

What seems inherently attractive is that energy flow as a universal process helps suppress entropy within increasingly ordered, localized systems evolving amidst increasingly disordered, global environments – a process that has arguably governed the emergence and maturity of our galaxy, our star, our planet, and ourselves. If correct, then energy itself is a principal mechanism of change in the expanding universe. And energy rate density is an unambiguous, objective complexity metric enabling us to gauge all organized systems in like manner, as well as to examine how over the course of time some systems evolved to optimally command energy and survive, while others apparently could not and did not.

The graph in Figure 21.1 encapsulates specifically our own evolutionary trek from big bang to humankind (Chaisson 2001, 2014b; Dick 2012). It empirically maps the evolutionary path, with neither purpose nor direction, of the Milky Way, Sun and Earth, as well as life and society. Other graphs of energy rate density would depict the rise of complexity on alien habited worlds, if we only knew of any. Each graph, for each extraterrestrial civilization, would likely differ, yet minimally as this is a semi-logarithmic scale; its events and phenomena are greatly compressed. All inanimate complex systems, such as galaxies, stars, and planets, populate only the lower parts of such curves; more complex systems, such as animate life forms and their societies and machines, reside in the upper parts. The general trend of rising complexity with the advance of cosmic evolution, which is presumed to be a universal process, likely pertains to all galactic civilizations, possibly causing in the above analogy each of their "bulbs" to illuminate – provided their energy-rate-density curves do not turn over or truncate as societal problems impair or destroy their continued development beyond rudimentary technological intelligence.

Close examination of energy rate density in Figure 21.1 for each type of individual complex system suggests a sharp rise for only limited periods of time, after which some of the curves begin flattening. Although caution is needed to avoid over-interpreting these data, some (but not all) systems do slow their rate of complexification; they seem to follow a classic, sigmoidal, S-shaped curve – much as microbes do in a petri dish while replicating with fixed food supplies or as human population is expected to plateau later this century. That is, energy rate densities for a whole array of physical, biological, and cultural systems first increase slowly and then more quickly during their individual evolutionary histories, eventually leveling off throughout the shaded area of Figure 21.1; if true, then the bold graph in Figure 21.1 that winds exponentially from lower left to upper right is probably the compound sum of multiple S-curves (Modis 2012). Viable, complex, surviving systems display no absolute decrease in their energy rate densities, rather merely lessened growth rates and S-shaped inflections as those systems apparently matured. Ultimately most systems, including unstable stars, stressed species, and inept civilizations, do collapse when they can no longer sustain themselves by optimally managing their energy budgets; such adverse fates are natural, common outcomes of cosmic evolution.

Might part of the "message" contained within the null "signal" be that successful, winning, complex systems, including intelligent life everywhere, are masters of continually increased energy usage – and those that cannot manage to do so are losers? Could this implicit message impact our society on Earth, including our fledgling candidacy for the Life Era?

The impact

The general trend of increased energy utilization in Figure 21.1 looks and feels like a cosmological imperative. Everlasting evolution and rising complexity may well be hallmarks of Nature, especially in a universe that expands at an accelerating rate (Chaisson 2001). There is no scientific reason to expect that the main, overall (bold) curve of energy rate density will halt its upward trend or ever turn over. Even if individual systems' (dotted) curves do plateau, such as for maturing galaxies, brainy species, or technical gadgets, some other ascendant complex system – such as a symbiosis of meat and machines – will likely vie for ever-higher energy rate density. Not everyone foresees such perpetual growth; some envision an end to high-tech devices lest societies self-destruct (Skrbina 2015), others infer limits to evolutionary complexity and a potential uniqueness for humankind (Conway Morris 2003, 2013).

All things considered and mindful of the extensive data comprising Figure 21.1, indefinitely rising energy rate densities imply a universal prerequisite for technological civilizations to endure. Societies capable of harnessing optimal amounts of energy consumption per capita – not too little as to starve them and not too much as to destroy them – potentially survive, if only locally and temporarily. If extraterrestrial intelligence is not as widespread as previously thought – and that seems to be what the lack of SETI success is relating to date – then the rarity of advanced beings implies short longevities probably owing to inability or unwillingness to adapt to larger energy use. A plethora of planetary issues might have caused them to succumb, including global warfare, climate change, asteroid collision, infectious disease, among many other natural or self-inflicted calamities capable of delivering too much or too little energy and thus terminating civilizations. Perhaps complexity does have its limitations after all.

Given that most living species are losers biologically, it is not unreasonable that most galactic civilizations might be losers culturally. They are, quite likely and naturally, cosmically selected out of the population of advanced civilizations, offering a pragmatic solution to the Fermi paradox. The Life Era is perhaps rarely populated by many technological civilizations at any one time, harking back to the few glowing bulbs in the vast chandelier analogy above. Such outcomes are neither special nor predicable. Frequent losers are disfavored while experiencing rapid disaster or slow extinction; they simply make a resource mistake or time their actions incorrectly, both outcomes failing to exploit energy budgets optimally. Rare winners survive favorably perhaps for no other reason than they manage to muddle through each new evolutionary step by effectively utilizing more energy per capita.

Thus, the impact of the "signal" and its "message" for us on Earth conceivably regards increased use of energy to power our civilization going forward. Yet a central issue confronts humanity here and now. How can our technological society get its act together, and quickly too, to foster increased energy use without adversely affecting the environment in which we live? That is, how can humankind's energy rate density continue to rise indefinitely atop Figure 21.1 without seriously harming our planetary abode? It's a practical matter and a timing issue; like much else in evolution writ large, survival at the threshold of the Life Era is all about energy viability and astute timing.

Society and its machines are currently driven upward and onward by economic growth, which is also an integral part of cultural evolution. The world economy is no different from any other open, complex system with its incoming energy and resources, outgoing products and wastes, and a distinctly non-equilibrium status. Much as for all complex systems, energy is central to the creation, growth, and operation (as well as demise) of any economy; neither too much nor too little but optimal amounts of energy utilization literally *drive* knowledge creation and product innovation. The bottom line – for this is economics, after all – makes clear that energy is the common currency of economies even more than money or self-interest. Today's most successful businesses are all about speed of production (including design and manufacture) as well as turnaround of new and better products; high-tech communications and intense social networking help to accelerate ideas, research, and development. And nothing speeds things along more than energy, which is at the heart of all complex systems' evolution.

Understandably, social scholars concerned about natural scientists treading on their turf will likely resist notions of non-equilibrium, market gradients, and non-linear dynamics, all of it implying economics (and politics) on the ragged edge of chaos. Yet if we have learned anything from cosmic-evolutionary analyses, realistic economies could be gainfully modeled using principles of complexity science and its underlying thermodynamics (Buchanan 2013; Chaisson 2014a; Gogerty 2014). Whereas orthodox economic modeling (with its supply–demand equilibrium and stable input–output harmony in the marketplace) spawns periodic crises among financial institutions and the nation-states that control them – a classic "heat death" of global markets and likely eventual collapse of technological civilization – non-equilibrium thermodynamics goes about its robust business of guiding energy flows through complex cultural systems, which for the case of human society is ourselves mostly within the vibrant and expansive cities of planet Earth.

Cities are as much a product of cosmic evolution as any star or life form, and their success and sustainability are closely tied to energy flowing through them. Energy use is already high in the cities where most people live; although cities occupy less than 1 percent of Earth's land area, they now house about 55 percent of humanity and account for about 70 percent of all global energy usage. Even more energy will likely be needed not only to lift developing nations out of poverty but also to increase the standard of living for the multitudes (about a million people per week) now flocking to cities during the greatest migration in human history. Urban energy metabolism has become a reality on planet Earth, and in today's energy-centered society efficiency gains often ironically dictate yet more energy use, with purported energy savings actually translating into higher consumption – which is why many people who buy cars with good mileage ratings typically drive more and those who are comfortable with smart gadgets tend to own more of them, ultimately often using just as much and sometimes more total energy. As cities grow and complexify, they hunger for more, not less, energy – both in absolute and per capita amounts; cities are more than the sum of their many residents as economy of scale saves little, if any, energy despite shared infrastructure (Bettencourt *et al.* 2007; Batty 2013; Fragkias *et al.* 2013). Cities exemplify cultural evolution at work in its most rapid and vigorous way to date, yet they are fundamentally no different than other complex systems described by cosmic evolution; humans cluster into cities much like matter assembles into galaxies, stars, and planets, or life itself organizes into bodies, brains, and society. The story is much the same all the way up the rising curve of Figure 21.1.

Humankind must raise its energy acumen – to think big and adapt broadly to what fundamentally drives human society. That driver is not likely information, the Internet, greed, or any other subjective factor that theorists and pundits often preach; rather, all complex systems, including cultured humanity and its smart gadgets, are root-based on energy. Cosmic evolution and its undeniable upward trend near the top of Figure 21.1 advise copious amounts of additional energy to power society, machines, cities, and the economy as well as their likely symbioses. The implication here is that our descendants' fate is not nearly dependent on more efficiency and less energy as it is on increased energy density. The cosmic-evolutionary narrative asserts that robust energy use, now and forevermore, is likely a cultural requirement if we are to enter the Life Era.

Yet an issue looms large and potentially fatal: cities are not only voracious users of energy; they are also the largest producers of entropy on the planet. Can humankind increase its energy budget, both in absolute terms and on a per capita basis, without degrading Earth's environment? The first law of thermodynamics is as inviolable as the second law; energy conservation demands that *all* energy utilized from non-renewable sources (not just that inefficiently wasted) eventually dissipates as heat. Alas, heat alone is pollution that can, even absent anthropogenic greenhouse gases, overheat our planet within a few centuries (Chaisson 2008, 2014a; Flanner 2009).

Solar energy is the only natural solution that permits us to circumvent the dilemma that society needs more energy yet cannot tolerate more of its inevitable heat byproduct. Unlike fossil fuels that are dug up on Earth, used on Earth, and warm the Earth internally, the Sun's daily rays land here externally and are already accounted for in our planet's thermal balance. Humankind and its machines can safely utilize more energy without being awash in heat only by adopting solar energy (including its renewable derivatives of wind, water, and waves), plenty of which is available to power economic growth indefinitely; a single hour's dose of incoming radiation from the Sun approximates that now used by civilization in a year. What's more, the Sun's energy seems the only seriously viable way that our technological society can avoid collapse, continue evolving, and ultimately be selected by Nature to endure perhaps indefinitely. This isn't the stance of an astrophysicist looking to the sky for solutions to earthly problems. It's an abbreviation of the single strongest scientific argument to create a solar economy as soon as possible to secure humanity's future well-being.

Such a global solar-based economy would still produce numerous goods and services, but the one vital resource – energy – that underlies all complex systems, including society itself, would no longer be subject to geopolitics, revolutions, or selfishness. Urban energy metabolism can become an earthly virtue, shepherding the structure and function of our cities and their residents without further degrading surrounding environments. Only our parent star can grant us the freedom to employ the needed additional energy required to survive – an idea pondered decades ago (Kardashev 1964) and now surmised as likely much the same everywhere: technological civilizations anywhere in the universe that manage to use starlight are naturally selected to endure, and those that don't, aren't. Again, it's all about energy and all about timing. Can civilizations execute both? Have any? Can we?

The sum

More than a half-century of searching for extraterrestrial intelligence has produced no contact whatsoever. Such a lack of signal, however, might actually constitute a "signal." At the least, many observers suspect that advanced technological civilizations are less abundant than earlier estimates of physicists and astronomers (but probably not most biologists). Astrobiology remains a useful context in which to study life in the universe, but currently there is no evidence that any intelligent entity has survived long in the Life Era, where resultant system complexity and required energy usage are likely essential, high, and sustained.

Given that ubiquitous change, energy flow, and rising complexity are hallmarks in the emergence of all ordered systems, it might well be that some galactic technological civilizations – perhaps nearly all of them, given the lack of extraterrestrial intelligence discovery to date – never manage to enhance optimally and expeditiously their energy rate densities. The "message" contained within the "signal" might reveal why extraterrestrial intelligence is apparently so scarce elsewhere in the universe. To survive long term, advanced civilizations need to use the energy of their parent star as quickly as technologically feasible – and those that don't are non-randomly removed from the population of galactic civilizations. In turn, that "signal's message" might also be telling us how to endure on Earth: embrace the abundant energy of the Sun – not merely use it or lose it, rather use it or lose.

Throughout all of Nature, if complex systems' energy usage is optimum – neither maximized nor minimized, rather within different ranges for different systems of different masses – then those systems have opportunities to survive, prosper, and evolve; if it's not, they are non-randomly eliminated. Civilizations are likely no different. Thus, my considered resolution of the Fermi paradox is mundane and boring. Most, perhaps all, technological societies, much like biological species, are naturally selected out of existence, not because of any dramatic inward failure or outward catastrophe, rather because they simply fail to manage their energy budgets smartly and quickly with the march of time.

Whether civilization on Earth endures or not – the choice is probably ours since technology is society's to wield as we collectively wish – the stars will shine and the galaxies will twirl, with or without sentient beings here or anywhere else. If we do manage to survive into the Life Era, the cosmic-evolutionary narrative can help empower humans and their descendants in countless ways to appreciate not only the importance of utilizing the boundless resource of our parent star, but also how wisely managed and optimally energized complex systems can safeguard the health, wealth, security, and perhaps destiny of humankind.

References

Batty, M. 2013. *The New Science of Cities*. Cambridge, MA: MIT Press.

Bettencourt, L. M. A., Lobo, L., Helbing, D., Kuhnert, C. and West, G. B. 2007. "Growth, Innovation, Scaling, and the Pace of Life in Cities." *Proceedings of the National Academy of Sciences* 104: 7301–7306. http://dx.doi.org/10.1073/pnas.0610172104.

Buchanan, M. 2013. *Forecast: What Physics, Meteorology, and the Natural Sciences Can Teach Us About Economics*. New York, NY: Bloomsbury.

Chaisson, E. 1987. *The Life Era: Cosmic Selection and Conscious Evolution*. New York, NY: Atlantic Monthly Press.

Chaisson, E. J. 2000. "Null or Negative Effects of ETI Contact in the Next Millennium." In *When SETI Succeeds*, edited by A. Tough, 59–60, Seattle: Foundation for the Future Proceedings.

Chaisson, E. J. 2001. *Cosmic Evolution: The Rise of Complexity in Nature*. Cambridge, MA: Harvard University Press.

Chaisson, E. J. 2003. "A Unifying Concept for Astrobiology." *International Journal of Astrobiology* 2: 91–101. http://dx.doi.org/10.1017/S1473550403001484.

Chaisson, E. J. 2008. "Long-term Global Heating from Energy Usage." *Eos Transactions of the American Geophysical Union* 89: 253–254. http://dx.doi.org/10.1029/2008EO280001.

Chaisson, E. J. 2014a. "Practical Applications of Cosmology to Human Society." *Natural Science* 6: 767–796. http://dx.doi.org/10.4236/ns.2014.610077.

Chaisson, E. J. 2014b. "The Natural Science Underlying Big History." *The Scientific World Journal* 2014, Article ID 384912. http://dx.doi.org/10.1155/2014/384912

Conway Morris, S. 2003. *Life's Solution: Inevitable Humans in Lonely Universe*. Cambridge: Cambridge University Press.

Conway Morris, S. 2013. "Life: the final frontier for complexity?" In *Complexity and the Arrow of Time*, edited by C. Lineweaver, P. Davies and M. Ruse, 135–161, Cambridge: Cambridge University Press.

Dick, S. J. 2012. "Critical Issues in History, Philosophy and Sociology of Astrobiology." *Astrobiology* 12: 906–927.

Dick, S. J. and Lupisella, M. L. (eds.). 2009. *Cosmos and Culture: Cultural Evolution in a Cosmic Context*. Washington, DC: NASA.

Flanner, M. G. 2009. "Integrated Anthropogenic Heat Flux With Global Climate Models." *Geophysical Research Letters* 36:L02801–L02805. http://dx.doi.org/10.1029/2008GL036465

Fragkias, M., Lobo, J., Strumsky, D. and Seto, K.C. 2013. "Does Size Matter? Scaling of CO_2 Emissions and US. Urban Areas." *PLoS ONE* 8:1–6. http://dx.doi.org/10.1371/journal.pone.0064727

Gogerty, N. 2014. *The Nature of Value*. New York, NY: Columbia University Press.

Kardashev, N. 1964. "Transmission of Information by Extraterrestrial Civilizations." *Soviet Astronomy* 8: 217–221.

Modis, T. 2012. "Why the Singularity Cannot Happen." In *Singularity Hypotheses*, edited by A. H. Eden *et al.*, 311–340, Berlin: Springer-Verlag.

Skrbina, D. 2015. *The Metaphysics of Technology*. New York, NY: Routledge.

Vidal, C. 2014. *The Beginning and the End*. Switzerland: Springer International Publishing.

Contributor biographies

Linda Billings is a consultant to NASA's Astrobiology and Near-Earth Object programs in the Planetary Science Division of the Science Mission Directorate at NASA Headquarters in Washington, DC. She also is Director of Communication with the Center for Integrative STEM Education at the National Institute of Aerospace in Hampton, Virginia. Dr. Billings earned her Ph.D. in mass communication from Indiana University. Her research interests include science and risk communication, social studies of science, and the rhetoric of science and space. Her papers have been published by the NASA History Division, *Space Policy, Acta Astronautica*, and *Advances in Space Research.* She has worked for more than 30 years in Washington, DC, as a researcher; communication planner, manager, and analyst; policy analyst; journalist; and consultant to the government. She is a contributing author to many books, including *First Contact: The Search for Extraterrestrial Intelligence* (New American Library, 1990). Dr. Billings was a member of the staff for the National Commission on Space (1985–1986), appointed by President Reagan to develop a long-term plan for space exploration. She served as an officer of Women in Aerospace (WIA) for 15 years, most recently as president (2003). She is the recipient of a WIA Lifetime Achievement Award (2009) and a WIA Outstanding Achievement Award (1991). She was elected a Fellow of the American Association for the Advancement of Science in 2009.

Eric J. Chaisson is an American astrophysicist at Harvard University in Cambridge, Massachusetts, where he has multiple appointments, most notably at the Harvard-Smithsonian Center for Astrophysics. His major interests are currently twofold. His scientific research addresses an interdisciplinary, thermodynamic study of physical, biological, and cultural phenomena, seeking to understand the origin and evolution of galaxies, stars, planets, life, and society, thus devising a unifying cosmic-evolutionary worldview of the universe and our sense of place within it writ large. His educational work engages master teachers and computer animators to create better methods, technological aids, and novel curricula to enthuse teachers, instruct students, and enhance scientific literacy from grade school to grad school. He teaches at Harvard an annual undergraduate course on natural science that combines both of these research and educational objectives, wherein he maintains a huge multi-media website:

"Cosmic Evolution: From Big Bang to Humankind" that contains a rich suite of astrobiology materials for both non-scientists (Introductory Track) and professional scientists (Advanced Track). Chaisson has published nearly 200 papers in professional journals, written a dozen books, and won several awards such as the B. J. Bok Prize (Harvard) for astronomical discoveries, the Smith-Weld Prize (Harvard) for literary merit, and the Kistler Award for increasing understanding of subjects shaping the future of humanity. He has also received scholarly prizes from Phi Beta Kappa, the American Institute of Physics, a Certificate of Merit from NASA for work on the Hubble Space Telescope, as well as fellowships from the Sloan Foundation and the National Academy of Sciences.

Carol E. Cleland is Professor of Philosophy at the University of Colorado. She arrived at CU Boulder in 1986, after having spent a year on a postdoctoral fellowship at Stanford University's Center for the Study of Language and Information. She received her Ph.D. in philosophy from Brown University in 1981 and her B.A. in mathematics from the University of California Santa Barbara in 1973. From 1998 to 2008 she was a member of NASA's Institute for Astrobiology (NAI). Professor Cleland specializes in philosophy of science, philosophy of logic, and metaphysics. Her current research interests are in the areas of scientific methodology, historical science, biology (especially microbiology, origins of life, the nature of life, and astrobiology), and the theory of computation. Cleland's published work has appeared in leading philosophy and science journals. She is co-editor (with Mark Bedau) of *The Nature of Life: Classical and Contemporary Perspectives from Philosophy and Science* (Cambridge University Press, 2010) and is currently finishing a book (*The Quest for a Universal Theory of Life; Searching for Life as We Don't Know It*), which is under contract with Cambridge University Press.

Brother Guy Consolmagno, SJ, president of the Vatican Observatory Foundation, is an astronomer and meteoriticist at the Vatican Observatory. A native of Detroit, Michigan, he earned undergraduate and masters' degrees from MIT, and a Ph.D. in Planetary Science from the University of Arizona, was a postdoctoral research fellow at Harvard and MIT, served in the US Peace Corps (Kenya), and taught university physics at Lafayette College before entering the Jesuits in 1989. At the Vatican Observatory since 1993, his research explores connections between meteorites, asteroids, and the evolution of small Solar System bodies, observing Kuiper Belt comets with the Vatican's 1.8-meter telescope in Arizona, and applying his measure of meteorite physical properties to understanding asteroid origins and structure. Along with more than 200 scientific publications, he is the author of seven popular books including *Turn Left at Orion* (with Dan Davis), and *Would You Baptize an Extraterrestrial?* (with Paul Mueller). He also has hosted science programs for BBC Radio 4, been interviewed in numerous documentary films, and writes a monthly science column for the British Catholic magazine, *The Tablet.* Dr. Consolmagno has been elected to the governing boards of a number of international scientific organizations, including serving as chair of the American Astronomical Society's Division for Planetary Sciences. In 2000, the small bodies nomenclature committee of the IAU named an asteroid, 4597 Consolmagno, in recognition of his work.

Steven J. Dick was most recently the 2014 Baruch S. Blumberg NASA/Library of Congress Chair in Astrobiology at the Kluge Center of the Library of Congress. He served as the Charles A. Lindbergh Chair in Aerospace History at the National Air and Space Museum from 2011 to 2012, and as the NASA Chief Historian and Director of the NASA History Office from 2003 to 2009. Prior to that he worked as an astronomer and historian of science at the US Naval Observatory in Washington, DC for 24 years, including three years on a mountaintop in New Zealand. He obtained his B.S. in astrophysics (1971), and M.A. and Ph.D. (1977) in history and philosophy of science from Indiana University. Among his books are *The Biological Universe: The Twentieth Century Extraterrestrial Life Debate and the Limits of Science* (Cambridge University Press, 1996); *Life on Other Worlds* (Cambridge University Press, 1998); (with James Strick) *The Living Universe: NASA and the Development of Astrobiology* (2004), and *Discovery and Classification in Astronomy: Controversy and Consensus* (Cambridge University Press, 2013). Dr. Dick is the recipient of the NASA Exceptional Service Medal, the Navy Meritorious Civilian Service Medal, the NASA Group Achievement Award for his role in NASA's multidisciplinary program in astrobiology, the NASA Group Achievement Award for the book *America in Space*, and the 2006 LeRoy E. Doggett Prize for Historical Astronomy of the American Astronomical Society. In 2012 he was elected a Fellow of the American Association for the Advancement of Science (AAAS). He has served as Chairman of the Historical Astronomy Division of the American Astronomical Society, as President of the History of Astronomy Commission of the International Astronomical Union, and as President of the Philosophical Society of Washington. He is a member of the American Astronomical Society, the International Astronomical Union, and a corresponding member of the International Academy of Astronautics. In 2009 the International Astronomical Union designated minor planet 6544 stevendick in his honor.

Iris Fry obtained her B.Sc. in Chemistry and M.Sc. in Biochemistry from The Hebrew University, Jerusalem; her M.A. in philosophy from Haifa University, with a thesis on "Immanuel Kant's concept of teleology"; and her Ph.D. in the History and Philosophy of Science from Tel Aviv University with a thesis on "L. J. Henderson's Fitness of the Environment for Life."

She has ten years' experience as a research assistant at the Department of Biological Chemistry, The Hebrew University, and a long teaching career at the Cohn Institute for the History and Philosophy of Science and Ideas at Tel Aviv University and at the Department of Humanities and Arts, Technion–Israel Institute of Technology in Haifa. She retired as of 2014.

Her main areas of interest are scientific, philosophical, and historical aspects of the origin of life; problems in the history and philosophy of biology; the historical debate on the plurality of worlds; philosophical aspects of astrobiology; and the relationship between intelligent design and Darwinism. Among her publications are *The Emergence of Life on Earth: A Historical and Scientific Overview*, Rutgers University Press (2000); "On the biological significance of the properties of matter: L. J. Henderson's theory of the fitness of the environment," *Journal for the History of Biology*, 29 (1996), 155–196; "The role of natural selection in the origin of life," in *Origins of Life and Evolution Biosphere*, 41 (2011), 3–16; "Is science metaphysically neutral?" *Studies in the History and Philosophy of Biology and the Biomedical Sciences*, 43 (2012), 665–673, "The Emergence of Life on Earth and the Darwinian Revolution," in Michael Ruse (ed.), *The Cambridge Encyclopedia of Darwin and Evolutionary Thought*, Cambridge University Press (2013), pp. 322–329; and "The Origin of Life: Scientific Solution versus Chance and Telos," (in preparation).

Robin W. Lovin is Director of Research at the Center of Theological Inquiry in Princeton, New Jersey, and Cary Maguire University Professor of Ethics emeritus at Southern Methodist University (SMU). He joined the SMU faculty in 1994, and served as Dean of Perkins School of Theology from 1994 to 2002. Dr. Lovin's most recent books are *Christian Realism and the New Realities* (Cambridge University Press, 2008) and *An Introduction to Christian Ethics* (Abingdon Press, 2011). He has also written extensively on religion and law and comparative religious ethics. In 2013, he held the Maguire Chair in Ethics and American History at the Library of Congress. He is a former Guggenheim Fellow, former president of the Society of Christian Ethics, and a member of the advisory board for the McDonald Centre for Theology, Ethics, and Public Life at Oxford University.

Mark Lupisella works on NASA's Human Spaceflight Architecture Team and manages Goddard's Advanced Exploration Systems portfolio for Human

Exploration. He has led or co-led a number of architecture and mission analysis efforts and recently worked on the NASA Technology Roadmap Team for human exploration systems. He previously worked on Mars surface science for the Human Spaceflight Architecture Mars Destination Operations Team, co-led the Cis-Lunar Destination Team, was a member of the Near-Earth Asteroid Working Group and was a participant in the Keck Institute for Space Studies Asteroid Retrieval Mission Study. He recently led a proposal to fly a miniaturized DNA sequencer on the International Space Station and beyond to Mars. Mark has also worked on the Hubble Space Telescope, wearable computing, cooperative robotics, and areas of astrobiology such as SETI, planetary protection, artificial life, and broader societal issues of astrobiology such as ethics and worldviews. He co-founded the Horizons Project which aims to explore mechanisms for long-term human survival, and has been developing methods for applying evolutionary dynamics to long-term strategic planning and organizational ethics. He has recently been developing and chairing an ethics committee for the satellite company Planet Labs. Mark was a panel member at the COSPAR workshop on "Developing a Responsible Environmental Regime for Celestial Bodies," was a participant in a Princeton workshop on the Ethics of Planetary Protection, and worked with the Secure World Foundation, helping formulate their space security program.

Mark has authored over 30 published works and is a co-editor of *Cosmos and Culture: Cultural Evolution in a Cosmic Context* with previous NASA Chief Historian, Steven Dick. He is working on a new book recently approved by Springer, tentatively titled *Philosophical and Policy Implications of Space Activities.* Mark has a B.S. in physics, an M.A. in philosophy (emphasis in Philosophy of Science and ethics), and a Ph.D. in evolutionary biology (Program in Behavior, Ecology, Evolution, and Systematics), from the University of Maryland at College Park, where he did his dissertation on modeling microbial contamination of Mars from human missions.

Jane Maienschein is Regents' Professor, President's Professor, and Parents Association Professor at Arizona State University, where she directs the Center for Biology and Society in the School of Life Sciences. She is also Adjunct Senior Scientist at the Marine Biological Laboratory in Woods Hole, Massachusetts. Maienschein specializes in the history and philosophy of biology and the way biology, bioethics, and biopolicy play out in society. Focusing on research in development, genetics, and cell biology, Maienschein combines detailed analysis of the epistemological standards, theories, laboratory practices, and experimental approaches, with study of the people, institutions, and changing social, political, and legal context in which science thrives. She loves teaching and is committed to public education about the life sciences and their human dimensions, having won the History of Science Society's Joseph Hazen Education Award and the Carnegie Foundation for the Advancement of Teaching (CASE) Professor of the Year award for Arizona. She has offered educational programs for federal and state judges through the Federal Judicial Center, served as a Congressional Fellow in the office of Congressman Matt Salmon during the 105th Congress, co-chaired the Gordon Conference on Science and Technology Policy in 2004, served as co-editor for the *Journal of the History of Biology*, and is co-chair for the Joint Caucus for Socially Engaged History and Philosophy of Science and co-organizer for the Digital HPS Consortium. Her most recent book is *Embryos Under the Microscope. The Diverging Meanings of Life* (Harvard University Press, 2014), and she is author of three other books and 11 (co-) edited volumes, along with many articles and essays.

Lori Marino is a neuroscientist and expert in animal behavior and intelligence who was on the faculty at Emory University for 19 years. She is internationally known for her work comparing the evolution of dolphin and whale brains with primate brains and has worked with the astrobiology and SETI community for two decades and currently advises the New Horizons Message Initiative. Presently she is the Executive Director of The Kimmela Center for Animal Advocacy and Science Director for The Nonhuman Rights Project. In both positions she applies scientific data and expertise to animal advocacy issues such as marine mammal captivity and the rights of non-human primates. Dr. Marino has authored over 100 peer-reviewed papers, book chapters, and magazine articles on marine mammal biology and cognition, comparative neuroanatomy, self-awareness in other animals, human–non-human animal relationships, and animal welfare and the captivity industry. Dr. Marino was recently featured as a National Geographic Innovator and appears in several films and television programs on the above topics, including the influential documentary Blackfish, about killer-whale captivity. She currently teaches online courses in Animal Behavior and Evolution.

Carlos Mariscal is a native of southern New Mexico and graduated from New Mexico State University in 2006. He had several brief careers as a newspaper editor, graphic designer, and math tutor before enrolling in a philosophy of biology Ph.D. program at Duke University working with Robert Brandon and Alex Rosenberg. While at Duke, he earned the James B. Duke Fellowship and the University Scholars Fellowship. He was subsequently awarded the Katherine Goodman Stern Dissertation Fellowship and a Summer Research Fellowship. He currently serves as the webmaster of the International Society for the History, Philosophy, and Social Studies of Biology (ISHPSSB), was a

Graduate Assistant to the University Scholars Program, and was the Assistant Director for the Center for the Philosophy of Biology. Carlos recently graduated from Duke and is presently a post-doctoral fellow at the Centre for Comparative Genomics and Evolutionary Bioinformatics in Halifax, Nova Scotia, supervised by W. Ford Doolittle of the Department of Biochemistry and Molecular Biology at Dalhousie University.

Michael A. G. Michaud, the author of more than 100 published works, was a US Foreign Service Officer for 32 years before turning full time to writing. During his diplomatic career, he served as Acting Deputy Assistant Secretary of State for Science and Technology, Director of the State Department's Office of Advanced Technology, and as Counselor for Science, Technology and Environment at the American embassies in Paris and Tokyo. He led US delegations in the successful negotiation of international science and technology cooperation agreements with the Soviet Union and Poland. He played an active role in negotiating a new United States–Soviet space cooperation agreement and in initiating United States–Soviet talks on outer space arms control. Michaud has been a leading figure in preparations for possible future contact with extraterrestrial intelligence. As chairperson of International Academy of Astronautics working groups, he coordinated the drafting of the Declaration of Principles Concerning Activities Following the Detection of Extraterrestrial intelligence, better known as the First SETI Protocol. His 2007 book *Contact with Alien Civilizations* (Springer) is a comprehensive overview of centuries-long debates about the probability and consequences of contact. His previous book was an informal history and analysis of the American pro-space movement. His publication record also includes more than 80 articles and journal papers. Michaud is a member of several professional organizations including the International Academy of Astronautics, the American Institute of Aeronautics and Astronautics, the American Astronautical Society, the American Association for the Advancement of Science, and the World History Association.

Margaret S. Race is a Senior Scientist at the SETI Institute (Mountain View, CA), who works with NASA on astrobiology, planetary protection, and risk communication. Currently, she is working on projects involving planetary protection and human Mars missions; the societal, ethical, and policy implications of space exploration and searches for ET life; and risk-preparedness and communications about hazardous asteroids. Over the past decade, she has served on numerous US and international studies of forward and back contamination on both robotic and human missions in the solar system. In addition to her research and analytical work, Dr. Race is actively involved in science education and outreach about astrobiology and science, technology, engineering, and mathematics (STEM) topics through the mass media, schools, museums, and public presentations. Dr. Race received her Ph.D. (Ecology) from the University of California at Berkeley, and her B.A. degree (Biology) and M.S. degree (Energy Management and Policy) from the University of Pennsylvania. Previously, she has been a professor and researcher at Stanford University (Human Biology Program; Center for International Security and Cooperation), Assistant Dean at UC Berkeley (College of Natural Resources), and Senior Science Policy Analyst at University of California Office of the President. She is the author of over 100 peer-reviewed papers, book chapters, and other publications.

Michael Ruse is the Lucyle T. Werkmeister Professor of Philosophy and the Director of the Program in the History and Philosophy of Science at Florida State University. He is a Fellow of the Royal Society of Canada, a Guggenheim Fellow, a Gifford Lecturer and the author or editor of over 50 books on the history and philosophy of the life sciences, with a special interest in the thought and influence of Charles Darwin. The recipient of numerous awards, including four honorary doctoral degrees, he is most proud of the fact that he is now in his fiftieth year of college teaching.

Susan Schneider is an associate professor at the University of Connecticut and a fellow with the American Council of Learned Societies and Australian National University. She works at the intersection of philosophy of mind, metaphysics, and cognitive science. She is the author of *The Language of Thought* (MIT Press), *The Blackwell Companion to Consciousness* (Wiley) and *Science Fiction and Philosophy* (Wiley). She is currently writing a book on the nature of the self and mind (forthcoming with Oxford).

Dirk Schulze-Makuch is a professor in the School of the Environment at Washington State University and an adjunct professor at the Beyond Center of Arizona State University. Currently he also holds a guest professorship at the Technical University in Berlin, Germany. Dirk obtained his Ph.D. from the University of Wisconsin-Milwaukee in 1996. Afterwards he worked as a senior project hydrogeologist at Envirogen, Inc. and took in 1998 a faculty position at the University of Texas at El Paso. During that time, he was also a summer faculty fellow at the Goddard Space Flight Center. Since 2004, Dr. Schulze-Makuch has been a faculty member at Washington State University. His interests are in astrobiology, planetary habitability, and medical microbiology. He has published seven books including *Life in the Universe: Expectations and Constraints*, which is in the second edition, more than 100 scientific papers, and has given more than 200 talks at scientific meetings. In 2010 he received the Friedrich-Wilhelm Bessel Award from the Humboldt Foundation for extraordinary achievements in the field of theoretical biology. The work of Dr. Schulze-Makuch has received much public attention and has been the subject of television programs on the BBC, the National Geographic, and the Discovery Channel, and of numerous articles in magazines such as *New Scientist, The Guardian*, and *Der Spiegel.*

Seth Shostak is the Senior Astronomer at the SETI Institute, in Mountain View, California. He has an undergraduate degree in physics from Princeton University, and a doctorate in astronomy from the California Institute of Technology. For much of his career, Seth conducted radio astronomy research on galaxies, publishing approximately 60 papers in professional journals. For more than a decade, he worked at the Kapteyn Astronomical Institute, in Groningen, The Netherlands, using the Westerbork Radio Synthesis Telescope. He also founded and ran a computer animation company. Seth has written more than 400 popular magazine and web articles on various topics in astronomy, technology, film, and television. He lectures on astronomy and other subjects at various academic venues, and gives approximately 60 talks annually at both educational and corporate institutions. Seth has been a Distinguished Speaker for the American Institute of Aeronautics and Astronautics. He also chaired the International Academy of Astronautics' SETI Permanent Committee for a decade. Frequently interviewed for radio and television, Seth is the host of a one-hour weekly radio program on astrobiology entitled "Big Picture Science." Seth has edited and contributed to nearly a dozen books. His first popular tome, *Sharing the Universe: Perspectives on Extraterrestrial Life*, appeared in March, 1998, followed by *Cosmic Company* in 2002. He has also co-authored an astrobiology text, *Life in the Universe*, which is now in its third edition, and his latest trade book is *Confessions of an Alien Hunter*. In 2004, he won the Klumpke-Roberts Prize for the popularization of astronomy.

John W. Traphagan is an anthropologist and professor in the Department of Religious Studies at the University of Texas at Austin where he is also Centennial Commission in the Liberal Arts Fellow. His research interests center on the relationship between science and culture, with an emphasis on exploration of how cultural ideas influence concepts of health and illness. He is the author of numerous articles and books related to Japanese culture, family, and religion, including *Taming Oblivion: Aging Bodies and the Fear of Senility in Japan* (SUNY Press, 2000) and *The Practice of Concern: Ritual, Well-Being, and Aging in Rural Japan* (Carolina Academic Press, 2004), and his most recent book, *Rethinking Autonomy: A Critique of Principlism in Biomedical Ethics* (SUNY Press, 2013). Traphagan has also pursued research related to SETI, which has been published in journals such as *Acta Astronautica* and he is currently writing a book entitled *Imagining Extraterrestrial Intelligence: SETI at the Intersection of Science, Religion, and Culture* (Springer).

Julian W. Traphagan is a student in the Department of Earth and Environmental Sciences at Lehigh University.

Douglas A. Vakoch is the Director of Interstellar Message Composition at the SETI Institute, as well as Professor of Clinical Psychology at the California Institute of Integral Studies. Dr. Vakoch researches ways that different civilizations might create messages that could be transmitted across interstellar space, allowing communication between humans and extraterrestrials even without face-to-face contact. He is particularly interested in how we might compose messages that would begin to express what it's like to be human. As a member of the International Institute of Space Law, Dr. Vakoch examines international policy issues related to sending such responses. He serves as Chair of the International Academy of Astronautics Study Group on Interstellar Message Construction. In addition to being a clinical psychologist (Ph.D., State University of New York at Stony Brook), he has formal training in comparative religion (B.A., Carleton College) and the history and philosophy of science (M.A., University of Notre Dame). Dr. Vakoch has published widely in psychology, space sciences, and environmental studies. He is the editor of several books, including *Communication with Extraterrestrial Intelligence* (SUNY Press, 2011), *Civilizations Beyond Earth: Extraterrestrial Life and Society* (Berghahn, 2011), *Psychology of Space Exploration: Contemporary Research in Historical Perspective* (NASA, 2011), *Feminist Ecocriticism: Environment, Women, and Literature* (Lexington Books, 2012), *Astrobiology, History, and Society: Life Beyond Earth and the Impact of Discovery* (Springer, 2013), *Archaeology, Anthropology, and Interstellar Communication* (NASA, 2014), and *Extraterrestrial Altruism: Evolution and Ethics in the Cosmos* (Springer, 2014). He serves as general editor of Springer's Space and Society series, as well as Lexington Books' Ecocritical Theory and Practice series.

Clément Vidal is a philosopher with a background in logic and cognitive sciences. He is co-director of the "Evo Devo Universe" community, founder of the "High Energy Astrobiology" prize and initiator of the "starivore hypothesis." He authored *The Beginning and the End: The Meaning of Life in a Cosmological Perspective* (Springer, 2014). To satisfy his intellectual curiosity when facing the big questions, he brings together many areas of knowledge such as cosmology, physics, astrobiology, complexity science, evolutionary theory, and philosophy of science.

Elspeth M. Wilson is a doctoral candidate in Political Science at the University of Pennsylvania, where she is writing a dissertation titled, "The Reproduction of Citizenship." In addition to being awarded Benjamin Franklin and Presidential Prize Fellowships to fund her graduate studies at the University of Pennsylvania, Wilson served as the Administrator for the Penn-Mellon Foundation Program on Democracy, Citizenship, and Constitutionalism from 2007–2011. She was also the recipient of a Teece Dissertation Research Fellowship for the summer of 2012, awarded by the School of Arts and Sciences to fund interdisciplinary empirical field research by graduate students in the sciences or social sciences, and was named the Penn-Mellon DCC Program's Boise Family Graduate Student Fellow for the 2012–2013 academic year. She received her M.A. from the University of Wisconsin-Madison, and her B.A. with Honors in Political Science from Columbia University. Wilson's research interests include bioethics, ethics and public policy, democratic theory, public health and social welfare, American political development, political philosophy, civil rights, and constitutional law.

Index